SEA SHELLS OF THE WEST INDIES

SEA SHELLS
of the
WEST INDIES

A Guide to the Marine Molluscs of the Caribbean

MICHAEL HUMFREY

A Taplinger World Wide Field Guide
TAPLINGER PUBLISHING COMPANY : NEW YORK

TO SUE, SEAN AND SHEENA

who always thought I could,
with my love

: : :

All the shells illustrated are from the author's collection.
The photographs were taken by Brian Motta and Neville Hylton,
and the shells arranged for photography by the author.

: : :

First published in the United States in 1975 by
TAPLINGER PUBLISHING CO., INC.
New York, New York

Library of Congress Catalog Card Number: 74-20213

ISBN 0-8008-7014-X

Contents

FOREWORD by Sir Maurice Yonge, CBE, FRS. 9

PREFACE 11

PART I Introduction to Molluscs 15

The Classes of Shells 17
The Living Animals 26

PART II The West Indian Marine Shells 39

Class Gastropoda – Snails 47
Class Pelecypoda – Bivalves 206
Class Cephalopoda – Squids, Octopuses, Nautiloids 287
Class Amphineura – Chitons 289
Class Scaphopoda – Tusk Shells 294

PART III Collecting, Cleaning and Storing Shells 297

PART IV The Value of Shells 323

BIBLIOGRAPHY 335

INDEX 339

List of Illustrations

1. The Classes of Shells *page* 64
2. Limpets: *Fissurellidae, Acmaeidae* 65
3. Turbans, Top Shells, Tegulas, Cyclostremes and Pheasant Shells: *Trochacea* 80
4. Nerites, Periwinkles and Prickly-winkles: *Neritacea, Littorinacea* 81
5. Turbans, Top shells, Turrets, Star Shells, Worm Shells, Sundials and Turciculas: *Trochacea, Cerithiacea* 96
6. Worm Shells, Sundials, Horn Shells, Modulus and Planaxis Shells: *Cerithiacea* 97
7. Ceriths, Wentletraps, Purple Sea Snails, Hoof Shells, Cap Shells, Slipper Shells and Cup and Saucer Shells: *Cerithiacea, Epitoniacea, Hipponicacea, Calyptraeacea* 112
8. Conchs and Carrier Shells: *Strombacea* 113
9. Cowries, Cyphomas and Simnias: *Cypraeacea* 128
10. Moon-shells and Baby's Ears: *Naticacea* 129
11. Bonnets, Tuns and small Helmets: *Tonnacea* 144
12. Helmets, Conchs, Tulip Shells, Tritons and Spindles: *Tonnacea, Strombacea, Buccinacea* 145
13. Hairy Tritons: *Tonnacea* 160
14. Hairy Tritons, Distorsios and Frog Shells: *Tonnacea* 161
15. Murex Shells: *Muricacea* 176
16. Murex Shells, Drupes, Aspellas, Dwarf Tritons and Coral Shells: *Muricacea* 177
17. Rock Shells, Dwarf Tritons, Cantharus Shells, Nassas, Phos Shells, Enginas, Dove Shells and Baily Shells: *Muricacea, Buccinacea* 192
18. Latirus and Vase Shells: *Buccinacea* 193
19. Olives and Mitres: *Volutacea* 208
20. Volutes, Nutmegs, Marginellas, Augers and Turret Shells: *Volutacea* 209
21. Cone Shells: *Volutacea* 224
22. Trivias, Turrets, Drillias, Melampus Shells, Pyrams, Bubble Shells, Sea Butterflies and others: *Cypraeacea, Volutacea, Opisthobranchia* 225
23. Arks, Bittersweets and Yoldias: *Arcacea, Nuculacea* 240
24. Mussels and Oysters: *Mytilacea, Pteriacea* 241

25. Thorny Oysters and Pen Shells: *Pectinacea, Pteriacea* 256
26. Scallops: *Pectinacea* 257
27. Scallops, Kitten's Paws and File Shells: *Pectinacea* 272
28. Oysters, Jingle Shells, False Mussels, Jewel Boxes and Carditas: *Ostreacea, Anomiacea, Dreissenacea, Lucinacea, Carditacea* 273
29. Lucinas, Diplodons and Cockles: *Lucinacea, Carditacea* 276
30. Venus Shells, Rupellarias, Cyclinellas and Dosinias: *Veneracea* 277
31. Tellins, Macomas and Strigillas: *Tellinacea* 284
32. Asaphis Shells, Sanguins, Semeles, Wedge Shells, Corbulas, Surf Clams, Mactras, Razor Shells, Martesias and Tagelus Clams: *Tellinacea, Solenacea, Mactracea, Myacea, Adesmacea* 285

Foreword

NATURALISTS are born not made. I confirm this well-worn statement out of long experience in University teaching – one 'made' professional zoologists of one type or another but never created a naturalist. These came, if they came to the University at all, ready made. They had been observing, most usually birds, insects of one type or another, or perhaps mammals or reptiles, since earliest childhood. They were among the fortunate minority of mankind born with a compelling interest in nature which, whatever their future calling might be, would enrich their entire lives.

Armed with this experience, because I had ceased University teaching before I met him, I was in no way surprised to find that the best conchologist in Jamaica held a high position (now indeed one of the highest) in the police. I was still less surprised when Mike Humfrey told me that his interest in shells went back to early childhood in Grenada. I then first heard about the book that he was writing on shells, which I encouraged him to complete and to which I am now so glad to write a brief foreword.

Interest in shells goes back to the dawn of civilisation. It rose to a high point in the years of exploration when all manner of exotic shells of bizarre form or striking beauty were brought from the tropical Indian and Pacific Oceans and from the Caribbean to enrich European collections. Later, in the nineteenth century these formed the subject matter of magnificently illustrated monographs. Then interest waned to be more recently revived since the world shrank with the advent of air travel and the shallow depths in which almost all molluscs live could be searched by all capable of aqualung diving.

The result has been a great increase of interest in shells and in shell collecting and the appearance of a new series of books many, as here, most beautifully illustrated by colour photography. As one who has worked on molluscs in all oceans for half a century, I cannot but welcome these developments especially when they permit the publication of books such as this which give the benefit of years of intense interest to those who search for shells around the margin of West Indian islands. I know, for one, how my studies of Jamaican molluscs have been aided by Mike Humfrey's knowledge just as the pleasure of my visits has been enhanced by his friendship.

I am certain that the growing interest felt by so many not only in shells but in the still more remarkable molluscan animals that produce them will be increased by this book. May it give them some portion of the pleasure to read and to use that I know it has given its author to write. But I make a final plea to this reader. As far as possible observe the animal and where and how it lives. Collect sparingly if this involves destroying the animal because however fascinating the shell the living creature is far more interesting. There is no process so utterly irrevocable as the extinction of a species, the end product of a process that began with the first appearance of life. And this book is about species.

C. M. Yonge

Preface

Shells have fascinated man from the dawn of his existence. They have fed him, served him as tools and currency, inspired his art, and in their exquisite diversity of form and colour satisfied his deepest aesthetic instincts. In more recent times, they have also led him to write about them.

This book has four major aims. The first is to provide a general illustrated guide to West Indian marine shells. The 'Caribbean Province' as it is called, extends from Bermuda to Brazil, and sustains many of the most beautiful and interesting shells in the world. The shells illustrated are widely distributed throughout this area.

The second purpose is to provide a census of all known Jamaican marine molluscs. Compared with the neighbouring islands of Cuba and Puerto Rico, very little has been written about Jamaican marine shells. C. B. Adams, the noted nineteenth-century American malacologist, made a brief study of Jamaican molluscs in 1843 and again in 1848, but there is practically nothing else of value available on the subject. This is all the more surprising because, from a conchologist's point of view, Jamaica is very favourably situated. The island lies roughly in the centre of the Caribbean Sea, and it is a meeting place for the molluscan fauna both of the Antilles and continental America.

As it applies to shells which occur at depths down to 50 fathoms (300 feet), the census is probably very nearly complete. For shells which are found only in deeper water, I make no such claim. Very little systematic dredging has been carried out in deep water off Jamaica, and when this is eventually done there is no doubt that my lists will have to be expanded.

The third purpose of the book is to examine the habitat of each of the 497 shells described and illustrated, and to list at least one locality where each species is known to occur. Most of the habitats described are Jamaican, but because molluscs have consistent habits, the information has a general Caribbean relevance. The geographical range of each species is also provided. I should mention that the depth ranges given for each species do not necessarily represent the maximum and minimum depths in which the various species occur. They are the result of personal observation, and true maximum depths of occurrence for some species may be well outside the ranges I have quoted.

In selecting which of the 850 species listed to describe and illustrate, I have been guided by two simple criteria:

(i) The adult species must measure more than 12 mm. in length (there are a few exceptions to this),

(ii) The species must occur where it can be collected by means available to most amateur collectors, i.e. in depths not greater than about 35 fathoms.

These arbitrary limits will not please everybody, but they seemed to me to be the most convenient places at which to draw the necessary line. The shells not described (i.e. those which do not satisfy the criteria) are listed under family headings at the end of the sections on each class, in the systematic account in Part II. For details of these species – many of which are exquisite in spite of their small size – one can consult the reference libraries of the better natural history muscums.

The fourth purpose of this book is to examine in detail the most efficient methods of collecting, cleaning and storing shells. When I started collecting seriously, I had no written guide and no one available to advise me. I learnt the hard way, and I regret it. The hard way is not the best way. In the section on collecting I have set out what I learnt by trial and error. Collecting is often hard physical work; the secret, if there is one, is to avoid unproductive effort.

The section on the classes of shells and the living animals deal with these subjects in broad outline. Molluscs are fascinating creatures, and if these chapters encourage the reader to study them in more detail they will have served their purpose. The brief bibliography at the end of this book may be useful in this connection.

The section on the commercial aspect of collecting is not intended to encourage shelling for money, but the fact is that some shells are valuable objects, and it is well that collectors should have an idea of the market value of what they possess. The history of shell collecting, and the wider subject of man's long association with shells, is of interest to most serious collectors, and there are several books which deal with these subjects in fascinating detail. It is this sort of knowledge which can provide a new dimension to collecting shells, and I have dealt briefly with some of the less familiar aspects.

It has been suggested that I should close this preface with a very brief account of my own association with shells, and I lack the modesty to spurn the chance. The very first occasion I can remember taking note of a shell was on my fourth birthday. I sat on a Grenadian beach and swallowed a worn *Thais rustica*. My family profess to have seen something symbolic in the act; my own recollection is that I had finished my cake and was still hungry.

Thereafter, I collected shells. I can remember some twelve years later, having amassed about 100 West Indian species I decided that there were no new shells to find. I recall as a schoolboy, collecting in the Grenadines – that chain of exquisite volcanic and coral islets which link Grenada with St Vincent – and finding a beach littered with species whose rarity and value I discovered only years later.

In 1957 I was 20 years old and resplendent in the uniform of an Officer in Her Majesty's Colonial Police Service. I stood in a market place in Uganda, on the shores of Victoria-Nyanza and at the source of the Nile. The nearest sea was 800 miles away, across another country. In a wooden tray, carelessly mixed with cheap combs and mirrors, was a Golden Cowry (*Cypraea aurantium*). I bought it for the price of two mirrors.

But the 'serious' phase of my collecting has been in Jamaica. I doubt whether there is a single strip of Jamaican coastline, accessible by road, which I have not visited. I know the mangrove swamps and the mud flats as well as the coral reefs and the incomparable beaches. I know all the unbelievable colours of the Caribbean Sea, so often reflected as a kind of permanent record in the shells it sustains.

I can neither define nor weigh the satisfaction and pleasure that my association with shells has brought. I am left with a vague feeling that I have taken much and contributed nothing. Perhaps this book will help settle the debt.

I wish to thank my friends Norman Copland, Ian Sutton-Brown and Ez Hamid for sharing with me their knowledge of West Indian marine molluscs. My constant stream of questions, spread over a period of several years, about depth, environment and incidence must often have seemed like some form of official interrogation. I hope that they will feel that their unfailing helpfulness has been put to a useful purpose.

To Sir Maurice Yonge I owe an immeasurable debt of gratitude, not only for his kindness in making the contacts which led to the publication of this book, but particularly for the encouragement he gave me at a time when I was far from certain that I had produced anything worth publishing.

I am most grateful also, to Dr Tucker Abbott for his valuable assistance in identifying a number of rare West Indian species. I very much doubt whether I would have been able to attempt this book had I not been able to refer to his own publications on American and Caribbean shells. Their influence on my own work will be obvious to anyone who has read them.

The photographs in this book were taken by Messrs. Brian Motta and Neville Hylton. I am most grateful to them for their expert work.

I also wish to thank my publishers for their help and advice, and especially for the encouragement they gave me by their evident enthusiasm for the book. May all writers have such considerate publishers.

Finally, and most of all, I thank my wife who not only put up with my constant withdrawals to the typewriter with patience and understanding, but also produced all the line drawings which appear in the book.

Kingston, J.M.H.
1974.

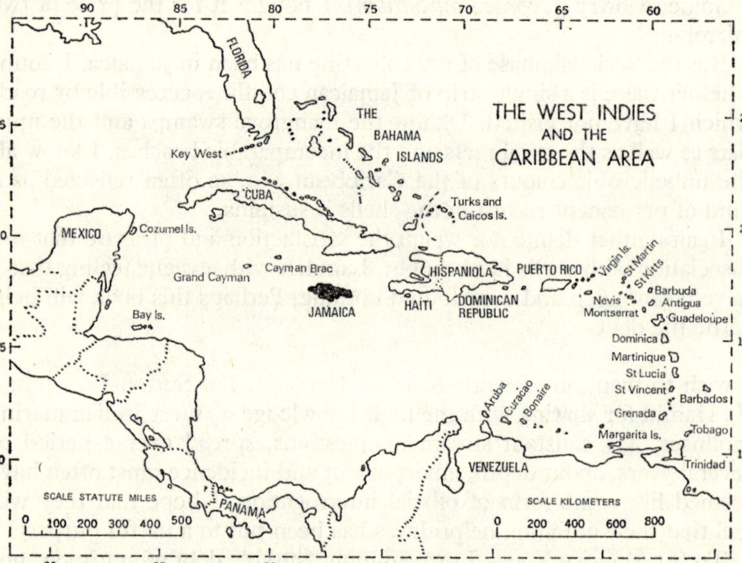

FIG. 1. Map of the West Indies

The Greater Antilles are Cuba, Jamaica, Hispaniola and Puerto Rico. The Leeward Islands extend from Barbuda south to Dominica. The Windward Islands extend from Martinique south to Grenada (excluding Barbados)

PART I

Introduction to Molluscs

The Classes of Shells

All shells belong to that ancient phylum of invertebrate animals known as molluscs. By any standards, they comprise one of the most remarkable groups of living creatures. There are more than 90,000 identified molluscan species, and they come in every conceivable shape, size and colour. They crawl, leap, swim, burrow and occasionally take to the air. Numerically they stand second only to the Arthropods; but in the scope of their diversity of form and habit they are unrivalled by any other animal group.

Molluscs have penetrated every habitat. Their ancestral forms developed in the sea, but in the long process of evolution they have adapted themselves successfully to life in fresh water and on land. They thrive in deserts and forests, lakes and rivers; they are found on the tops of mountains and in the perpetual night of the abyssal sea floor.

There is no standard molluscan shape, but their diversity of form is imposed on a uniform and unmistakable pattern. In spite of their obvious external differences, the basic structure of the oyster and the octopus is very similar.

The evolution of the phylum has produced six distinct major classes, and every known mollusc belongs to one of them. They are:

(i) The Snails (*Gastropoda*), with about 80,000 species in the sea, in fresh water and on land,
(ii) The Bivalves (*Pelecypoda*), with about 15,000 species in the sea and in fresh water,
(iii) The Squids, Octopuses and Nautiloids (*Cephalopoda*), with about 800 marine species,
(iv) The Chitons (*Amphineura*), with about 600 marine species,
(v) The Tusk Shells (*Scaphopoda*), with about 200 marine species,
(vi) The Gastroverms (*Monoplacophora*), with 5 marine species.

Until recently, the *Monoplacophora*, which are blind, limpet-like, dubiously segmented molluscs, were known only from fossil specimens and were thought to have been extinct for many million years. The discovery of a living species in 1952 off the west coast of Mexico excited the interest of malacologists all over the world, and upset many widely held theories on the form of the archetypal mollusc. They have not yet been found in the Caribbean Sea, however, and do not really fall within the scope of this book.

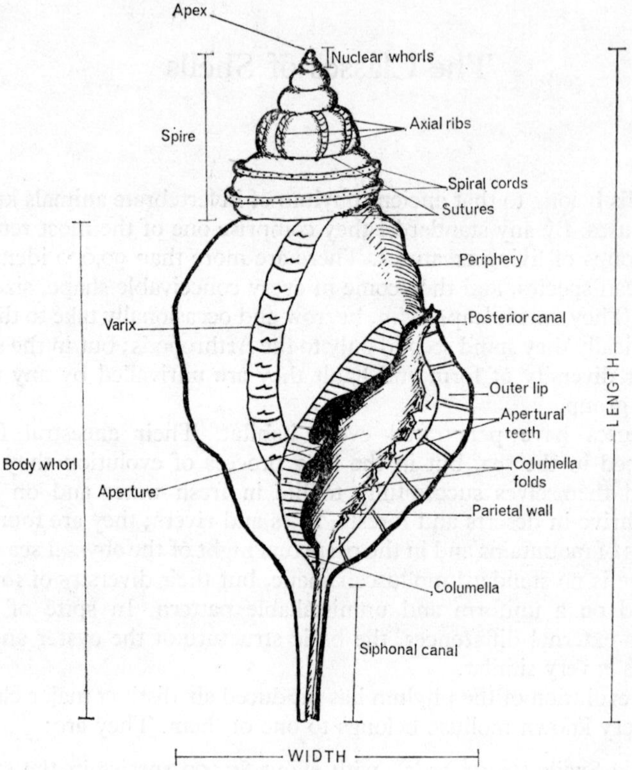

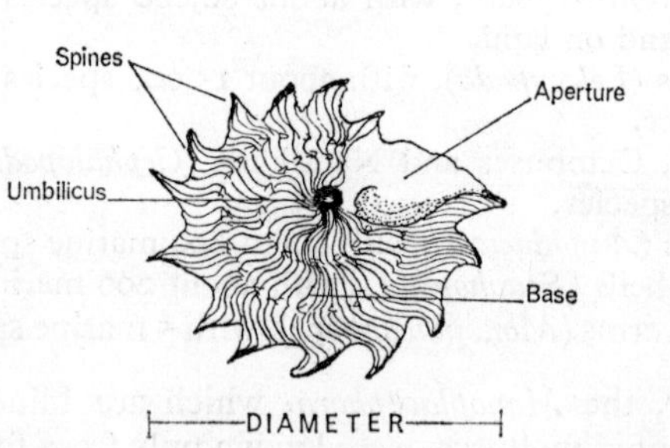

FIG. 2. Parts of the gastropod shell. *Above*, side view. *Below*, ventral view. *Right*, various gastropod opercula: (a) Paucispiral and calcareous (*Nerita peloronta*), (b) Multispiral and corneous (*Cittarium pica*), (c) Paucispiral and corneous (*Littorina nebulosa*). (d) Sickle-shaped and corneous (*Strombus gigas* – reverse side), (e) Concentric and corneous (*Cassis tuberosa*)

The other five classes are all represented in the Caribbean area, and together they comprise the approximately 1,100 marine species known to occur in the West Indies. The following sections of this chapter provide a brief synopsis of the main features of the shells of each class.

Snails *Gastropoda*

The members of this class, with a few notable exceptions, have single spiral shells. Essentially, a gastropod's shell is a geometrically increasing cone coiled on its own axis, but the most beautiful and diverse shapes have evolved from this principle. Gastropod shells range in size from minute snails of less than one twenty-fifth of an inch, to giants two feet long, and they have exploited every conceivable combination of colours. A few genera have abandoned the spiral form in their adult stage, and the common Limpets (*Fissurellidae*) are familiar examples. The main features of a gastropod shell are shown in fig. 2.

About 75% of the known West Indian marine shells belong to this class, and nearly all are 'right handed', i.e. coil to the right when viewed with the apex nearest the eye. Occasionally one finds freak 'left handed' examples of normally 'right handed' shells, but these are rare and consequently highly prized by collectors. (It has been estimated that one of every 5,000 shells of the Queen Conch (*Strombus gigas*) is 'left handed'.)

The exterior surface of many gastropod shells is covered by a tough

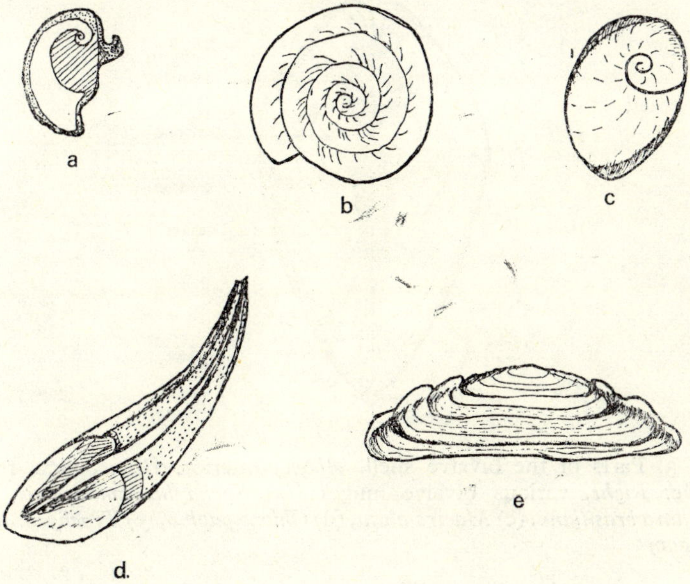

leathery material known as the periostracum, which is in fact an integral part of the shell itself. The shell's underlying colour pattern is not generally visible until the periostracum is removed.

Many gastropods, like the Turban Shells (*Turbinidae*), are equipped

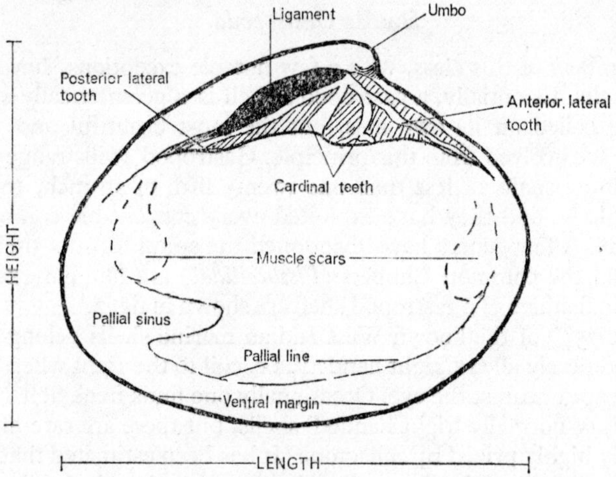

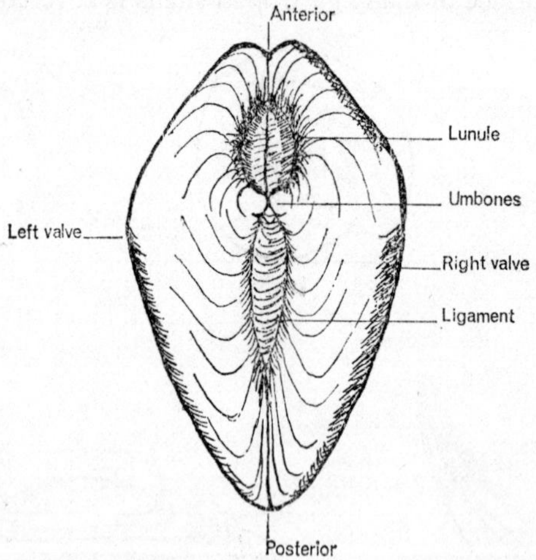

FIG. 3. Parts of the bivalve shell. *Above,* interior. *Below,* view from above. *Right,* various bivalve hinges: (a) *Spondylus americanus,* (b) *Anadara brasiliana,* (c) *Mactra alata,* (d) *Chione paphia,* (e) *Trachycardium magnum*

with an operculum, which is a horny or calcareous appendage that serves as a kind of trap door with which the animal can seal the entrance to the shell when it retracts inside. In some genera, like the familiar Conchs (*Strombidae*), the operculum has become modified primarily for use in locomotion, and is of limited help in closing the aperture. In all cases, the operculum is attached to the animal's foot, and is the last part retracted into the shell. Various operculums are shown in fig. 2.

A few advanced marine genera have lost their external shells entirely during the long period of evolution. In the West Indies, these are typified by the superfamily *Aplysiacea*, which includes the common Sea Hare (*Aplysia protea*), a mollusc which shares with most members of the Squid class the ability to emit a cloud of opaque ink when disturbed.

Shell-less molluscs are not usually of interest to amateur collectors, and the beautiful animals are virtually impossible to preserve attractively.

Bivalves *Pelecypoda*

The shells of this class are composed of paired, and usually inequilateral, valves which are joined by a hinge at one end and closed voluntarily by the mollusc. The form of the hinge varies significantly from family to family, and is a very useful aid to identification. In most species the elastic ligament which secures the hinge is quickly broken when the animal dies, and the beachcombing collector is usually presented with only one half of the complete shell. The main features of a bivalve shell are shown in fig. 3.

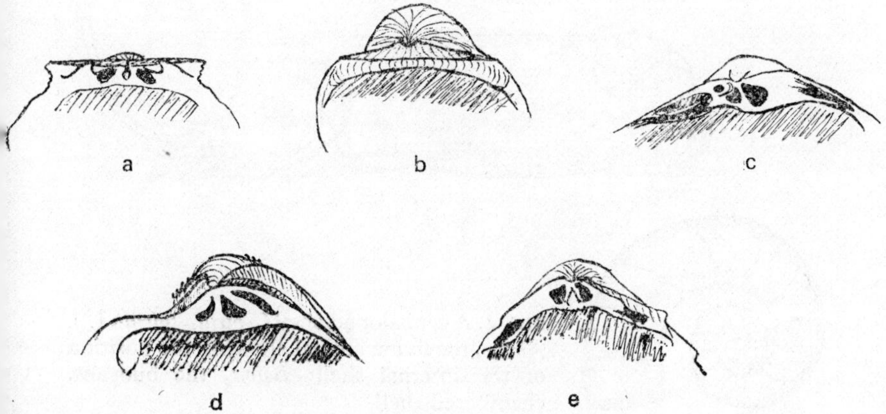

There is very marked diversity of colour, shape and sculpture among marine bivalves, but the basic double valve form remains constant, for unlike the snails, bivalves have concentrated on relatively few structural patterns. All species produce a periostracum.

In the Pacific, the legendary Giant Clam (*Tridacna gigas*) attains a length of four feet and a weight of five hundred pounds. In the West Indies, where this particular genus is not represented, the largest bivalve measures about six inches, and at the other extreme, few are smaller than half an inch. Bivalves account for about 20% of all West Indian marine molluscs, and include some of the most delicately beautiful shells found in these waters.

Squids, Octopuses and Nautiloids *Cephalopoda*

Most members of this highly specialised class have lost their shells in the course of evolution, and the Common Octopus (*Octopus vulgaris*) is an example familiar to most West Indians. A few families have retained a vestigial shell internally, and the white, porous 'Cuttlebone' occasionally found on Caribbean beaches and commonly used in parrot cages, is in fact a functional remnant of the shell of a Cuttlefish (*Sepia*). Only the primitive Nautiloids of the Indo-Pacific Ocean have retained an external shell, which is in the form of a plane spiral with the interior divided by septa into gas filled chambers.

The one cephalopod shell commonly found on West Indian beaches is the Common Spirula (*Spirula spirula*), which is the internal shell of a small squid-like animal which lives at great depths far out in the

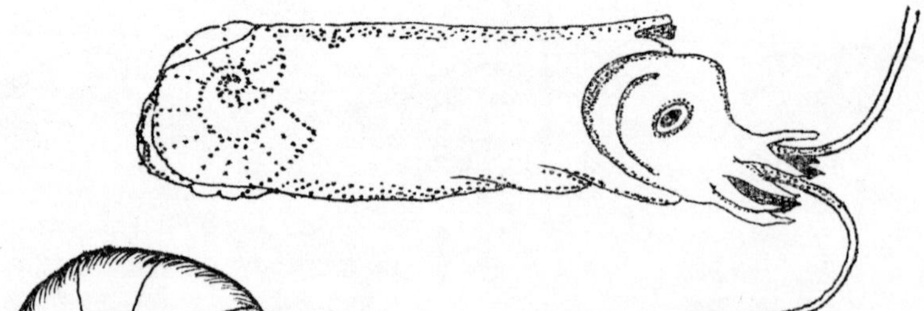

FIG. 4. A cephalopod shell (*Spirula spirula* L.). *Above,* the living animal showing the position of the internal shell; *Below,* the buoyant, chambered shell

Atlantic Ocean (see fig. 4). When the mollusc dies and decomposes, the delicate chambered shell rises to the surface and is frequently washed ashore.

The only other shell of this class normally found in the Antilles is the rare and beautiful octopod, Paper Nautilus (*Argonauta argo*), which is not a shell in the usual sense, but in fact is the delicate egg case of the female Argonaut. The very closely related Brown Paper Nautilus (*Argonauta hians*) also occurs in the Atlantic Ocean, but apparently it has rarely been recorded in West Indian waters.

Chitons *Amphineura*

The class *Amphineura* consists of two very distinct groups. One is the *Polyplacophora*, which includes the familiar chitons; the other is the *Aplacophora*, a specialised group of aberrant, worm-like molluscs without shells. The latter does not fall within the scope of this book.

A chiton's shell consists of eight interlocking V-shaped plates bound together by a leathery peripheral girdle. There is very little variation in

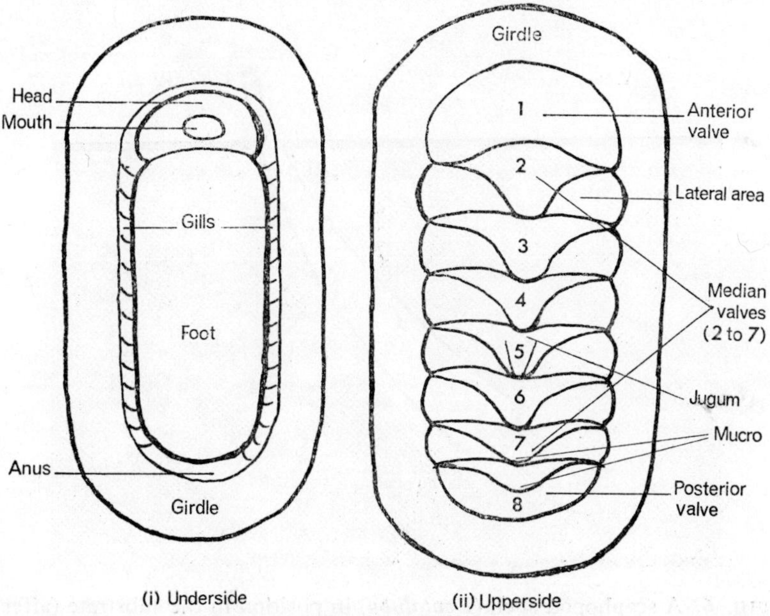

FIG. 5. Parts of the chiton shell

basic shape among the genera which range in size from about $\frac{1}{3}$ to 10 inches, and they are quite distinct from other molluscs. Most West Indian species occur in the inter-tidal zone, and they are relatively easy to collect. The main features of a chiton's shell are shown in fig. 5.

When the animal dies and the girdle decomposes, the light shelly plates separate and may wash ashore. In their individual state they bear no resemblance to a complete chiton's shell, and they are frequently sold to gullible visitors as 'sharks' teeth'.

With few exceptions, chitons have made modest use of colour and sculpture. The shells of most species are in combinations of green and brown; a few show traces of bright pink. The colour of the peripheral girdle does not necessarily match that of the valves and in some species the girdle is decorated with tufts of bristle, and may even envelop the entire shell.

Tusk Shells *Scaphopoda*

This is a small, distinctive class, the shells of which are shaped like miniature elephant tusks, open at both ends. The majority of these animals live at considerable depths, but a few small, white and rather

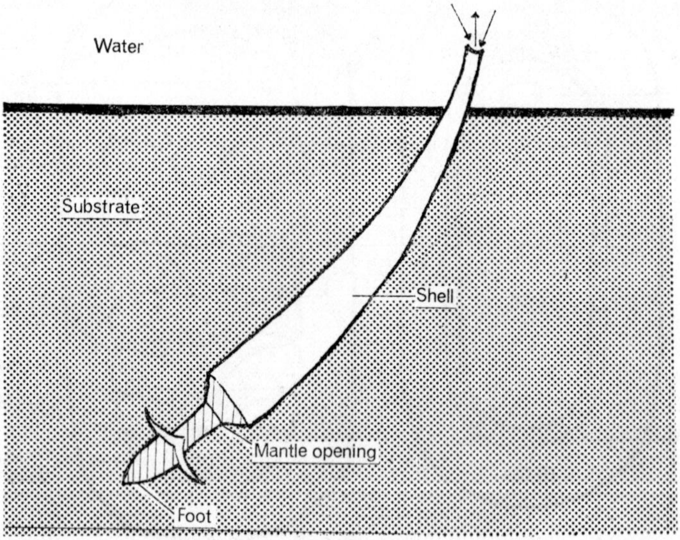

FIG. 6. A scaphopod shell (*Dentalium*) in position in the substrate (after Morton)

plebeian species occur in shallow water in the West Indies. The largest measures just over two inches. Fig. 6 shows a shell in position in the substrate.

Tusk Shells are perhaps the least complex of all molluscs, and their simple shells, with the characteristic apical notch in *Dentalium*, show very little variation in basic shape.

The Living Animals

Fossil remains from the Cambrian period – about five hundred million years ago – clearly indicate that molluscs were a well developed and flourishing group of animals even at that distant time. Their ancestral forms lived at periods probably far more remote. They belong, therefore, to one of the oldest groups of animal life. (If one assumes that the first mollusc appeared on earth 24 hours ago, then man arrived within the last 90 seconds.)

Their evolution has been incredibly slow. At Bowden, in the eastern end of Jamaica, there are several extensive beds of well preserved fossil shells. One can recognise there species apparently identical with those which occur to-day in the sea just fifty yards away. Only the most detailed comparison of the fossil and living shells will reveal the morphological differences between them. Yet they are believed to be separated in time by a gulf of sixty million years.

In an evolutionary sense, molluscs are plastic material, and their shells have slowly evolved into an amazing variety of shapes and sizes, and in some cases disappeared altogether. In spite of this, the basic structure of the animals themselves has remained constant, and the modifications characteristic of each class are far more superficial than they appear.

Essentially a typical mollusc's body consists of two regions. The first is anterior, muscular, and responsible for locomotion, the capture of food, and retraction into the shell. It includes the animal's foot, and also its head in those classes equipped with one. (Snails, chitons and cephalopods have a distinct head, tusk shells an indistinct one, and bivalves none at all.) In general, it is the part of the animal which can be thrust out of the shell. The second region is posterior, and forms a soft visceral mass which is entirely non-muscular. It is responsible for such basic functions as respiration, reproduction, the digestion of food, and the circulation of the blood. (The most obvious variants from this general pattern are the cephalopods. Most cephalopods have lost their shells, and the edge of the foot has been modified into a series of prehensile tentacles.)

Enclosing the visceral mass is the mantle. This membrane envelops all the soft parts like a loose skin, and is responsible for the secretion of the shell, and in some cases much of the respiratory activity. It is capable of extraordinary modification, and can be adapted to serve a variety of functions in the same animal. The possession of a mantle (particularly the presence of the gills in the mantle cavity between the membrane and the visceral mass) and a flexible foot, are two of the most important characteristics which link all members of the molluscan phylum.

Molluscs as a group cover a wide span in the development of their nervous systems. In the case of the most advanced examples, the cephalopods, the brain and sense organs are nearly as complex as those of the lower vertebrates; at the other extreme, the chitons possess a nervous system as primitive as that of the flatworms.

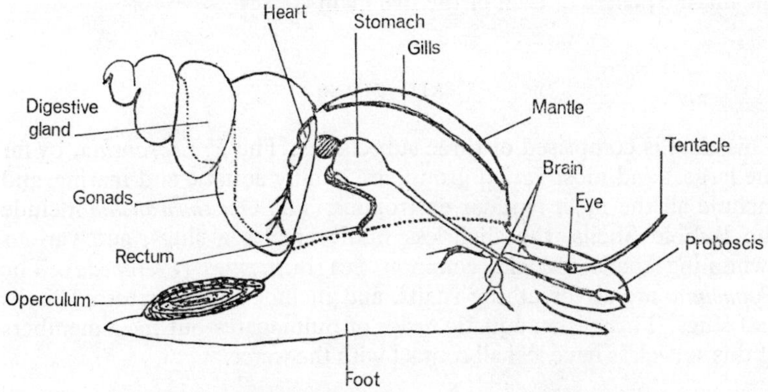

FIG. 7. Diagram of a gastropod animal (after Zim)

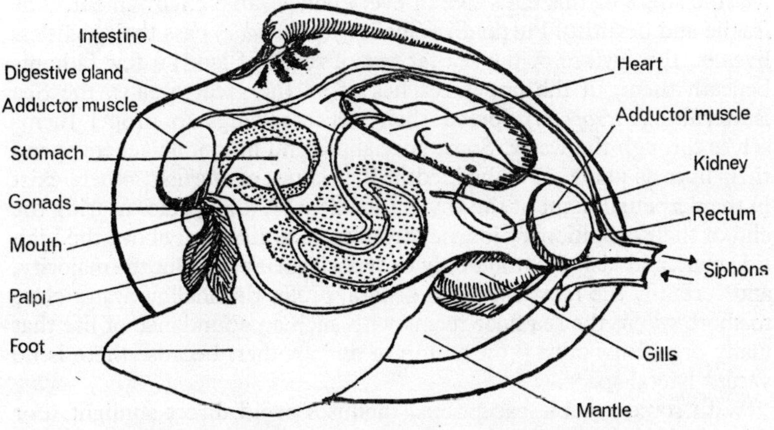

FIG. 8. Diagram of a bivalve animal (after Zim)

Most gastropods and cephalopods are equipped with a pair of eyes, and in the squids and octopuses these are highly developed and very efficient. Tusk shells, nearly all chitons, and most bivalves are sightless, a condition directly related to the way in which they obtain their food. Many molluscs have a keen sense of 'smell', and the cephalopods in particular possess a well developed and easily demonstrated tactile sense. Diagrams of gastropod and bivalve animals are shown in figs. 7 and 8.

The habits of living molluscs are of special interest because they show just how diverse can be the behaviour patterns of animals within the same phylum. The following sections deal briefly with the behaviour of the marine genera of each of the five main classes.

SNAIL CLASS

This class is comprised of three sub-classes. The *Prosobranchia*, by far the largest and most varied group, are mainly aquatic and marine, and include all the most familiar gastropods. The *Opisthobranchia* include the Bubble Shells, the shell-less nudibranch sea slugs, and various swimming snails like the common Sea Butterflies (*Pteropoda*). The *Pulmonata* are air-breathing snails, and include most terrestrial shells and slugs. There is an aquatic order of pulmonates but most members of this sub-class have lost all contact with the water.

(i) *Environment*

Marine shells of this class live in every conceivable environment. The fragile and beautiful Purple Sea Snails (*Janthinidae*) pass their sightless lives on the surface of the sea far out of reach of land; a few fathoms beneath them, in the upper reaches of all the great oceans, the Sea Butterflies (*Pteropoda*) spread their delicate wings to propel themselves through the water. Some specialised and immobile genera spend their lives as parasites embedded in the tissues of starfish; others exist in the perpetual night of the abyssal sea floor. A few species nearing the end of their evolutionary journey from the sea, live well above the high tide mark and visit the water only to deposit their eggs. But the majority, and certainly the most colourful genera, prefer the shallow water close to shore where the sea floor teems with such an abundance of life that many organisms must grow on top of one another, because there is no vacant lateral space.

With some notable exceptions, molluscs avoid direct sunlight. For this reason, most shallow water gastropods conceal themselves during the day. They lie hidden under rocks and slabs of broken coral, partially buried in the sand and mud of the sea floor, or wedged into crevices in the reefs, waiting for darkness to invade their world and spur them to the task of hunting food

Most species confine themselves exclusively to a particular environment, and one learns by experience which genera will be found only in the mud of the mangrove swamps, and which hide in the coral of the open sea. A few active and hardy species confound the general rule by occurring in every imaginable environment, and the ubiquitous Apple Murex (*Murex pomum*) is a well known Caribbean example.

Nearly all gastropods (and chitons) move by means of a continuous succession of muscular contractions which pass over the sole of the foot from the posterior end. These waves of movement are clearly visible through the glass of an aquarium tank when a snail obligingly scales the side.

(ii) *Feeding habits*

Gastropods are nearly as diverse in their manner of feeding as in the shape of their shells. Some are ferociously carnivorous and will often attack their own kind; others are harmless vegetarians and eat nothing but algae; the remainder are omnivorous, and will indiscriminately eat practically anything. Contrary to popular opinion, size is not an important factor in this matter. The powerful Queen Conch (*Strombus gigas*), which may weigh more than five pounds, grazes peacefully on the algae which grow on the eel grass blades, while some voracious species of Murex less than one inch long, prey on bivalves two or three times their own size.

The mouth of most snails, often borne on a retractable snout, contains a flexible tongue-like ribbon set with rows of teeth. (See fig. 9.) This is the radula, and the mollusc eats by rasping it against the flesh or vegetable

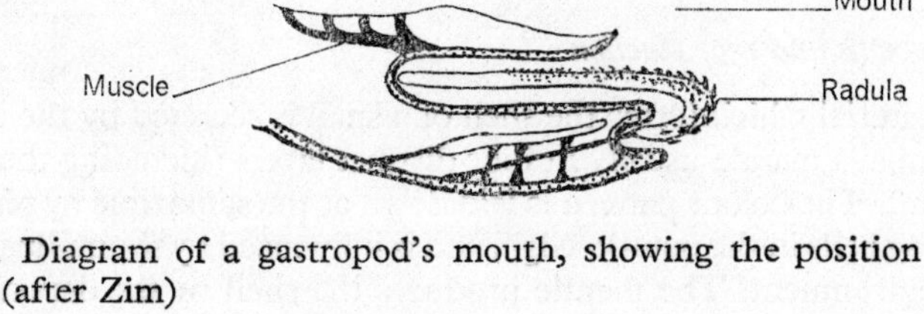

FIG. 9. Diagram of a gastropod's mouth, showing the position of the radula (after Zim)

matter it has selected for its meal. The teeth are constantly replaced as they wear out. Many carnivorous gastropods can bore a hole in the shell of their victims by ingenious use of the radula, in conjunction with the simultaneous secretion of an acid solution, to penetrate the victim's shell in the selected spot. The bivalves one finds on the beaches conveniently equipped with small holes for stringing on a necklace, were all the succulent meals of predatory gastropods.

Several families of this class have developed an interesting and dangerous modification of the standard radula pattern. One of these is the aristocratic family of Cone Shells (*Conidae*), a beautiful and much coveted group. In this family the radula is modified in such a way that the teeth are miniature harpoons which can be ejected from the snail's snout and jabbed into the victim's flesh. This is accompanied by the simultaneous ejection of a highly toxic liquid, in much the same manner as the bite of a poisonous snake. In the Pacific, there are three or four species of large Cone Shell whose sting can be fatal to a man. No West Indian Cone is known to be dangerous, but the very uncommon Atlantic Agate Cone (*Conus ranunculus*) attains a length of four inches, and a sting by a Cone of this size is a painful experience. It is as well to remember that most, if not all, *Conidae* have the same venomous apparatus, and deserve to be treated with caution while they remain alive.

(iii) *Respiration*

Gastropods breathe by passing a current of water over their gills, which extract the oxygen from the water and dissolve the waste gases. In a very few species respiration takes place in the skin of the mantle. Most snails are equipped with a trunk-like siphon which projects from the anterior end of the shell, and serves to conduct the water to the gills. In the species which burrow under the substrate, the siphon is thrust up through the sand to supply the animal with a constant flow of water.

A few advanced amphibious genera (*Basommatophora*) are pulmonates, and possess a rudimentary lung which may operate both in andout of the water.

(iv) *Growth and reproduction*

The material which forms the shell of a snail is secreted by the edge of the animal's mantle usually in four distinct layers (including the periostracum). The colour pattern is laid down at the same time by pigment-secreting cells in the mantle-edge, and is influenced by the mollusc's diet and environment. The mantle produces the shell by the deposition of calcium carbonate in an organic substance known as conchiolin. The secretion hardens immediately it is laid and comes in contact with the water. Growth usually takes place in stages, often interspersed with long resting periods, and the lines of growth are clearly visible in most gastropods. In many genera, the 'varices', or built up edges of each growth stage, remain in place permanently, and in the Murex shells for example, the varices are the spiny protuberances which appear at regular intervals on the whorls. The mantle is also capable of repairing damage incurred by the lip of the shell, and with marvellous dexterity, it will even recreate much of the original colour pattern on the repaired section.

The process of reproduction among gastropods is both varied and complicated. Most are of separate sexes, but many marine genera possess male and female organs simultaneously, and in some families like the Slipper Shells (*Crepidula*), each individual begins life as a male and then develops female characteristics when it is fully grown.

After the liberation and meeting of sperm and egg (whether from the same or different individuals), and the consequent fertilisation, most female marine gastropods place the fertilised eggs in specially prepared capsules, and either release them to float freely or attach them to some solid object. In many cases the individual protective capsules are joined to form long, narrow ribbons. In a very few families, which include the Cowries, the female protects the eggs.

In the more primitive species, the eggs usually hatch free-swimming larval trochophores, which develop after a short period into second-stage veligers. Both trochophores and veligers become members of the sea's plankton, and those veligers which survive this precarious existence eventually sink to the bottom and are rapidly transformed into minute replicas of their parents. In the more advanced gastropods, the free-swimming trochophore stage is omitted, and the eggs hatch veligers; in a few genera, the entire larval stage is passed within the egg capsules, which finally break open to release perfectly developed young snails. (Gastropods provide no after-care for their progeny, but it is interesting to note that at certain times of the year, about half the adult *Murex brevifrons* found in Kingston harbour have 'babies' clinging to their shells.)

All larval gastropods undergo a twisting process known as 'torsion', which brings the mantle cavity to the front of the body and displaces the visceral organs through 180 degrees. Torsion is quite separate from spiral coiling and its value to the mollusc remains a matter of speculation.

BIVALVE CLASS

This class is comprised of only two sub-classes. The *Protobranchia* is a small marine group of three primitive but highly diversified families. All other bivalves are *Lamellibranchia*, and there are many fresh water species.

(i) *Environment*

Marine bivalves are commonly dismissed as prosaic creatures cemented to the sea floor and so condemned to a sedentary life. While this is true of some genera, the class as a whole is surprisingly active, and lives in widely diverse environments which extend from the intertidal area to the substrate six miles deep. Some bivalves are accomplished, if clumsy, swimmers, and one family (*Tridacnidae*) is believed to 'farm' its own food in the folds of its mantle.

Most species use their distensible, muscular foot to bury themselves just below the sea floor in muddy or sandy substrate, with their siphons extended to the surface to supply them with a constant flow of water (see fig. 10). These are all capable of movement. Others, like the Mussels and Pearl Oysters (*Mytilacea* and *Pteriacea*), which ensure the survival of their species by sheer numbers, usually attach themselves by a fibrous byssus to rocks and wharf piles, but even they are capable of slow movement by manipulation of their byssus. A few specialised groups, like the Date Mussels (*Lithophaga*), drill their way into coral or limestone, and the notorious Shipworms (*Teredo* and *Bankia*) burrow deep into wooden pilings and the planks of ships. In the days of sail, before the advent of steel hulls, the activities of Shipworms created havoc among wooden vessels, and Admiral Nelson, who was once stationed in Jamaica, considered them a greater threat to his ships than the French.

The Scallops (*Pectinidae*) – perhaps the most familiar of all molluscan groups – 'swim' in a clumsy but effective fashion by rhythmically opening and closing their valves, which produces a kind of water jet-propulsion. The scallop's mantle usually directs the jet of water to the rear of the hinge, so that the mollusc appears to be moving in reverse, but *Pectinidae* can travel surprisingly quickly over short distances, as many disappointed collectors will confirm. In the West Indies, many of the common scallops live in shallow water on the eel grass beds, and at a diver's approach they will flutter off in different directions like a cloud of disturbed butterflies.

A bivalve's large adductor muscles, which are attached to the inner surface of each valve and contract to close the shell, are unusually

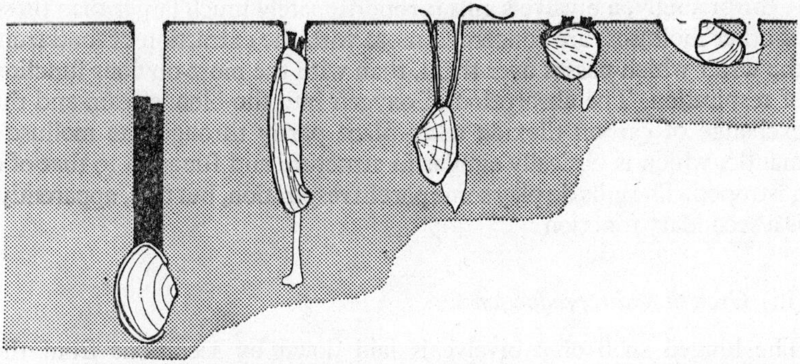

FIG. 10. Bivalves in position in the substrate. From the left, *Mya, Ensis, Tellina, Cardium, Nucula* (after Morton)

powerful in the larger species. An adult Thorny Oyster (*Spondylus americanus*), a truly sessile bivalve which is cemented to some firm surface, can exert painful pressure on a finger caught between its valves, and some of the stories of Pacific divers being trapped and drowned by the Giant Clam (*Tridacna gigas*) may well be true. (An experiment conducted on the Great Barrier reef some years ago proved that the plastic leg of a tailor's dummy could be released from the grip of a Giant Clam only by destroying the leg.) The closing of the shell when a foreign object is thrust inside is, of course, a reflex defence mechanism. Bivalves do not bite people.

(ii) *Feeding habits and respiration*

Most bivalves live on minute organisms which occur in suspension in the sea. The water containing this food is taken into the mollusc's body, generally through the siphon, which is a flexible, tubular appendage leading to the gills, or through a slit in the mantle lobes in those species without siphons. The particles of food are removed from the water by the gills, and are then passed by a mucus secretion to the mouth which, in this class, lacks a radula. The water is voided through the exhalant siphon or slit. A few bivalves use their siphons to suck up food from the sea floor, and display a certain selectivity about what they eat. These are classified as 'deposit feeders'.

The length of a bivalve's siphons usually indicates the depth at which it normally lives within the substrate. This is also reflected in the size of the pallial sinus on the interior of the valves – a fact which helps to determine the original habits of fossil bivalves.

Surprisingly, a bivalve's gills (proportionately much larger than those of other molluscs) are concerned more with the extraction of food from the water which passes over them, than with the more obvious function of respiration. The absorption of oxygen into the bloodstream and the exchange of carbon dioxide takes place partly through the mollusc's mantle, which is basically similar in structure and function to that of a gastropod. The gills do play some part in respiration, but this, apparently, is a secondary function.

(iii) *Growth and reproduction*

The hinged shell of a bivalve is laid down by secretions from the mantle-edge in very much the same way as that of a gastropod (see fig. 11). The lines of growth are usually visible and play a large part in

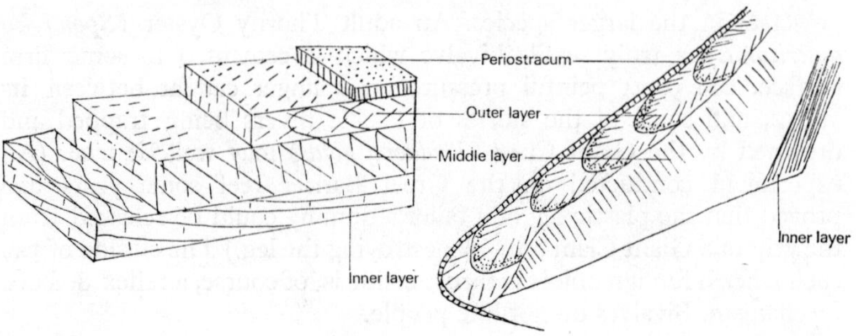

FIG. 11. Structure of a bivalve shell (*Tellina*). *Left*, diagrammatic form; *Right*, cross section of *Tellina tenuis* showing the protruding end of the periostracum (after Truman)

the sculpture of the shell's surface. The elastic ligament, which keeps the valves together, is an exclusively bivalvian product and is not represented in any form in the other molluscan classes.

Sexually, the bivalves share the strange habits of the gastropods, but their reproductive system, by contrast, is relatively simple. The majority are of separate sexes, but a few species are hermaphroditic, and some – in the same way as the *Crepidula* snails – begin life as males and ultimately become females. One group of Oysters is even more enterprising, and undergoes a constant series of sex changes, sometimes starting the year as males, spending the summer as females, and reverting to male status in time for Christmas.

With few exceptions, eggs and sperm are shed directly into the water where they meet and fertilisation takes place. Bivalve eggs, like those of the more primitive gastropod families, hatch trochophores which develop into veligers, and then settle to the bottom to become minute replicas of the shells which produced them.

SQUID CLASS

There are only two living sub-classes of *Cephalopoda.* They are the *Nautiloidea*, which includes the few surviving species of the primitive Nautilus, and the *Coleoidea* which is comprised of the squids, cuttlefish and octopuses. The cephalopods, by any standards, are the most advanced molluscan class, and the intelligence of the octopus, in particular, is superior to that of many of the lower vertebrates. All cephalopods can swim, and a few genera take to the air like flying fish – a facility which astonished the members of the famous Kon-Tiki expedition in the Pacific some years ago. Cephalopods are the most delicately specialised of all molluscs.

The fact that these clever, active animals are closely related to such primitive, sedentary creatures as the Oysters is not easy to accept. Nevertheless, the two share a common ancestry, and certain basic anatomical characteristics, which place them unmistakably in the same phylum. Several species of octopus occur in West Indian waters, but contrary to the well embroidered legend, the largest of these rarely exceeds a length of three feet in radial spread, and none presents a danger to man.

Octopuses are characterised by the modification of the molluscan foot into eight tentacles, equipped with rows of powerful suckers. As a group, they are shy and retiring, and usually live in holes in the reefs or on the sea floor. These 'nests' are easily identified by the piles of discarded shells which surround them. The octopus normally travels along the bottom on its tentacles, but when alarmed it can propel itself backwards at considerable speed by ejecting a stream of water from its siphon. The animals are carnivorous, and subsist chiefly on bivalves and small crustacea. Like most cephalopods, they possess both a formidable horny beak and the characteristic molluscan radula.

Squids and cuttlefish differ from octopuses in several ways, mainly in the possession of two long tentacular arms as well as eight shorter ones. Squid are not bottom dwellers, and spend most of their lives suspended in the water. Like octopuses, both make use of 'jet-propulsion' to travel over short distances with astonishing speed, and employ this facility to capture fish. They are masters of the art of colour change and camouflage.

The legendary Giant Squid (*Architeuthis harveyi*), which occurs off the eastern seaboard of the United States and Canada, attains a length of fifty-five feet. These enormous molluscs are known to give battle to whales, and no doubt inspired many of the sea monster fables of the days of sail.

Many genera of *Cephalopoda* live in the abyssal depths of the sea, from where they are occasionally brought to the surface in the dredges of scientific research vessels. With minor exceptions, the sexes of this class are separate. Unlike most other molluscs, there is no free-swimming larval stage and the embryo emerges fully developed from the egg.

The ancestors of the present cephalopods possessed large and remarkably beautiful shells. Most of the genera living to-day, like the octopuses and squids, have lost theirs in the long process of evolution, but it is interesting to note that the process is still incomplete in the squid and cuttlefish, which retain a small internal remnant.

Fotunately, three primitive members of the class still produce an external shell, and the largest, the beautiful Pearly Nautilus (*Nautilus pompilius*), is well known and admired all over the world. Unfortunately for West Indians, this species occurs only in the Pacific Ocean. Nautiloids are one of the few cephalopod groups which do not possess the characteristic sack of black ink.

As mentioned above, the two cephalopod shells usually found in West Indian waters (excluding the vestigial shells of the squid and cuttlefish) are the Common Spirula (*Spirula spirula*) and the exquisite Paper Nautilus (*Argonauta argo*). The Paper Nautilus, which is in fact the egg case of the pelagic female Argonaut, is formed of calcified conchiolin. This 'shell' is a recent adaptive development, and was not produced by the ancestral Argonaut. The very few collectors who have discovered a perfect specimen of the Paper Nautilus on a West Indian beach must be excused for claiming divine guidance.

CHITON CLASS

The two sub-classes which comprise the *Amphineura* have already been mentioned. They are the non-littoral, shell-less, worm-like *Aplacophora*, and the more familiar *Polyplacophora* which includes the chitons.

Generally speaking, chitons are not popular with collectors, and tend to be sketchily represented even in the more comprehensive museum collections. This is probably due to the undeniable fact that they are the least attractive of the molluscan classes, and at the same time require more effort to prepare for a collection.

Chitons are primitive members of the molluscan phylum. The animals are essentially similar to gastropods, but there are basic and obvious differences in the shells of the two classes. Chitons are the only molluscan class to possess a flexible shell. If the animal is removed from the surface on which it is resting, it will immediately curl up into a tight ball much like a disturbed armadillo.

A chiton's powerful foot can secure it so tightly to the rock surface that nothing short of the destruction of the shell will remove it. The more common West Indian species live in the inter-tidal zone, often on rocks exposed to the full blast of the open sea, and this ability to cling on under stress is therefore essential to the mollusc's survival. A number of chitons are equipped with tactile organs and primitive 'eyes', which are set in pits on the exposed surface of the shell. Unlike most gastropods, chitons do not possess tentacles or a siphon. The animals breathe by lifting a small section of the girdle, at any convenient point, to form a narrow tunnel for the intake of water. After passing over the gills, the water is voided at the posterior end.

A number of West Indian species also occur in shallow water, usually on the underside of smooth rocks. All chitons require a hard, firm surface on which to attach themselves, and consequently they are rarely found in open sandy or muddy areas. One common exception is a small species which lives on the valves of one of the big Pen Shells, which occur in large numbers in sandy places.

Most chitons are herbivorous and use their well developed radulas to graze on the algae which cover the rocks on which they live. The sexes are separate in this class, and there is a free-swimming veliger stage only in some species. In others, the immature chitons remain under the protection of their mother's shell, until they have developed sufficiently to take care of themselves.

TUSK SHELL CLASS

Most tusk shells live in deep water, but there are a few small West Indian species which occur in shallow, sandy areas, usually in protected bays or inlets. The animal lives buried obliquely in the sand with the narrow posterior end of the shell projecting just above the surface of the sea floor. The foot, which is situated at the opposite end, can be extended out of the shell to draw the mollusc deeper into the sand.

Tusk shells, which have no true gills, breathe by inhaling a current of water which passes over the transverse folds in the lining of the mantle, which in this class is modified to form a complete tube. The extraction of oxygen and exchange of carbon dioxide take place in much the same way as in the *Pelecypoda*.

The molluscs do not extract food from this current of water, but use the adhesive heads of their long tentacles, which are grouped near the mouth at the buried end, to grasp microscopic organisms from the substrate. All species possess a large radula, but lack eyes.

Scaphopods start life with an embryonic shell consisting of two minute valves. These soon fuse to form a tube, and the mantle then secretes the characteristic tubular shell as in the other molluscan classes. The sexes are separate in *Scaphopoda.*

To pioneering Americans tusk shells were familiar as the West Coast equivalent of 'wampum', one of the American-Indian media of exchange. West Indian species rarely exceed a length of two inches, however, and lack the attractive colours present in some Pacific genera. Their attachment to the sea floor is tenuous in shallow water, and after a storm the beaches may be covered with thousands of their strange little shells, torn from their anchorage by the force of the waves.

PART II

The West Indian Marine Shells

A systematic account of West Indian marine molluscs

This systematic account includes the scientific and colloquial names, detailed descriptions, and remarks on the habitat and areas of occurrence of 497 West Indian marine shells. The known geographical range of each species is given immediately below its scientific name. (The territorial limits of the Caribbean fauna are shown in fig. 12.)

Information on the occurrence of each species *in Jamaica* is provided under a separate sub-heading following the description, and it should be noted that the details of each mollusc's habitat apply generally throughout the West Indies. For collectors in Jamaica, the informa-

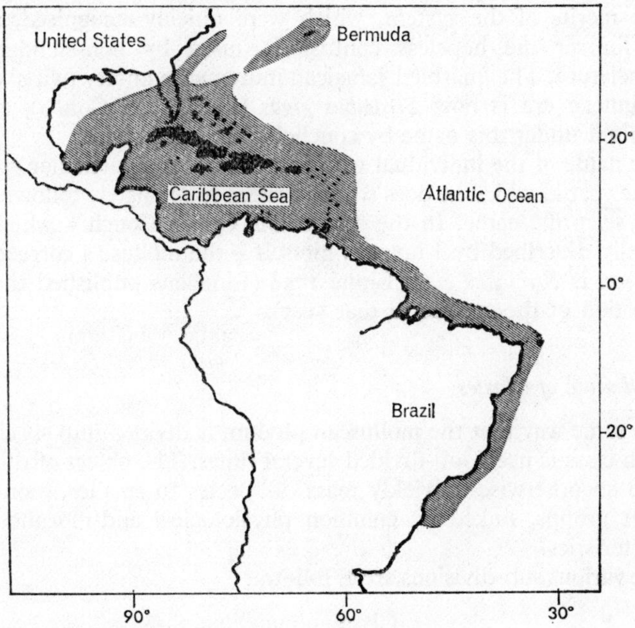

FIG. 12. The Caribbean Province. The shaded area defines the territorial limits of the Caribbean fauna

tion provides an accurate guide to at least one locality where each of the 497 species is known to occur.

At the end of the sections on each of the five classes, there are lists under family headings (and without descriptions) of 353 additional species which are known to occur in Jamaica and elsewhere in the West Indies. Nearly all of these are shells less than ½ inch in length. These lists, plus the shells fully described, provide a more or less complete census of identified Jamaican marine molluscs, and include a very large percentage of all West Indian sea shells.

Nomenclature

In 1758 the great Swedish naturalist Carl Linnaeus (later von Linné) perfected the 'bi-nomial system' for giving every known species of mollusc an individual name. In this system every species, in fact, is given two names. The first represents the mollusc's genus, and the second its particular species. The language used for the purpose was Latin, and as Linnaeus had hoped, the system gained universal acceptance and has been applied to the classification of all living things.

The merits of the system, which were quickly recognised, swept away forever the hopeless confusion caused by non-standardised nomenclature. The 'marbled Jamaican murex with knotty twirls' of the pre-Linnean era is now *Strombus gigas* (the Queen Conch), and is recognised under this name by conchologists everywhere.

The name of the individual who first describes a particular species, and the year in which he does so, are written immediately following the shell's scientific name. In the case of the Queen Conch – which was originally described by Linnaeus himself – the mollusc's correct title, therefore, is *Strombus gigas* Linné 1758 (Linnaeus published the first description of the species in that year).

Sub-Division of Classes

In the same way that the molluscan phylum is divided into six classes, so each class is itself sub-divided several times. The object of this is to reduce an otherwise unwieldy mass of species to smaller, more convenient groups, linked by common physiological and morphological characteristics.

The various sub-divisions are as follows:

Class
Order
Superfamily

Family
Genus
Species

Each of these divisions (except Superfamily), may be prefixed by the word 'sub', which has the effect of further dividing the categories.

Identification

Most shells can be correctly identified simply by comparing them with an illustration in a book such as this one, and then ascertaining that the written description fits exactly.

Unfortunately, there are some species which are not so easily disposed of. The most troublesome shells in this respect are those small white bivalves which share a very similar external shape and sculpture – and there are many of them. The points of difference in these cases are most obvious in the hinge construction and the internal muscle scars, and it is these aspects, together with the shape of the pallial sinus and pallial line, which should be examined and compared when difficulty arises.

The other common cause of confusion is provided by juvenile shells, which in many species bear no obvious resemblance to their eventual adult forms, and which are often assumed to be separate species. One of the best illustrations of this is the case of the young Conchs (*Strombidae*) which far more closely resemble Cone shells (*Conidae*) than adults of their own species – until they finally mature and produce the characteristic flaring lip. The point to remember is that once produced, the external form of the early portions of a shell cannot be altered by the mollusc (there are a few unimportant exceptions to this). When a shell is suspected of being an immature specimen of a particular species, it should be compared with the early portion of a well preserved adult of that species. If they are both in fact the same species, this comparison should quickly make it apparent.

When in spite of all efforts, doubts about identification persist, assistance can usually be obtained either from expert collectors or from Museums. There are a number of unwritten rules on seeking help in this way:

(i) Don't send shells for identification without the prior approval of the people concerned,
(ii) Don't send photographs in place of the shells themselves,
(iii) Ensure that the shells you do send are accompanied by all the relevant data available.

Localities

To avoid the constant repetition of Parish names after commonly used Jamaican localities (mentioned in connection with the distribution of species), it should be noted that the following places are in the parishes indicated. Reference to any detailed map of the Island will provide their exact locations, which are shown approximately in fig. 13.

Bull Bay – St Andrew (6 miles east of Kingston)
Cow Bay – St Thomas (just west of Yallahs)
Old Harbour Bay – St Catherine (due south of Old Harbour)
Orange Bay – Portland (3 miles east of Buff Bay)
Port Henderson – St Catherine (across the harbour from Port Royal)
Priory – St Ann (2 miles west of St Anns Bay)
Prospect – St Thomas (between Morant Bay and Port Morant)
Robins Bay – St Mary (4 miles west of Annotto Bay)
The Bogue Islands – St James (1 mile west of Montego Bay; now known as Montego Freeport)

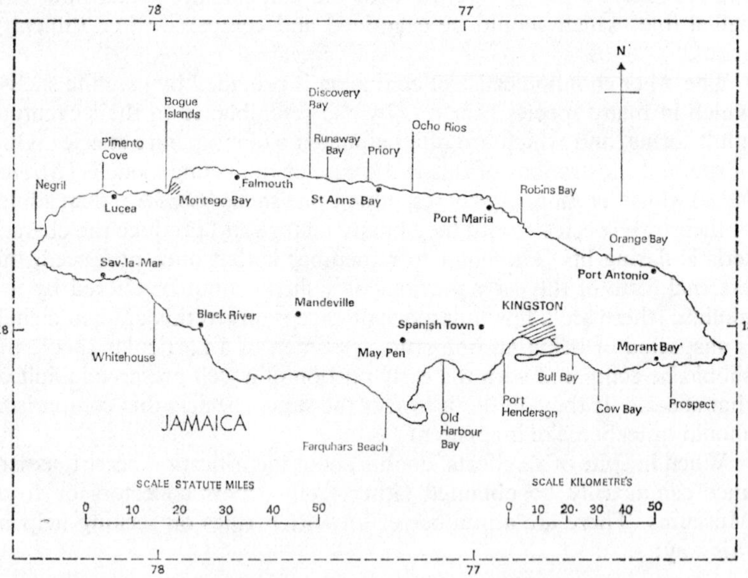

FIG. 13. Map of Jamaica showing the collecting localities most commonly referred to in the text (all of which are accessible by road)

Measurements

Measurements in this book, with occasional exceptions, are given in feet and inches. Most English speaking readers will be more familiar with these units than the metric scale. In this context, the following information may be useful:

1 inch=25.37 millimetres
1 fathom=6 feet=1.83 metres
1 (statute) mile=5,280 feet=1.61 kilometres

The measurement of any shell in this book, unless otherwise stated, is its largest dimension, and may be described as its height, length, or diameter. When very precise measurements are required, it is better to quote them in the metric scale.

Conchological Terms

To permit reasonably concise descriptions, it is necessary to use certain conchological terms. All of these are quite straightforward and can be found in any dictionary. The following glossary includes most of the commonly used terms:

Anterior – in front of; the forward end
Axial – following the direction of the axis
Body whorl – the last whorl of the shell
Calcareous – made of hard calcium carbonate
Callus – a thick deposit of smooth shell material
Cancellate – lines of sculpture, usually crossing each other
Columella – the central column around which the gastropod's shell coils
Dorsal – belonging to the back
Habitat – the place where an organism generally lives
Lamella – a thin plate or ridge
Maculated – irregularly marked
Nodulose (or Nodulous) – with small knobs
Operculum – a gastropod's 'trap door' which may close the shell when the animal retracts inside it
Parietal wall – (in gastropods) the area on the whorl near the columella and opposite the outer lip
Periphery (of a whorl) – the edge of the whorl
Plication – a fold
Posterior – opposite of anterior
Reticulate – network of crossed lines
Spiral – (in the context of a gastropod's sculpture) going in the direction of the turning of the whorls

Suture – a spiral line or groove, where one whorl comes in contact with another
Tubercles – small, raised projections
Umbilicus – the hole in the centre of some gastropod shells
Varix – (plural: varices) a thickened rib on the surface of a gastropod's shell marking the end of a particular growth stage
Ventral – opposite of dorsal
Whorl – a complete turn of a gastropod's shell

THE GASTROPODA

SUB-CLASS
PROSOBRANCHIA

Order Archaeogastropoda

SUPERFAMILY PLEUROTOMARIACEA

FAMILY PLEUROTOMARIIDAE

This is the coveted family of ancient and primitive Slit Shells, whose distinctive, top-shaped shells are characterised by a long axial slit on the periphery of the body whorl. The mollusc discharges excrement through the slit to avoid contamination of the gill plumes.

Pleurotomariidae were extremely common several million years ago, but there are now only about 15 recognised living species. These are all found in deep water in temperate and tropical seas.

All Slit Shells are thought to be vegetarian, but some species are known from less than ten specimens, and much research remains to be carried out on the entire family.

Genus *Perotrochus* P. Fischer 1885
Subgenus *Entemnotrochus* P. Fischer 1885

PEROTROCHUS ADANSONIANUS Crosse and Fischer 1861
Adanson's Pleurotomaria **pl. 1**
Cuba, Jamaica and the Lesser Antilles
Description: Diameter 3 to 7 inches. Colour cream, with large irregular patches of red or maroon; the apex and shoulders of the early whorls are bright yellow. Sculpture of coarsely beaded spiral threads. The slit on the periphery of the body whorl extends from the aperture half way around the shell. There is a large, deep umbilicus, and the operculum is thin, brown, corneous and multispiral.
Occurrence: Two live specimens of this most rare and valuable of West Indian shells were recovered off Discovery Bay by the retractable claw of a two-man submersible while this book was in proof (1972). They were collected on coral rubble in 350 and 500 feet respectively, on the steeply sloping ocean side of the reef at the mouth of the bay. These are the first Jamaican records of this species.

Note: These rather sluggish molluscs eject a dense cloud of opaque white mucus from the top of the slit when disturbed. Experiments to determine what they eat were inconclusive.

FAMILY FISSURELLIDAE

This is the family of Keyhole Limpets, whose distinctive cap-shaped shells are characterised by a small hole at the apex, or a slit on the anterior slope. There are several hundred species, with a world-wide distribution in shallow tropical and temperate seas; the largest, from California, attains a length of 5 inches.

All *Fissurellidae* are vegetarian, and most lack an operculum. They cling tenaciously to some firm surface. The shells are widely used throughout the West Indies in the manufacture of shell artifacts.

SUBFAMILY EMARGINULINAE

Genus *Emarginula* Lamarck 1801

EMARGINULA PHRIXODES Dall 1927

Ruffled Rimula **pl. 2**

Eastern U.S. and the West Indies.

Description: Length $\frac{1}{3}$ inch. Colour white. Sculpture of radial ribs and concentric cords, with a slit at the base of the anterior slope. This is a delicate and beautiful little shell.

Occurrence: Uncommon at Orange Bay and Ocho Rios in 45 feet on coral reefs, and possibly in the same environment all along the north coast. Dead shells are rarely washed ashore.

EMARGINULA PUMILA A. Adams 1851

Pigmy Emarginula **pl. 2**

Bermuda, south-east Florida to Brazil.

Description: Length $\frac{1}{3}$ inch. Colour white, occasionally yellowish. Variable in shape, with an oval or squarish base. Sculpture of radial ribs and irregular concentric cords. There is a slit at the base of the anterior slope. Flatter and thinner than *E. phrixodes*.

Occurrence: Relatively common from 6 to 500 feet near coral reefs at Negril, Discovery Bay and Orange Bay. The dead shells are often found in drift sand.

Genus *Hemitoma* Swainson 1840

Subgenus *Hemitoma* Swainson 1840

HEMITOMA OCTORADIATA Gmelin 1791
Eight-ribbed Limpet **pl. 2**
South-east Florida and the West Indies.
Description: Length about 1 inch. Colour cream or grey. Sculpture of eight nodulose ribs with smaller riblets between them. Oval base with ribs projecting beyond it; the apex is inclined backwards.
Occurrence: On and under rocks exposed to the open sea, at Prospect, Runaway Bay and in similar areas all around Jamaica. It is an abundant species, and beach specimens are common everywhere.

HEMITOMA EMARGINATA Blainville 1825
Emarginate Limpet **pl. 2**
Florida Keys and the Greater Antilles.
Description: Length $\frac{3}{4}$ to $1\frac{1}{4}$ inches. Colour white or grey. Shell high and rather narrow, with a reticulate sculpture; apex inclined backwards. The base is higher at both ends than in the centre.
Occurrence: Not uncommon on coral reefs in 20 to 60 feet at Discovery Bay, off the Bogue Islands, and in Ocho Rios harbour. Beach specimens are rare.
Forms: There is a broad, flattened form with more delicate sculpture, which is often taken as a separate species. It occurs at Negril and in Ocho Rios harbour.

SUBFAMILY DIODORINAE

Genus *Diodora* Gray 1821

Subgenus *Diodora* Gray 1821

DIODORA LISTERI Orbigny 1853
Lister's Keyhole Limpet **pl. 2**
Southern Florida and the West Indies
Description: Length $1\frac{1}{2}$ to 2 inches. Colour grey, cream or white, often with darker rays. Sculpture of alternating large and small ribs, crossed by concentric cords. Orifice black and keyhole in shape. The base is oval with small paired teeth around the margin.
Occurrence: Very common on rocks from 1 to 10 feet, at Port Henderson, Port Antonio, and in rocky areas all around Jamaica.

DIODORA CAYENENSIS Lamarck 1822

Cayenne Keyhole Limpet **pl. 2**

South-east U.S. to Brazil.

Description: Length 1½ to 2 inches. Colour variable, but generally dull brown rays on a cream background, often with short black radial lines near the apex. Sculpture of many thin ribs, crossed by numerous fine concentric cords. Every fourth rib is larger. The orifice is keyhole shaped, and the posterior end noticeably convex. Apex slopes forward and is set nearer the anterior end.

Occurrence: As for *D. listeri*, but it is not as common. Beach specimens are found all around the Island, and perfect examples occur in the 'rivers' which connect the lagoons with the sea at Port Royal.

DIODORA DYSONI Reeve 1850

Dyson's Keyhole Limpet **pl. 2**

Florida, the Bahamas and south to Brazil.

Description: Length ½ to ¾ inch. Colour cream with 8 dotted black rays. Apex set slightly towards the anterior end.Orifice small, triangulate, usually bounded by a thin black line except at the posterior end. Sculpture of many ribs crossed by numerous, thin concentric cords.

Occurrence: Uncommon on and under rocks in about 6 feet at Runaway Bay, Prospect, and in rocky areas all around the Island. Dead shells are sometimes found in beach drift.

DIODORA SAYI Dall 1899 Say's Limpet **pl. 2**

South-east Florida and the Greater Antilles to Brazil.

Description: Length about 1 inch. Colour white or cream, sometimes with faint olive rays: surface dull. Sculpture of numerous fine, close-set radiating ribs crossed by many concentric threads. Apex directed forward: anterior slope short, posterior slope long. The shell resembles *D. cayenensis*.

Occurrence: This is an uncommon species in Jamaica. It occurs under rocks in about 20 feet at Negril and in Ocho Rios harbour, but beach specimens are rarely found.

DIODORA VIRIDULA Lamarck 1822

Green Keyhole Limpet **pl. 2**

Florida south to Trinidad.

Description: Length ¾ to 1 inch. Colour, white ribs on a greenish background; orifice blue-black. Sculpture of numerous, rather coarse radiating ribs. Apex set well forward towards the anterior end.

Occurrence: Relatively common in the lower inter-tidal zone, on rocks, at Prospect, Robins Bay and Negril. It is an attractive species which is often washed ashore.

DIODORA MINUTA Lamarck 1822

Dwarf Keyhole Limpet **pl. 2**

South-east Florida and the West Indies.

Description: Length about ½ inch. Colour white, with many ribs which are partly or entirely black and form radiating triangles. Sculpture of numerous radiating ribs crossed by concentric threads. The apex is at the anterior third of the shell.

Occurrence: Relatively common on and under rocks in about 10 feet at Port Antonio, Runaway Bay and along the southern side of the Palisadoes. Dead shells are common on the beaches at Negril.

DIODORA VARIEGATA Sowerby 1862

Variegated Keyhole Limpet

Bahamas to the northern Lesser Antilles.

Description: Similar to *D. minuta* (the preceding species), but usually mostly black in colour, with the pigment both on the ribs and the spaces between them. The shell tends to be smaller and more depressed than *minuta*.

Occurrence: Relatively common on rocks in about 15 feet at Prospect, Negril and in Ocho Rios harbour. Dead shells are found in beach drift along the southern side of the Palisadoes.

DIODORA ARCUATA Sowerby 1862

Arcuate Limpet **pl. 2**

Bahamas, Florida south to Trinidad

Description: Length about ⅓ inch. Colour white with rays of tan or light brown. Sculpture of minute scales. This species is similar to *D. variegata*, but higher and with a smaller and more anterior orifice.

Occurrence: Uncommon in about 30 feet in Ocho Rios harbour. Beach specimens are found at Negril after stormy weather.

Genus *Lucapina* Sowerby 1835

LUCAPINA SUFFUSA Reeve 1850

Cancellate Fleshy Limpet **pl. 2**

South Florida and the West Indies.

Description: Length 1 to 1½ inches. Colour a pale pink or mauve, with the large orifice generally blue-black. Sculpture of many ribs crossed by numerous, thin concentric cords. The shell has a low profile with the apex sloping forward. An orange mantle envelops the shell.

Occurence: Relatively uncommon in 5 to 25 feet in the vicinity of coral reefs at Negril and in Ocho Rios harbour. Beach specimens are generally rare.

LUCAPINA SOWERBII Sowerby 1835
Sowerby's Fleshy Limpet **pl. 2**
South-east Florida to Brazil.
Description: Length $\frac{3}{4}$ inch. Colour, dirty white to light brown, with 7 to 9 irregular rays of a darker brown. Sculpture of about 60 alternating large and small ribs, crossed by 10 to 12 thin concentric cords.
Occurrence: Uncommon in about 20 feet in Ocho Rios harbour. We found a single specimen in a few inches of water in the 'river' which connects the lagoon with the sea nearest Port Royal, but it is rare at this depth.

LUCAPINA PHILIPPIANA Finlay 1930
Philippi's Keyhole Limpet **pl. 2**
Florida, Bahamas and south to the Lesser Antilles.
Description: Length $\frac{1}{2}$ to $\frac{3}{4}$ inch. Colour cream or white, irregularly speckled with brown. Sculpture of about 38 radiating alternately larger and smaller ribs. Shape similar to *L. sowerbii*, but it is smaller, thinner, narrower and less depressed.
Occurrence: Rare in about 30 feet, under rocks, in Ocho Rios harbour. It may also occur at Orange Bay in deeper water.

LUCAPINA AEGIS Reeve 1850 **pl. 2**
Florida, the Bahamas and the Greater Antilles.
Description: Length $\frac{3}{4}$ to $1\frac{1}{2}$ inches. Colour variable, usually light brown or olive green, with 8 broken darker rays. Sculpture of about 40 radiating ribs separated by deep grooves, and numerous concentric laminae.
Occurrence: This is a rare species in Jamaica. It occurs in about 45 feet in Ocho Rios harbour, and beach specimens have been reported from Negril.

Genus *Lucapinella* Pilsbry 1890

LUCAPINELLA LIMATULA Reeve 1850 **pl. 2**
North Carolina south to the Virgin Islands.
Description: Length about $\frac{3}{4}$ inch. Colour white, yellow, brown or pink, usually with 6 broad darker rays. Sculpture of alternately large and small ribs at the anterior end; at the posterior end there are 3 riblets between each pair of primary ribs. There are numerous concentric laminae.
Occurrence: Uncommon in about 40 feet in Ocho Rios harbour and at Orange Bay. Beach specimens have occasionally been found at Negril.

SUBFAMILY FISSURELLINAE

Genus *Fissurella* Bruguière 1789

Subgenus *Cremides* H. and A. Adams 1854

FISSURELLA NODOSA Born 1778

Knobby Keyhole Limpet **pl. 2**

Lower Florida Keys and the West Indies.

Description: Length 1 to 1½ inches. Colour white or light grey. Sculpture of 20 to 22 very strong, nodulous ribs. Interior with incised lines running from the base of each rib to the apex. Orifice contracted in the centre. Exterior of shell frequently disfigured by calcareous growths.

Occurrence: Very common on rocks exposed to the open sea in the inter-tidal zones at Runaway Bay, Prospect and Negril. Beach specimens are common in rocky areas all around the Island.

FISSURELLA BARBADENSIS Gmelin 1791

Barbados Keyhole Limpet **pl. 2**

Bermuda, south-east Florida and the West Indies.

Description: Length 1 to 1¾ inches. Colour very variable, generally with irregular lines or patches of pink or purple on a white background. Interior with broad concentric rings of varying shades of green. Sculpture of many large and small irregular, nodulous ribs, with a few very faint concentric cords.

Occurrence: As for *F. nodosa*, but it is a more abundant species. It is widely used in the manufacture of shell artifacts throughout the West Indies. Dead shells are found on most beaches.

FISSURELLA ROSEA Gmelin 1791

Rosy Keyhole Limpet **pl. 2**

South-east Florida and the West Indies to Brazil.

Description: Length 1 to 1½ inches. Colour consists basically of a series of alternating whitish and dark pink rays, but there are many variations. Sculpture similar to *F. barbadensis* (the preceding species), but much less pronounced; the shell is thinner, flatter and narrower at the anterior end.

Occurrence: Relatively common on and under rocks in the inter-tidal zone at Prospect, Runaway Bay and Negril. Dead shells are found on many beaches all around the Island.

FISSURELLA ANGUSTA Gmelin 1791

Pointed Keyhole Limpet **pl. 2**

Bahamas, lower Florida Keys and south to Guyana.

Description: Length ¾ to 1 inch. Colour variable, generally with 9 or 10 slightly nodulose white ribs on a red background. Sculpture similar to *F. barbadensis*, but not as coarse. The shell is low and pointed at the anterior end.

Occurrence: As for *F. nodosa*, but it is the least common of the genus in Jamaica. Beach specimens are found occasionally at Negril.

Subgenus *Clypidella* Swainson 1840

FISSURELLA FASCICULARIS Lamarck 1822

Wobbly Keyhole Limpet **pl. 2**

South-east Florida and the West Indies.

Description: Length ¾ to 1¼ inches. Colour dark pink with the rib interspaces a dirty white. Interior light green, with the orifice outlined in pink. Sculpture of numerous flat ribs. The profile of the shell is much lower than other West Indian *Fissurellidae*, and both ends are elevated.

Occurrence: Common on and under rocks, in the inter-tidal areas at Prospect, Discovery Bay, and in rocky localities all around Jamaica. Beach-worn specimens tend to lose their distinctive pink colour.

Subgenus *Fissurella* Bruguière 1789

FISSURELLA NIMBOSA Linné 1758

Rayed Keyhole Limpet **pl. 2**

Jamaica (?), Puerto Rico and south to Venezuela.

Description: Length 1½ to 2 inches. Colour, purplish-red or dark brown rays on a buff background. Sculpture of many smooth ribs crossed by numerous concentric growth lines. Orifice large and oblong. The shell is heavy and elevated.

Occurrence: Reported in 1 to 6 feet on rocks exposed to the open sea at Drax Hall Estate, St Ann. We have not found this species in Jamaica, and the single record is open to question. It is common in the Southern Caribbean.

SUPERFAMILY PATELLACEA

FAMILY ACMAEIDAE

This is an attractive family of about 400 species which resemble the Keyhole Limpets (*Fissurellidae*) in many respects, but lack a hole at the apex of the shell. The sexes are separate.

Most of the Jamaican species occur in the inter-tidal area. All members of the family have a remarkably powerful foot, and cling stubbornly to the surface on which they are resting. *Acmaeidae* live in temperate and tropical seas, and all the genera are vegetarian.

Genus *Acmaea* Eschscholtz 1830

Subgenus *Collisella* Dall 1871

ACMAEA ANTILLARUM Sowerby 1831

Antillean Limpet **pl. 2**

Southern Florida and the West Indies.

Description: Length $\frac{3}{4}$ to 1 inch. Colour variable, usually with a number of reddish-brown radial lines on a white background. Exterior smooth; the shell is oval, thin and rather flat. Interior white with a brown callus at the centre.

Occurrence: An abundant species on rocks exposed to the open sea, in inter-tidal areas at Runaway Bay, Port Henderson, and in rocky areas all around Jamaica. Beach specimens are common everywhere.

ACMAEA PUSTULATA Helbling 1779

Spotted Limpet **pl. 2**

Bermuda, south-east Florida and the West Indies.

Description: Length $\frac{1}{2}$ to 1 inch. Colour white, liberally spotted with red or orange. These spots are usually in radial lines from the apex to the edge of the shell. Sculpture of many axial ribs, crossed by fine concentric threads. Interior white, usually with an orange or yellow callus in the centre. The shell is very variable in height.

Occurrence: As for *A. antillarum*, but it prefers smooth surfaces free from dense marine growths. It is a very common and delicately beautiful shell.

Forms: There is a thin, smooth, narrow red and orange form (adapted for life on *thalassia* blades) known as *A. pustulata* form *pulcherrima* Guilding. It is less widespread than typical *pustulata* but locally common, especially at Negril (pl. 2).

ACMAEA JAMAICENSIS Gmelin 1791
Jamaica Limpet **pl. 2**
South-east Florida and the West Indies.
Description: Length about ¾ inch. Colour black, with 5 or 10 raised white ribs. Interior white with a brown or black callus.
Occurrence: Common on smooth rocks, in the inter-tidal area, all around Jamaica. It is abundant at Discovery Bay and Negril.

ACMAEA LEUCOPLEURA Gmelin 1791
Dwarf Suck-on Limpet **pl. 2**
South-east Florida and the West Indies.
Description: Diameter about ½ inch. Height very variable. Colour black and white, in alternating rays. The black rays divide near the shell's base. Interior white with the callus sometimes brown or black. The exterior of the shell is smooth.
Occurrence: Relatively common on rocks exposed to the open sea at Runaway Bay, Port Antonio, and in rocky areas all around Jamaica. It sometimes lives well above the high tide line. Abbott (*Caribbean Seashells*) reports that it is also commonly found attached to the live West Indian Top Shell (*Cittarium pica*).

ACMAEA CUBENSIS Reeve 1855 Cuban Limpet **pl. 2**
Bahamas and the West Indies.
Description: Diameter about ½ inch. This shell is very similar to the preceding species (*A. leucopleura*), but it is higher and has more numerous and less clearly defined rays. The callus is usually brown and extends over most of the interior. It is probably a form of *A. leucopleura*.
Occurrence: As for *A. leucopleura*, and it is equally common.

SUPERFAMILY TROCHACEA

FAMILY TROCHIDAE

This is a family of exceptionally beautiful shells, characterised by their uniform conical shape, pearly interiors, and thin, multispiral operculums. The sexes are separate.

There are over 1,000 littoral and deep-sea species, which are found mainly in temperate and tropical waters. The genus *Calliostoma* includes the most delicate and attractive members of the family. All *Trochidae* are believed to be vegetarian (but see note on *Calliostoma javanicum*).

Genus *Cittarium* Philippi 1847

CITTARIUM PICA Linné 1758 West Indian Top Shell **pl. 5**
South-east Florida and the West Indies.
Description: Diameter 3 to 4 inches. Colour dirty white, with large irregular patches of purplish-black. Shell strong and heavy, with a deep umbilicus. Aperture round and pearly within; operculum circular, multispiral, and dark greenish-red in colour. The animal is tough but edible, and makes an acceptable meal when properly prepared.
Occurrence: This is undoubtedly the most common of the larger West Indian gastropods. It occurs in great numbers on and under rocks in inter-tidal areas, usually exposed to the open sea, everywhere around Jamaica. The very large specimens are sometimes found in deeper water. Port Henderson, Prospect, and Port Antonio are all excellent collecting localities for this species.

Genus *Tegula* Lesson 1832

TEGULA FASCIATA Born 1778
Smooth Atlantic Tegula **pl. 3**
South-east Florida and the West Indies.
Description: Diameter $\frac{1}{2}$ to $\frac{3}{4}$ inch. Colour extremely variable, generally dark brown or reddish, mottled with irregular white patches; many specimens are entirely light brown. There is often a dotted red line on the periphery. The whorls are smooth and rounded, and there is a deep umbilicus. There are two white teeth at the base of the columella. Operculum brown, horny and multispiral. The shell is strong, and many of the colour variations are unusually attractive.
Occurrence: Abundant in 1 to 6 feet under rocks at Runaway Bay, Prospect, Port Antonio, and in rocky areas all around the Island. This is a favourite shell of the small hermit crabs, which occupy it after the death of the mollusc.

TEGULA LIVIDOMACULATA C. B. Adams 1845
West Indian Tegula **pl. 3**
Key West and the West Indies.
Description: Diameter $\frac{1}{2}$ to $\frac{3}{4}$ inch. Similar in shape to *T. fasciata*, but lighter, and the whorls are more angular and sculptured with low spiral cords. Colour greenish-grey to light brown, with small dark brown mottlings. Umbilicus deep and furrowed; columella with one large

tooth and several small, indistinct ones below it. Operculum brown, horny and multispiral.
Occurrence: As for *T. fasciata*, with which it is often found living under rocks, in shallow water.

TEGULA EXCAVATA Lamarck 1822 Green Tegula **pl. 3**
Caribbean area.
Description: Length about ¾ inch, and the same in width. Colour variable, but usually either light brown lines on a greyish background, or white patches on a dark grey background. Sculpture of fine spiral cords and oblique growth lines. Base concave; umbilicus deep, and surrounded by a green or pearly ring. Operculum thin, light brown and multispiral.
Occurrence: Very common on and under rocks in the lower inter-tidal areas along the south side of the Palisadoes, at Robins Bay, and in rocky localities all around the Island. It prefers areas directly exposed to the open sea.

Genus *Calliostoma* Swainson 1840

CALLIOSTOMA JAVANICUM Lamarck 1822
Chocolate-lined Top Shell **pl. 3**
Lower Florida Keys and West Indies.
Description: Height ¾ to 1½ inches. Colour light brown with several incised, spiral, chocolate-brown lines. Exposure to the sun fades both lines and background. Base, light olive with several fine, spiral lines. Umbilicus deep; the shell is light and relatively fragile. This is a remarkably beautiful species, and is only infrequently collected in an undamaged condition.
Occurrence: Very uncommon in 25 to 70 feet in reefs at Discovery Bay, the Bogue Islands, Ocho Rios harbour, and other areas off both the north and south coasts. This is generally considered a rare species, but recent commercial dredging at Ocho Rios and Discovery Bay has revealed that it occurs in relatively large numbers, often close to shore. It prefers to live well within the reefs and is consequently difficult to obtain. Damaged specimens are very occasionally washed ashore.
Note: Live shells have been found grazing on the coral *Favia fragum*, though they may be feeding primarily on the zooxanthellae in the superficial coral tissues. This very untypical behaviour is now being investigated, for Trochids were thought to be exclusively vegetarian.

CALLIOSTOMA JUJUBINUM Gmelin 1791

Jujube Top Shell **pl. 3**

Bahamas, lower Florida Keys and the West Indies.

Description: Height 1 to 2 inches. Colour variable, but generally dark or light brown, maculated with white. Occasional specimens are a rich, dark red, and a few have purple on the rounded periphery of the whorls. Sculpture of many beaded threads; umbilicus deep, narrow and smooth-sided. The adult shell is relatively thick and strong.

Occurrence: This species lives in much the same environment as *C. javanicum*, and the two are occasionally collected together. It is a shallower water species, however, and not infrequently washed ashore at the Bogue Islands. Recent dredgings in Ocho Rios harbour and at the mouth of the Salt River (Clarendon), indicate that it is common there in about 25 feet.

CALLIOSTOMA PULCHRUM C. B. Adams 1850

Beautiful Top Shell **pl. 3**

South-east U.S., Gulf of Mexico and the West Indies.

Description: Height $\frac{1}{2}$ inch. Colour light reddish yellow, maculated with patches of white and orange-brown. Sculpture of two thin spiral cords above the suture; these cords are white with red-brown dots. There are 6 or 7 very fine, beaded spiral threads. Aperture pearly; no umbilicus. This is a delicately beautiful species which merits close examination under a hand lens.

Occurrence: Relatively common in 6 to 20 feet under rocks in sandy areas at Negril and in Ocho Rios harbour. It is uncommon at Runaway Bay, and also at Old Harbour Bay on the south coast (where its colours tend to be muted).

CALLIOSTOMA SARCODUM Dall 1927 **pl. 3**

Bahamas south to Barbados.

Description: Height $\frac{5}{8}$ inch. Colour light brown or tan, with occasional small, faint white maculations. On the base, usually two or more of the many spiral lines have dots of a darker colour. Whorls fat and rounded, with numerous finely beaded spiral cords; there are more of these on the base than the periphery. Aperture pearly; no umbilicus. The shell is solid and strong, and has a polished sheen.

Occurrence: The very few specimens we have found all came from dredgings which involved the wholesale removal of sections of coral reef at Discovery Bay and in Ocho Rios harbour. The depth range was 20 to 40 feet, and the shell may live deep inside the reefs. It is a rare and choice species not previously recorded from Jamaica.

Subgenus *Turcicula* Dall 1881

TURCICULA IMPERIALIS Dall 1881
Imperial Turcicula **pl. 5**
Cuba and Barbados.
Description: Length about 2½ inches. Colour grey, with a beautiful nacreous interior. Sculpture of 3 revolving rows of close-set tubercles, and a series of short, sharp ribs immediately below the suture. There are a number of spiral threads on the base which become nodulous towards the aperture. The outer lip is reflected, and the shell is light but strong.
Occurrence: This appears to be a very rare shell indeed. The only previous records we can trace are two small specimens collected by the Blake Expedition off Cuba in about 1,200 feet. Our single specimen was collected dead on the beach at Seven Man's Bay, Barbados, which greatly extends the range and suggests that it also occurs in relatively shallow water. It has not yet been found in Jamaican waters, and is included here as a matter of interest.

FAMILY TURBINIDAE

This is a large family of 500 species, most of which are confined to tropical seas. The shells are pyramidal or top-shaped, with beautiful, nacreous interiors. The sexes are separate.

Many *Turbinidae* are highly coloured, and most possess a characteristic, hard, limy operculum which distinguishes them from the *Trochidae*. The family is exclusively vegetarian.

SUBFAMILY LIOTIINAE

Genus *Arene* H. and A. Adams 1854

ARENE CRUENTATA Mühlfeld 1829 Star Arene **pl. 3**
South-east Florida and the West Indies.
Description: Diameter ¼ to ½ inch. Colour white, with small, fairly regular red or brown patches on the upper surface of the whorls. There are 4 to 6 whorls, bearing two rows of triangular spines on the periphery; the bottom row is generally much smaller. Sculpture of many minute axial threads. Umbilicus deep and bordered by 3 spiral beaded cords. Aperture circular and pearly within; operculum horny, circular

and multispiral. Adult and perfect specimens are strikingly attractive shells.
Occurrence: Uncommon from 1 to 20 feet on the underside of rocks at Runaway Bay, Montego Bay, Prospect, and in many other rocky areas all around Jamaica. The dead shells are often claimed by small hermit crabs which bring them into shallower water.
Forms: A. vanhyningi Rehder 1939 is a form with underdeveloped spines. Our specimens are from Runaway Bay.

Genus *Cyclostrema* Marryat 1818

CYCLOSTREMA CANCELLATUM Marryat 1818
Cancellate Cyclostreme **pl. 3**
South-east Florida and the West Indies.
Description: Diameter ½ inch. Colour white. Whorls angular, with 15 to 17 axial ribs. There are 3 spiral cords on the periphery. Umbilicus broad and deep; operculum thin and multispiral. The shell is light but strong. It is a delicate, beautiful and coveted species.
Occurrence: In 30 to 60 feet (and probably deeper) under rocks, at Robins Bay and in Ocho Rios harbour. It has also been found on the beach at Negril. It is correctly considered a rare species in Jamaica, but the commercial dredgings at Ocho Rios indicate that it may occur locally in fair quantity.

SUBFAMILY TURBININAE

Genus *Turbo* Linné 1758

Subgenus *Marmrostoma* Swainson 1829

TURBO CASTANEA Gmelin 1791 Chestnut Turban **pl. 3**
South-east U.S., Texas and the West Indies.
Description: Length ¾ to 1½ inches. Colour very variable: generally a combination of white with large patches of grey, orange or reddish-brown. Sculpture of many nodulous spiral cords; aperture round and pearly within. Operculum thick and calcareous, with a large purplish-brown marking on a white background. There is no umbilicus (cf. *T. filosus*).
Occurrence: Very common in 1 to 4 feet along the northern side of the Palisadoes on blades of eel grass, and in the same environment at Port Antonio and Negril. Dead shells and operculums often surround the holes of octopuses, and the dead shell is a favourite home of the small hermit crab.

Forms: There are no accepted Jamaican forms, but the specimens which live in deeper water (down to 25 feet) tend to have a rougher sculpture and a more clearly defined pattern of alternating patches of white and orange. Occasional examples from Ocho Rios harbour bear a narrow, bright orange spiral band.

TURBO FILOSUS Wood 1828 Threaded Turban **pl. 3**

Dry Tortugas south to Guadeloupe.

Description: Length about 1 inch. Colour variable, but usually a rayed mottling of dull orange and dark brown, with about 12 smooth, lighter coloured (occasionally red) spiral cords. In general shape it is very similar to *T. castanea*, but it is easily separated from it by the presence of a partially closed umbilicus. The operculum is similar to that of *castanea.*

Occurrence: Until recently this was considered a comparatively rare shell. However, commercial dredging at Discovery Bay has revealed that it occurs there in great quantity in coral reefs, at depths of 15 to 45 feet. It has also been found in similar environments in several areas from Montego Bay to Port Antonio. We have seen no beach specimens.

Forms: There is a very uncommon and striking bright yellow and orange colour form, and we have occasionally found juvenile shells which are entirely red.

Subgenus *Taeniaturbo* Woodring 1928

TURBO CANALICULATUS Hermann 1781
Channelled Turban **pl. 5**

Lower Florida Keys and the West Indies.

Description: Length 2 to 3½ inches. Colour tan, variegated with white, orange, brown and olive green. Area around columella often pearly green. Sculpture of 16 to 18 strong, smooth, spiral cords; there is a deep channel just below the suture, and no umbilicus. The early whorls are usually pinkish. It is a solid, graceful and exceptionally attractive shell.

Occurrence: In 50 to 200 feet on mud or sand bottoms all along the south coast, and in Ocho Rios harbour and at Port Antonio on the north coast. In Kingston harbour, Old Harbour Bay, Alligator Pond and Whitehouse (Westmoreland), excellent specimens are commonly carried into fishpots by large hermit crabs, and later thrown out on the beaches by the fishermen. We have not yet seen a living specimen, and it is probable that the hermits bring the dead shells up from deeper water. We have found one small dead shell washed ashore on the southern side of the Palisadoes.

Genus *Astraea* Röding 1798

Subgenus *Lithopoma* Gray 1850

ASTRAEA CAELATA Gmelin 1791
Carved Star Shell **pl. 5**
South-east Florida and the West Indies.
Description: Diameter 2 to 3 inches, and not quite as high. Colour dirty white, mottled with emerald green, pink and brown. Sculpture of very coarse, oblique, scaly ridges, with rough axial cords on the lower part of each whorl. Base with several coarse spiral cords; there is no umbilicus. Operculum, thick, white and finely pustulose. The colour and much of the sculpture of this shell is often concealed by calcareous growths, particularly in the adult stages.
Occurrence: In 1 to 6 feet on and under rocks at Runaway Bay, Prospect, and in rocky areas all around Jamaica. It prefers areas exposed to the open sea, and is the most common member of the genus in Jamaica.

ASTRAEA TUBER Linné 1758 Green Star Shell **pl. 5**
South-east Florida and the West Indies.
Description: Diameter 1½ to 2 inches, and nearly as high. Colour a mottled combination of green, white and reddish-brown, with the base always lighter than the rest of the shell. Sculpture of coarse, oblique ridges, interspersed with very fine ridges. Base relatively smooth; no umbilicus. Operculum white with a raised, pustulose comma-shaped callus on the exterior. The shell is very thick and solid.
Occurrence: In 1 to 8 feet, in rocky areas all around Jamaica. It is abundant on and under rocks in certain areas at Prospect, where we counted over 100 adults living in an area 6 feet square. Dead shells are very common on most beaches facing the open sea.

Subgenus *Australium* Link 1807

ASTRAEA PHOEBIA Röding 1798
Long-spined Star Shell **pl. 5**
South-east Florida and the West Indies.
Description: Diameter 2 to 3½ inches; low in profile. Colour yellowish brown, with traces of green in the early whorls; very young shells have a golden appearance. Easily distinguished by the flat, triangular spines at the periphery of the whorls, and the axial cords of small, oblique

Plate 1 THE CLASSES OF SHELLS

Scale: reduced by 25%

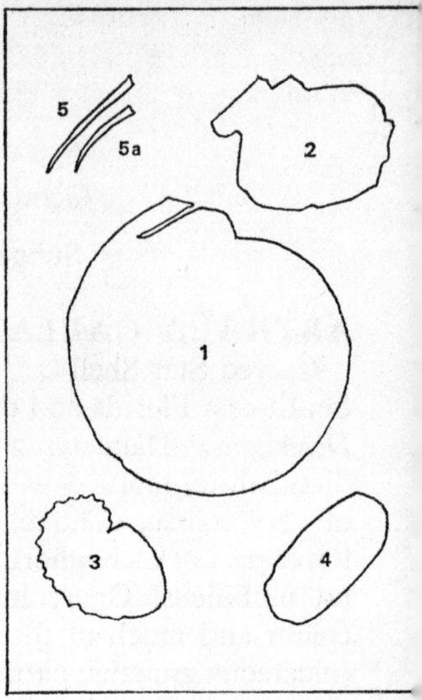

1. Class **Gastropoda:** Snails — 47

   Perotrochus adansonianus Crosse and Fischer Adanson's Pleurotomaria. Note the partially retracted living animal visible through the peripheral slit.

2. Class **Pelecypoda:** Bivalves — 228

   Lyropecten nodosus Linné Lion's Paw Scallop

3. Class **Cephalopoda:** Squids, Octopuses and Nautiloids — 288

   Argonauta hians Lightfoot Brown Paper Nautilus

4. Class **Amphineura:** Chitons — 291

   Chiton tuberculatus Linné Common West Indian Chiton

5. Class **Scaphopoda:** Tusk Shells — 294

   Dentalium eboreum Conrad Ivory Tusk Note hole in centre pierced by a carnivorous gastropod.

   5a, *Dentalium didymum* Watson Flattened Tusk — 295

Plate 2 LIMPETS

Scale: natural size

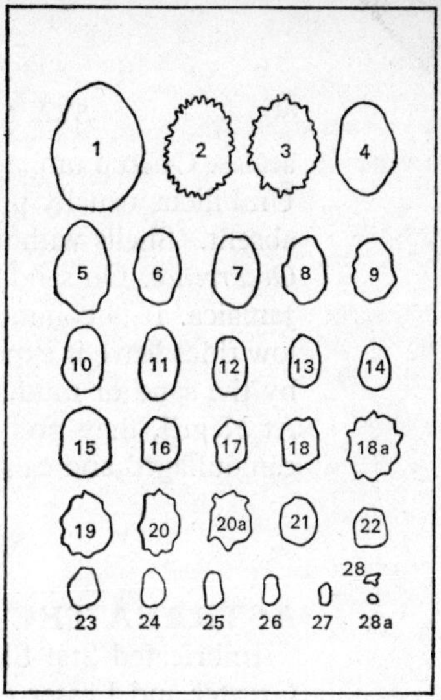

1. **Fissurella nimbosa** Linné 54
Rayed Keyhole Limpet
See text for note on the need for confirmation that this species occurs in Jamaica.

2. **Fissurella nodosa** Born 53
Knobby Keyhole Limpet

3. **Fissurella barbadensis** 53
Gmelin Barbados Keyhole Limpet

4. **Diodora listeri** Orbigny 49
Lister's Keyhole Limpet
Note that red patches are coralline growths.

5. **Fissurella rosea** Gmelin 53
Rosy Keyhole Limpet

6. **Diodora cayenensis** Lamarck 50
Cayenne Keyhole Limpet

7. **Diodora viridula** Lamarck 50
Green Keyhole Limpet

8. **Fissurella angusta** Gmelin 54
Pointed Keyhole Limpet

9. **Fissurella fascicularis** 54
Lamarck Wobbly Keyhole Limpet

10. **Lucapina sowerbii** Sowerby 52
Sowerby's Fleshy Limpet

11. **Lucapina suffusa** Reeve 51
Cancellate Fleshy Limpet

12. **Diodora sayi** Dall Say's 50
Limpet

13. **Lucapina philippiana** Finlay 52
Philippi's Keyhole Limpet

14. **Lucapina aegis** Reeve 52

15. **Acmaea antillarum** Sowerby 55
Antillean Limpet

16. **Acmaea pustulata** Helbling 55
Spotted Limpet

17. **Acmaea pustulata pulcherrima** Guilding 55
This is the narrow form of **16,** adapted for life on blades of *Thalassia* eel grass. Length and range identical with true *pustulata*.

18. **Acmaea jamaicensis** Gmelin 56
Jamaica Limpet
18a, ventral view.

19. **Hemitoma octoradiata** 49
Gmelin Eight-ribbed Limpet

20. **Hemitoma emarginata** 49
Blainville Emarginate Limpet
20a, the flattened, more delicately sculptured variety.

21. **Acmaea leucopleura** 56
Gmelin Dwarf Suck-on Limpet

22. **Acmaea cubensis** Reeve 56
Cuban Limpet
This is probably a form of **21,** not a true species.

23. **Emarginula pumila** 48
A. Adams Pigmy Emarginula
Note the anterior marginal slit.

24. **Diodora dysoni** Reeve 50
Dyson's Keyhole Limpet

25. **Lucapinella limatula** Reeve 52
File Fleshy Limpet

26. **Diodora minuta** Lamarck 51
Dwarf Keyhole Limpet

27. **Diodora arcuata** Sowerby 51
Arcuate Limpet

28. **Emarginula phrixodes** Dall 48
Ruffled Rimula
28a, dorsal view.

S.W.I.—E

scales. Operculum smooth and white with a small area of light brown. Umbilicus usually present, but often partially closed, and occasionally absent. (Shells without an umbilicus tend to have a higher profile.)
Occurrence: On sand flats with eel grass, in inter-tidal areas, all around Jamaica. It is common at Priory, and Runaway Bay, where unusually low tides leave it exposed on the sand flats. The shells live just covered by the sand or mud, and can be turned up by judicious use of a rake. At Negril, they prefer somewhat deeper water. They are always well camouflaged and can easily be overlooked.

Subgenus *Calcar* Montfort 1810

ASTRAEA TECTA Solander 1786
Imbricated Star Shell **pl. 5**
Greater and Lesser Antilles.
Description: Diameter 1 to 1½ inches, and about the same in height. Colour variable, generally greenish-orange with white peripheral tubercles, but occasionally mostly white. Young shells are sometimes solid orange or pearly green. Sculpture of numerous strong ribs running obliquely across the whorls. Base with a number of weak spiral cords. The operculum is similar to that of *A. tuber*.
Occurrence: In 1 to 8 feet at Port Royal (harbour side), particularly on the walls of the pool at the Police Training School; also on rocks at Negril, Priory, and in similar areas all along the north coast. It is a relatively uncommon species, but has a wide distribution.

FAMILY PHASIANELLIDAE

The West Indian members of this widely distributed family are small, polished, and brilliantly coloured. The unusual beauty of the colour patterns of all the local species merits close examination under a hand lens. The operculums are white, solid and shelly. The family is vegetarian and lives in shallow water.

Unlike other gastropods, Pheasant Shells (as they are called) progress by moving alternately the left and right sides of their foot, which appears to be split longitudinally down the centre.

SUBFAMILY TRICOLIINAE

Genus *Tricolia* Risso 1826

TRICOLIA AFFINIS C. B. Adams 1850
Polka-dot Pheasant **pl. 3**
Lower Florida Keys and the West Indies.
Description: Length $\frac{3}{8}$ inch. Colour rose or light orange-brown, with many small dots of red and white; there are often zigzag axial red bars and streaks of white. Umbilicus slit-like. The shell is smooth, light and delicate.
Occurrence: On *thalassia* blades in 4 to 15 feet in Kingston harbour, at Negril, Runaway Bay and similar areas all around Jamaica. It is uncommonly found alive, but can often be collected, with a hermit crab occupant, at low tide in pools in the eel grass flats along the north coast.
Note: The four smaller members of this genus are listed (but not described) at the end of this section on gastropods. They are essentially similar in shape and colour to *T. affinis*.

SUPERFAMILY NERITACEA

FAMILY NERITIDAE

There are several hundred members of this widely distributed mainly littoral family, which are notable for their diversity of habitat. Nerites are found in lakes and rivers as well as the sea, and some species even live in trees.

The family is characterised by the absence of an umbilicus, and, in *Neritinae*, by the small projecting arm on the shelly operculum (which is used for muscular attachment). All *Neritidae* are vegetarian, and they usually occur in large colonies.

SUBFAMILY NERITINAE

Genus *Nerita* Linné 1758

NERITA PELORONTA Linné 1758 Bleeding Tooth **pl. 4**
Bermuda, south-east Florida and the West Indies.
Description: Length 1 to $1\frac{1}{2}$ inches. Colour very variable, usually with

irregular red or black markings on a white, yellow, red or orange background. Parietal area stained blood-red. There are 2 or 3 white teeth. Sculpture of low spiral cords, which are more clearly defined on the earlier whorls. The operculum is dark brown and orange.
Occurrence: Abundant on rocks just above and below the high tide mark, at Port Henderson, Port Antonio, and in rocky areas exposed to the open sea all around the Island. The beauty of this shell is often overlooked because it is so commonly and easily collected.

NERITA VERSICOLOR Gmelin 1791
Four-toothed Nerite **pl. 4**
Bermuda, southern Florida and the West Indies.
Description: Length $\frac{3}{4}$ to 1 inch. Colour variable, but usually white more or less variegated with black and red. Sculpture of many smooth spiral cords. Parietal area white with 4 prominent teeth; operculum greyish. Occasional shells are all black.
Occurrence: As for *N. peloronta*, with which it is usually found.

NERITA TESSELLATA Gmelin 1791
Tessellated Nerite **pl. 4**
Bermuda, Florida to Texas and the West Indies.
Description: Length $\frac{1}{2}$ to $\frac{3}{4}$ inch. Colour white with black lines or speckling; many shells are virtually all-black. Sculpture of many low spiral cords; parietal area white with two minute teeth. The operculum is black.
Occurrence: Abundant on and under rocks in the inter-tidal area at Port Henderson, Port Antonio, and in rocky areas all around Jamaica.

NERITA FULGURANS Gmelin 1791
Antillean Nerite **pl. 4**
Bermuda, south-east Florida and the West Indies.
Description: Length $\frac{3}{4}$ to $1\frac{1}{4}$ inches. Colour yellowish white, with irregular maculations of black. Sculpture of many low spiral cords; parietal area yellowish, with two small teeth. The aperture of the shell is wider than *N. tessellata*, which it otherwise resembles, and the teeth are more prominent. The operculum is yellowish-grey.
Occurrence: Relatively common in the inter-tidal area of the mangrove swamps on the harbour side of the Palisadoes. It is not a widespread species and is confined to fairly sheltered, brackish water. We have found the occasional live specimen with a Caribbean Oyster (*Crassostrea rhizophorae*) cemented to its back.

Genus *Puperita* Gray 1857

Subgenus *Puperita* Gray 1857

PUPERITA PUPA Linné 1767 Zebra Nerite **pl. 4**
South-east Florida and the West Indies.
Description: Length ½ inch. Colour white with black stripes, giving a 'zebra pattern' effect; interior light yellow, with the operculum much the same colour. The shell is smooth and without sculpture.
Occurrence: Abundant in rock pools, usually above the high water mark, at Old Pera (St Thomas), Port Antonio, and in many other rocky areas facing the open sea. Its distribution is not as wide as most other members of the family.

PUPERITA TRISTIS Orbigny **pl. 4**
The Antilles.
Description: Length about ⅜ inch. Colour white overlaid with an extensive network of fine black lines. Aperture yellowish. The shell is smooth but not polished. This species is very similar to *P. pupa*, but the spire is more extended, and the pattern of black lines is quite different.
Occurrence: This species is locally common, but not very widely distributed in Jamaica. Like *P. pupa*, it occurs in rock pools, usually above the high water mark, in large colonies. Our specimens are from Prospect and Falmouth.
Note: Numerous intergrades suggest that this and the preceding shell (*P. pupa*) are in fact the same species.

Genus *Neritina* Lamarck 1816

Subgenus *Vitta* Mörch 1852

NERITINA VIRGINEA Linné 1758 Virgin Nerite **pl. 4**
Bermuda, Florida to Texas, and the West Indies.
Description: Length ¼ to ¾ inch. Colour: patterns endlessly variable, generally in combinations of white, black, red, yellow, olive or purple, with stripes, dots or irregular markings. The parietal area is usually yellowish, and the surface of the shell is smooth and polished. There are many small, irregular teeth. Operculum generally black. Many of the colour patterns are unusually beautiful, and these shells merit careful examination.
Occurrence: Common on broad mud flats in 6 to 24 inches at Port

Henderson, and abundant just east of Falmouth. They prefer protected, brackish water and can usually be found wherever these conditions exist.

NERITINA MELEAGRIS Lamarck 1822 Scaly Nerite **pl. 4**
The Antilles, central and northern South America.
Description: Length ½ inch. Colour brownish olive to bluish grey or purple, with the pattern in the form of scales outlined in white. Shell smooth and polished. Columella edge with a variable number of irregular teeth – generally not more than 10.
Occurrence: Abundant on beds of eel grass in 6 to 24 inches in Kingston harbour, and many other protected, brackish-water areas all around Jamaica. Beach specimens are very common.

NERITINA CLENCHI Russell 1940 Clench Nerite **pl. 4**
Florida, the Antilles and Central America.
Description: Length ¾ to 1 inch. Colour not unlike *N. virginea*, but less variable. The parietal area opposite the few minute teeth is bordered with orange-yellow. The shell is smooth and polished; it is larger and less globose than *N. virginea*. Operculum black to pink.
Occurrence: This can really be classified as a fresh water species, occasionally found in brackish water. Dead shells found in the sea have almost certainly been washed down by the rivers. It is common in the stream running through Frenchman's Cove Hotel (Portland) and in many other Jamaican rivers.

NERITINA PIRATICA Russell 1940 Netted Nerite **pl. 4**
Antilles, Central to South America.
Description: Length about ¾ inch. Colour brownish yellow to olive green, with black zigzag axial lines forming a reticulated network. Columella edged with a number of minute irregular teeth. It is an exceptionally beautiful species.
Occurrence: This is a locally common brackish-water species with a very limited distribution. It occurs in a few inches of water in the mangrove swamps of Dawkins Pond, west of the causeway in Kingston harbour. We have found beach specimens at Orange Bay (Hanover).

NERITINA PUNCTULATA Lamarck 1816
Spotted Nerite **pl. 4**
Antilles and Central America.
Description: Length ¾ to 1 inch. Colour generally purplish-pink, with numerous, suboval whitish spots. Columella area white, sometimes with very faint irregular teeth. Shell smooth but dull, with a low spire.

Occurrence: Reasonably common in the Richmond-Llandovery River (St Ann), and in many other rivers throughout Jamaica. This is a fresh water species, and is included here only because its shell is often found on beaches near the mouths of rivers. As in most members of the genus, the arrangement of the colour pattern produces a strange and attractive three dimensional effect.

NERITINA RECLIVATA Say 1822 Olive Nerite **pl. 4**
Florida, Texas and the West Indies.
Description: Length about $\frac{3}{4}$ inch. Colour brownish-green or dark olive, usually with numerous fine axial lines of black-brown or lavender. Shell smooth and polished; parietal area yellowish.
Occurrence: Relatively common in shallow, brackish water at Port Henderson, where beach specimens are often claimed by hermit crabs. It has also been reported from Negril and Falmouth.

SUBFAMILY SMARAGDIINAE

Genus *Smaragdia* Issel 1869

Subgenus *Smaragdia* Issel 1869

SMARAGDIA VIRIDIS VIRIDEMARIS Maury 1917
Emerald Nerite **pl. 4**
Bermuda, south-east Florida and the West Indies
Description: Length $\frac{1}{4}$ to $\frac{1}{3}$ inch. Colour usually bright green or yellowish, sometimes with small white or purple markings. Deeper water specimens sometimes lack the green or yellow base colour, and have only purplish-white markings on a white background. Shell smooth and polished; parietal area with 7 to 9 minute teeth. The bright green colour of most specimens makes this an easy species to spot in drift sand, in spite of its small size.
Occurrence: A very common species on eel grass in 1 to 10 feet at Negril, Ocho Rios, and in shallow sandy areas all around Jamaica.

Order Mesogastropoda

SUPERFAMILY LITTORINACEA

FAMILY LITTORINIDAE

Members of this family are small, rather drab inhabitants of the littoral zone. Most Caribbean genera live above the high tide mark.

The eggs of several species hatch inside the oviducts, in contrast to the usual gastropod process. The family is characterised by the typical horny, paucispiral operculum, and the absence of an umbilicus. *Littorinidae* are all vegetarian and the sexes are separate.

Genus *Littorina* Férussac 1822

Subgenus *Melaraphe* Menke 1828

LITTORINA ZICZAC Gmelin 1791 Zebra Periwinkle **pl. 4**
Bermuda, Southern Florida, Texas and West Indies.
Description: Length about ½ inch. Colour light or dark grey, with oblique wavy or zigzag lines of purplish brown. Aperture and columella varying shades of brown. Operculum dark brown. Shell with very fine, lightly incised spiral threads.
Occurrence: Abundant on the groins, within and just above the intertidal area, on the southern side of the Palisadoes, and in rocky, exposed areas all around Jamaica.
Note: Recent research suggests that what were previously considered juvenile examples of this shell may be the sibling species *L. floccosa* and *L. lineolata* (see *Seashells of North America* by R. T. Abbott).

LITTORINA NEBULOSA Lamarck 1822
Cloudy Periwinkle **pl. 4**
West Indies.
Description: Length ½ to ¾ inch. Colour usually bluish-white, often with pale red-brown speckling; operculum pale brown. Sculpture of weak spiral lines on all but the first two or three whorls. The shell is light but strong. This species is very similar in many respects to *L. angulifera*, but is more globular, less colourful and smaller in size.
Occurrence: This is one of the less common of the Jamaican *Littorinidae*. It occurs on rocks and wooden pilings, just above the water, in Kingston harbour, Prospect, Discovery Bay and Falmouth, but rarely

in large numbers. Excellent examples can be found on the jetty near the Government Explosives Store at Port Henderson.
Forms: L. nebulosa tessellata is a form with a regular checked pattern of red-brown on a dirty white background (pl. 4). Our specimens come from Negril, where intergrades are not uncommon.

Subgenus *Littoraria* Gray 1834

LITTORINA ANGULIFERA Lamarck 1822
Angulate Periwinkle **pl. 4**
Bermuda, southern Florida and West Indies.
Description: Length 1 to 1½ inches. Colour variable, generally reddish or purplish brown, with darker speckling. The earlier whorls tend to be darker. Operculum pale brown; columella purple in some specimens. Sculpture of many fine spiral threads, except on first two whorls. The shell is thin but strong.
Occurrence: Nearly always above high water mark, generally attached to wood rather than rock. Very common on mangrove roots on the harbour side of the Palisadoes, and in similar areas all around Jamaica.

Subgenus *Neritrema* Récluz 1869

LITTORINA MELEAGRIS Potiez and Michaud 1838
Spotted Periwinkle **pl. 4**
Southern Florida and the West Indies.
Description: Length ¼ inch. Colour light to dark brown or purple, with numerous white spots more or less in spiral rows. The shell has a very thin periostracum, and a brown operculum.
Occurrence: In the inter-tidal zone in crevices on moss covered rocks, often exposed to the open sea. It is common on the rock tables on the southern side of the Palisadoes, and in many similar areas around Jamaica.

LITTORINA MESPILLUM Mühlfeld 1824
Dwarf Periwinkle **pl. 4**
Florida, Key West and the Caribbean.
Description: Length ¼ inch. Colour white to brown; some specimens are covered with rows of very small black or dark red spots. Aperture glossy brown with a narrow umbilicus. Periostracum thin and brown.
Occurrence: In rock pools from the high tide mark to several feet above

it. It is locally abundant – as at Prospect and Old Pera (St Thomas) – but for some unexplained reason is found in relatively few areas around the Jamaican coast.

Genus *Nodilittorina* von Martens 1897

Subgenus *Echinolittorina* Habe 1956

NODILITTORINA TUBERCULATA Menke 1828

Common Prickly-winkle **pl. 4**

Bermuda, southern Florida and the West Indies.

Description: Length ½ to ¾ inch. Colour dull grey with lighter tubercles; columella dark brown. Sculpture of several spiral rows of fairly sharp nodules. Operculum dark brown and paucispiral.

Occurrence: On rocks just above the high tide mark at Port Henderson and Runaway Bay, and in rocky areas all around Jamaica. It is a very common species, often found in association with the following species (*Echininus nodulosus*).

Genus *Echininus* Clench and Abbott 1942

Subgenus *Tectininus* Clench and Abbott 1942

ECHININUS NODULOSUS Pfeiffer 1839

False Prickly-winkle **pl. 4**

South-east Florida and the West Indies.

Description: Length ½ to 1 inch. Very similar in most respects to *Nodillitorina tuberculata*, but it has a broader and squarer base, more prominent nodules, and the operculum is multispiral.

Occurrence: As for *N. tuberculata*. They are occasionally found side by side, but *nodulosus* prefers higher ground.

Genus *Tectarius* Valenciennes 1833

Subgenus *Cenchritis* von Martens 1900

TECTARIUS MURICATUS Linné 1758

Beaded Periwinkle **pl. 4**

Bermuda, Lower Florida Keys and West Indies.

Description: Length ¾ to 1 inch. Colour light brown, grey or white. Interior light brown; operculum dark brown and paucispiral. Sculpture of several spiral rows of white beads; more or less umbilicated. Shell strong and solid with a thick lip.

Occurrence: Abundant on rocks and trees from 1 to 30 feet above the high tide mark at Buff Bay, Port Henderson, and in rocky areas all around Jamaica. One of the most common and accessible of Jamaican gastropods.

SUPERFAMILY RISSOACEA

FAMILY HYDROBIIDAE

These are small, cylindrical, essentially land-dwelling shells, characterised by their corneous paucispiral operculums. The family is thought to be vegetarian.

In the genus *Truncatella*, the first 4 or 5 whorls are shed when the snail attains maturity, and this accounts for the shell's truncated appearance.

Genus *Truncatella* Risso 1826

TRUNCATELLA PULCHELLA Pfeiffer 1839

Beautiful Truncatella **fig. 14**

South-east U.S. and the West Indies.

Description: Length about $\frac{1}{3}$ inch. Colour white to light amber. Sculpture of 17 or more poorly developed axial costae. Operculum paucispiral and corneous.

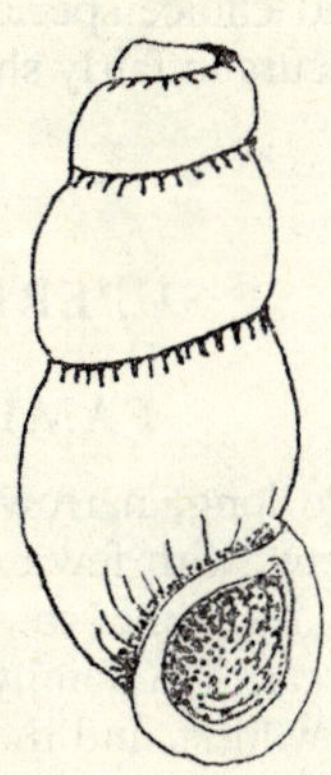

FIG. 14. *Truncatella pulchella*

Occurrence: Members of this genus are essentially land dwellers, though their shells are frequently found in drift sand. *T. pulchella* is reasonably common just above the high water mark along the harbour side of the Palisadoes, and in many other similar areas. The other Jamaican members of the genus are listed at the end of this section on gastropods.

FAMILY VITRINELLIDAE

These are small, delicate, circular shells, generally either opaque-white or transparent. Many species possess an unusually beautiful sculpture, which can be clearly seen only with the aid of a hand lens. Most West Indian members of the family are very uncommon and prefer moderately deep water. Operculums are corneous and multispiral.

Genus *Cyclostremiscus* Pilsbry and Olsson 1945

Subgenus *Ponocyclus* Pilsbry 1953

CYCLOSTREMISCUS BEAUI Fischer 1857

Beau's Vitrinella **pl. 3**

South-east U.S. and the West Indies.

Description: Diameter about $\frac{1}{2}$ inch. Colour chalky white. There are 4 rounded whorls sculptured with several smooth, low spiral cords which have numerous finer threads between them. Umbilicus broad and deep. Shell thin and light, but relatively strong.

Occurrence: In about 40 feet, on a sandy or muddy bottom, in Kingston harbour, Discovery Bay and at Treasure Beach (St Elizabeth). This is a rare and choice species in Jamaica. Its known distribution suggests that it occurs in fairly sheltered water right around the island.

SUPERFAMILY CERITHIACEA

FAMILY TURRITELLIDAE

These are long, narrow, shells with sharp apexes and many whorls. They occur, with few exceptions, on muddy bottoms.

There are some 50 species, widely distributed in temperate and tropical seas. The family is characterised by thin corneous operculums with few whorls, and the habit, in some species, of producing detached later whorls which are drawn out in an irregular way.

All species are ciliary feeders.

Genus *Turritella* Lamarck 1799

TURRITELLA EXOLETA Linné 1758

Eastern Turret Shell **pl. 5**

Southern Florida and the West Indies.

Description: Length 2 to 3½ inches. Colour whitish, sometimes with flame-like splotches of dark or very faint brown. The whorls are ridged at the top and bottom, and the intervening area is concave. There are scale-like axial lamellations on the whorls which become more pronounced towards the aperture.

Occurrence: In 100 feet off Rackhams Key (near Port Royal); in 40 feet off the Bogue Islands; and in 30 to 160 feet at Discovery Bay and in many other sandy or muddy areas around Jamaica. Perfect adult shells are still choice items, but commercial dredgings on the north coast have recently brought up hundreds of small or damaged specimens.

Note: A live specimen in an aquarium was seen to travel backwards, i.e. with the apex foremost.

TURRITELLA VARIEGATA Linné 1758

Variegated Turret Shell **pl. 5**

The West Indies.

Description: Length 3 to 4½ inches. Colour pale brown with reddish or purple-brown variegations. The whorls are nearly smooth and the suture is moderately indented. Occasional specimens are found with rounded whorls.

Occurrence: This is a common species in many West Indian islands. Surprisingly, it does not occur in Jamaica. It is found in shallow water, on mud or sand, mainly in protected areas, and beach specimens are quite common in Puerto Rico and Antigua. It is included here as a matter of interest.

Genus *Vermicularia* Lamarck 1799

VERMICULARIA KNORRI Deshayes 1843

Knorr's Worm Shell **pl. 5**

South-east U.S. to the West Indies.

Description: Length up to 4 inches. Colour light brown except for the early whorls, which are usually whitish and evenly coiled; the later whorls become detached and irregularly drawn out. Operculum light brown, circular and concave. Shell light but strong.

Occurrence: In sponges at about 45 feet at Cow Bay and in sponge-supporting areas all around Jamaica. Damaged shells are quite commonly washed ashore.

VERMICULARIA SPIRATA Philippi 1836

West Indian Worm Shell **pl. 5**

South-east Florida and the West Indies.

Description: Essentially similar to the preceding species (*V. knorri*), but thicker and with distinctive sharp spiral cords. The early whorls are dark in colour.

Occurrence: As for *V. knorri*, but this is a far less common species. Beach specimens are occasionally found at Negril, but rarely in perfect condition.

FAMILY ARCHITECTONICIDAE

Members of this family are compact, tightly coiled shells with a wide and deep umbilicus. There are some 40 species which occur in most tropical seas, often concealed in the substrate.

The family is characterised by the horny, pill-like operculum present in most species. *Architectonicidae* are attractive shells, and favourites with collectors everywhere. Many species live in association with corals.

Genus *Heliacus* Orbigny 1842

HELIACUS CYLINDRICUS Gmelin 1791

Cylinder Sundial **pl. 6**

Lower Florida Keys and West Indies.

Description: Height about ½ inch, and the same in diameter. Colour light reddish brown to very dark grey, usually with dotted white lines above the suture and on the periphery. Rounded whorls sculptured with many minutely beaded spiral cords. Umbilicus deep and narrow with a crenulated border. The very young shells lack the conical shape of the adults, and are sometimes mistakenly classified as a separate species.

Occurrence: In 10 to 40 feet on *Palythoa* coral at Bull Bay, in Ocho Rios harbour and in many areas around Jamaica. Fair specimens are quite often washed ashore at Negril, and excellent examples, inhabited by small hermit crabs, can be collected there in very shallow water.

HELIACUS BISULCATUS Orbigny 1845

Orbigny's Sundial **pl. 6**

South-east U.S. to the West Indies.

Description: Diameter ¼ to ½ inch. Colour light yellowish brown. The shell is flat and the whorls are sculptured with revolving, minutely beaded cords. There are two rows of larger beads on the periphery. Umbilicus deep and wide with a crenulated border; operculum conic. *Occurrence:* In 35 to 40 feet on muddy bottoms at the Bogue Islands and Port Royal. It is occasionally found in drift sand, particularly in areas where deep water occurs close to the beach. It is an uncommon species.

HELIACUS INFUNDIBULIFORMIS Gmelin 1791
Channeled Sundial **pl. 6**
West Indies.
Description: Diameter about $\frac{5}{8}$ inch. Colour light or dark brown, sometimes with lighter streaks on the last whorl. Whorls rounded, separated by a deep channel and sculptured by minutely beaded, spiral cords. Umbilicus wide and deep. In large specimens, the final whorl is occasionally detached from the preceding one.
Occurrence: Uncommon in about 25 feet at Discovery Bay and Orange Bay, either on, or close to, coral reefs. Small specimens occasionally turn up in drift sand.

Genus *Architectonica* Röding 1798

ARCHITECTONICA NOBILIS Röding 1798
Common Sundial **pl. 5**
South-east U.S. to the West Indies.
Description: Diameter 1 to 2 inches. Colour cream with orange-brown spots and light purple markings. Sculpture of 4 or 5 low spiral cords which may or may not be beaded; umbilicus deep, and bordered by a strongly crenulated cord. Operculum light brown, thin and paucispiral. The shell is quite heavy and strong.
Occurrence: In sand at about 35 feet at Montego Bay, Kingston harbour and Ocho Rios, and off Bull Bay. Elsewhere in the West Indies this is considered a moderately common species, but it is a choice and much sought after shell in Jamaican waters.

Genus *Philippia* Gray 1847

Subgenus *Psilaxis* Woodring 1928

PHILIPPIA KREBSI Mörch 1875 Kreb's Sundial **pl. 6**
South-east U.S. to the West Indies.
Description: Diameter $\frac{1}{4}$ to $\frac{1}{2}$ inch. Colour tan, with irregular light brown markings, particularly at the periphery and below the suture. The shell is smooth and glossy except for two microscopic spiral

Plate 3 TURBANS, TOP SHELLS, TEGULAS, CYCLOSTREMES, PHEASANT SHELLS

Scale: reduced by 20% (See also **Pl. 5**)

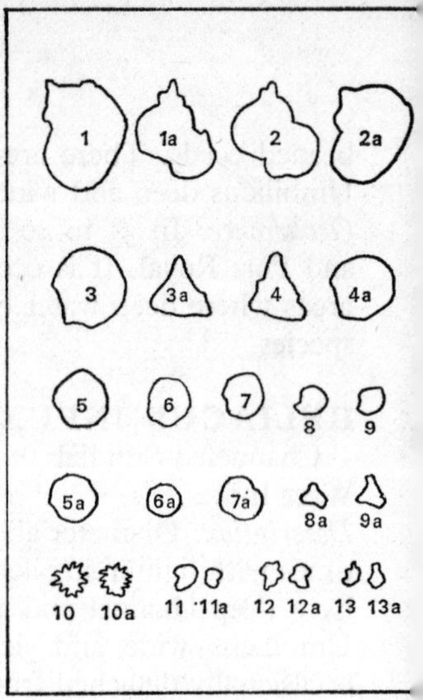

1. **Turbo castanea** Gmelin Chestnut Turban 61
Typical colour variety. **1a,** side view. Note distinctive operculum.

2. **Turbo filosus** Wood Threaded Turban 62
Side view of the rare yellow form. **2a,** dorsal view. Typical colour variety.

3. **Calliostoma javanicum** Lamarck Chocolate-lined Top Shell 58
3a, side view.

4. **Calliostoma jujubinum** Gmelin Jujube Top Shell 59
Side view of typical colour variety. **4a,** dorsal view. Purple and black colour variety.

5. **Tegula lividomaculata** C. B. Adams West Indian Tegula 57
5a, ventral view.

6. **Tegula excavata** Lamarck Green Tegula 58
6a, ventral view.

7. **Tegula fasciata** Born Smooth Atlantic Tegula 57
Red colour variety. **7a,** ventral view.

8. **Calliostoma sarcodum** Dall 59
8a, side view.

9. **Calliostoma pulchrum** C. B. Adams Beautiful Top Shell 59
9a, side view.

10. **Arene cruentata** Mühlfeld Star Arene 60
10a, ventral view.

11. **Cyclostremiscus beaui** Fischer Beau's Vitrinella 76
11a, ventral view.

12. **Cyclostrema cancellatum** Marryat Cancellate Cyclostreme 61
12a, ventral view.

13. **Tricolia affinis** C. B. Adams Polka-dot Pheasant 67
13a, side view.

Plate 4 NERITES, PERIWINKLES, PRICKLY-WINKLES

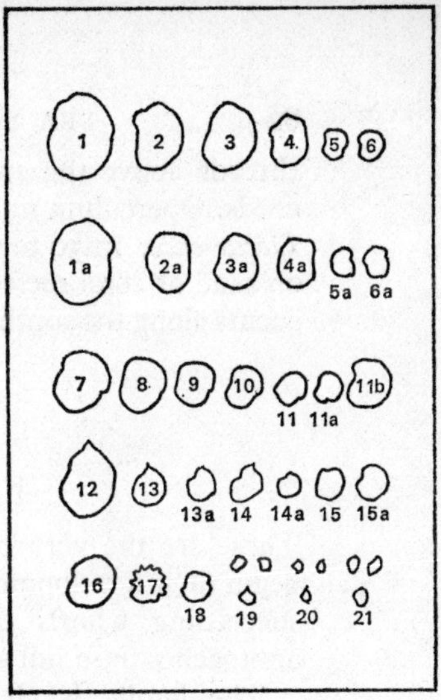

Scale: natural size

1. Nerita peloronta Linné 67
Bleeding Tooth
1a, ventral view. Note the characteristic parietal stain.

2. Nerita versicolor Gmelin 68
Four-toothed Nerite
2a, ventral view.

3. Nerita fulgurans Gmelin 68
Antillean Nerite
3a, ventral view.

4. Nerita tessellata Gmelin 68
Tessellated Nerite
4a, ventral view.

5. Puperita puper Linné 69
Zebra Nerite
5a, ventral view.

6. Puperita tristis Orbigny 69
6a, ventral view.

7. Neritina reclivata Say 71
Olive Nerite

8. Neritina piratica Russell 70
Netted Nerite

9. Neritina clenchi Russell 70
Clench Nerite

10. Neritina punctulata 70
Lamarck Spotted Nerite

11. Neritina virginea Linné 69
Virgin Nerite
11a and 11b, colour variation.

12. Littorina angulifera 73
Lamarck Angulate Periwinkle

13. Littorina ziczac Gmelin 72
Zebra Periwinkle
13a, sibling species. See text note under *Littorina ziczac.*

14. Littorina nebulosa 72
Lamarck Cloudy Periwinkle
14a, *L.n. tessellata* philippi. Colour form of **14.**

15. Neritina meleagris 70
Lamarck Scaly Nerite
15a, colour variation.

16. Tectarius muricatus 74
Linné Beaded Periwinkle

17. Echininus nodulosus 74
Pfeiffer False Prickly-winkle

18. Nodilittorina tuberculata 74
Menke Common Prickly-winkle

19. Littorina mespillum 73
Mühlfeld Dwarf Periwinkle

20. Littorina meleagris 73
Potiez and Michaud
Spotted Periwinkle

21. Smaragdia viridis viridemaris Maury 71
Emerald Nerite

threads above the suture. Umbilicus bordered by two beaded spiral cords. Operculum multispiral.

Occurrence: In 30 to 60 feet in Ocho Rios harbour and at Orange Bay on sand or coral reefs. Locally it is quite common, and it probably also occurs along the south coast. We have rarely found a beach specimen.

FAMILY VERMETIDAE

These are the very common and familiar Worm Shells which usually begin their development with regularly coiled, Turitella-like shells. The succeeding whorls detach and grow haphazardly, as the mollusc approaches the adult stage.

The family is characterised by circular apertures and corneous operculums. Most species are ciliary feeders.

Genus *Petaloconchus* H. C. Lea 1843

Subgenus *Macrophragma* Carpenter 1856

PETALOCONCHUS IRREGULARIS Orbigny 1842

Irregular Worm Shell **pl. 6**

Southern Florida and the West Indies.

Description: Length 1 to 2½ inches, depending on the tightness of the coiling. Colour brown. The shells are worm-like, much coiled and bear heavy longitudinal and spiral cords. They live in closely packed groups.

Occurrence: Attached to rocks in 3 to 20 feet at Prospect, Negril, and in many areas all around Jamaica. A common species, frequently washed ashore, usually in clusters of many individual shells.

PETALOCONCHUS ERECTUS Dall 1888

Erect Worm Shell **pl. 6**

South-east Florida and the West Indies.

Description: Length 1 to 2 inches. Colour white or cream. Similar in most respects to *P. irregularis*, but smoother, smaller in diameter, and generally less coiled.

Occurrence: As for *P. irregularis*, but not as common. Detached shells are occasionally found, in about 6 feet, on the reefs at Discovery Bay.

PETALOCONCHUS MCGINTYI Olsson and Harbison 1953 McGinty's Worm Shell **pl. 6**

Florida, Jamaica and Puerto Rico.

Description: Length 2 to 3 inches. Colour pinkish brown. The shell is solitary and attached by the earliest whorls; the terminal tube rises from the centre. Sculpture of growth lines and very faint, pustulose spiral threads.

Occurrence: On coral reefs at Discovery Bay and Negril in about 10 feet of water; on wharf pilings in Kingston harbour. Sections of the shell are commonly broken off and washed ashore. They resemble straight pink tubes, and are relatively common on the beach at Negril.

PETALOCONCHUS FLORIDANUS Olsson and Harbison 1953 Florida Worm Shell

Florida and the West Indies.

Description: Length about 1 inch. Colour brown. The shell is irregularly coiled and attached to some object along the whole length of one side. The surface is coarsely reticulate. The shells are usually found in tightly packed colonies, but may occasionally be solitary.

Occurrence: As for *P. mcgintyi.* It is a common species, and beach worn clusters are washed ashore on most beaches all around Jamaica.

Genus *Serpulorbis* Sasso 1827

SERPULORBIS DECUSSATA Gmelin 1791 Decussate Worm Shell **pl. 6**

South-east U.S. to the West Indies.

Description: Length about 1 inch. Colour brown or yellowish. Sculpture of prominent longitudinal ridges.

Occurrence: Commonly found attached to solid objects in 2 to 8 feet at Negril and in many other areas around Jamaica. Beach specimens are very common.

SERPULORBIS RIISEI Mörch 1862 Riise's Worm Shell **pl. 6**

West Indies.

Description: Length $1\frac{1}{2}$ to 2 inches. Colour cream or light brown above, dark brown or purple below. Sculpture similar to *S. decussata,* but coarser.

Occurrence: As for *S. decussata.* It is equally common, and worn purplish tube sections are found on most beaches.

FAMILY SILIQUARIIDAE

These are distinctive worm-like shells, irregularly coiled and unattached throughout. The family is characterised by the unusual, narrow axial slit in the centre of the whorls, which extends for the entire length of the shell. All species are ciliary feeders.

Genus *Siliquaria* Bruguière 1789

SILIQUARIA ANGUILLAE Mörch 1861
Slit Worm Shell **pl. 6**
West Indies.
Description: Length 1 to 4 inches. Colour white; later whorls may occasionally have a brownish tinge. Early whorls smooth; later whorls spinose. The shell is easily identified by the narrow slit in the whorls, which extends for the length of the shell. The whorls are irregularly coiled and separate. The shell is light but surprisingly strong.
Occurrence: Reasonably common from 35 feet in most protected bays along the north coast. It lives in mud or soft sand, and commercial dredgings at Ocho Rios and Montego Bay have shown that it can be locally abundant.
Forms: There is a smoother, more delicate form (possibly a separate species) which occurs in the Blue Hole (Portland). Sculpture of numerous fine growth lines.

FAMILY PLANAXIDAE

These are small, imperforate shells with sharp spires. They prefer very shallow water, and are characterised by oval, paucispiral operculums. The family is vegetarian.

Genus *Planaxis* Lamarck 1822

PLANAXIS NUCLEUS Bruguière 1789
Black Atlantic Planaxis **pl. 6**
South-east Florida and the West Indies.
Description: Length ½ inch. Colour dark brown or black; periostracum grey-black. Sculpture of widely spaced grooves, deeply etched on the

base, behind the outer lip and below the suture. The outer lip is thick and grooved within. The adult shell is strong and solid.
Occurrence: At Robins Bay, Sandy Bay (Hanover) and Cow Bay in the inter-tidal area, on smooth rocks. Locally this is a common species, but its areas of occurrence are quite widely separated. It is found on the north and south coasts, in both exposed and protected habitats.

FAMILY MODULIDAE

This family contains a single genus with about 12 species. Only two of these occur in the western Atlantic, and both are found in West Indian waters.

The shells are small, solid and circular, with round apertures. In both local species there is a sharp tooth at the base of the columella. *Modulidae* are vegetarian and prefer quiet, shallow water.

Genus *Modulus* Gray 1842

MODULUS MODULUS Linné 1758 Atlantic Modulus **pl. 6**
Florida to Texas and the West Indies.
Description: Length about ½ inch. Colour variable, but generally greyish white with irregular brown flecks. Sculpture of strong radiating ribs and numerous spiral cords; outer lip with a number of sharp ridges within. There is a prominent tooth where the columella terminates. The colour variations of this shell may cause some difficulty in identification; the basic shape and sculpture always remain constant however. Shells from deeper water tend to be larger and all white.
Occurrence: Common on eel grass in 1 to 6 feet on the harbour side of the Palisadoes, and in this relatively protected type of environment all around Jamaica. The living shell is often covered with algae, which effectively camouflages it on the *thalassia* blades. Dead shells are commonly adopted by hermit crabs.

MODULUS CARCHEDONIUS Lamarck 1822
Angled Modulus **pl. 6**
Greater Antilles to northern South America.
Description: Length about ½ inch. Colour variable, generally greyish purple to light orange, with dotted spiral cords. The base of the shell is always of a lighter colour, with brown dots on all the spiral cords. Sculpture of several small spiral cords, with a sharply angled periphery.

Its columella tooth is always white, and the interior of the aperture is ridged. The shell is not unlike *M. modulus*, but can easily be separated from it by the sharp periphery and lack of axial ribs.
Occurrence: Reasonably common on algae in 4 to 16 feet at Port Henderson, Rocky Point (Clarendon), and in many other areas where there is a mud or soft sand bottom conducive to the growth of the larger algae. Dead shells are found along the causeway in Kingston harbour.

FAMILY POTAMIDIDAE

Members of this family are small, elongate shells with convex whorls. All the Jamaican species live on inter-tidal mud flats, and generally occur in large colonies. Most *Potamididae* are detritus feeders.

The shells are characterised by thin, horny multispiral operculums with the nucleus at the centre.

SUBFAMILY POTAMIDINAE

Genus *Cerithidea* Swainson 1840

CERITHIDEA COSTATA da Costa 1778
Turret Horn Shell **pl. 6**
Florida and the West Indies.
Description: Length ½ inch. Colour light to dark brown. Sculpture of thick, prominent ribs (25 to 30 on the penultimate whorl).
Occurrence: Abundant at Port Henderson on the inter-tidal mud flats, and washed up on the beach. It is not a widespread species and apparently requires relatively sheltered, brackish water.
Forms: Cerithidea costata turrita Stearns is a less common form with 15 to 20 axial ribs on the penultimate whorl. It is also found at Port Henderson.

CERITHIDEA BEATTYI Bequaert 1942
Beaded Horn Shell
Bermuda, Bahamas, Puerto Rico, Virgin Islands and Trinidad.
Description: Length about 1 inch. Colour variable, generally light brown with a light ring on the lower part of each whorl; occasionally all white. Sculpture much like *C. costata*, but 2 or 3 spiral grooves divide the vertical ribs into tubercles. According to Abbott (*Caribbean Sea Shells*) this may be a form of *C. costata*.
Occurrence: As for *C. costata*, but it is less common.

CERITHIDEA PLICULOSA Menke 1829
Plicate Horn Shell **pl. 6**
Texas, Louisiana and the West Indies.
Description: Length about 1 inch. Colour, varying shades of brown. Sculpture of prominent former varices, and many vertical ribs and spiral threads. Base of shell with several spiral ridges.
Occurrence: As for *C. costata*, with which it is often found living. Dead shells are found all along the causeway in Kingston harbour.

SUBFAMILY BATILLARIINAE

Genus *Batillaria* Benson 1842

BATILLARIA MINIMA Gmelin 1791 False Cerith **pl. 6**
Southern Florida and the West Indies.
Description: Length $\frac{1}{2}$ to $\frac{3}{4}$ inch. Colour varies from all greyish white to all black, with innumerable combinations of both colours (often in rings). Sculpture of coarse axial protuberances and uneven spiral threads. In some localities as many as 20% of the shells may be deformed in some way. Operculum brown and multispiral.
Occurrence: An extraordinarily abundant species on inter-tidal mud flats, in brackish water, everywhere around Jamaica. In some places, at low tide the exposed areas are thickly carpeted with these shells. Port Henderson and the swamps near Gunboat Beach (Palisadoes) are two accessible places where they can be collected.

FAMILY CERITHIIDAE

Members of this family live in shallow water in tropical seas all over the world. There are 300 known species, all of which are detritus feeders.

The shells are characterised by the deep channel at the front of the aperture, and the thin, horny, paucispiral operculum. Most of the Jamaican species are locally abundant.

Genus *Cerithium* Bruguière 1789

CERITHIUM LITTERATUM Born 1778
Stocky Cerith **pl. 7**
Bermuda, South-east Florida and the West Indies.
Description: Length about 1 inch. Colour white, maculated with light

orange, brown or black; occasional specimens may have a purplish cast to the aperture. Sculpture of many spiral threads, and 9 to 12 nodules in a spiral row just below the suture. The prominence of the nodules varies greatly, even in species from the same locality. Operculum thin and reddish brown.

Occurrence: Very common on weed-covered rocks in 1 to 150 feet at Discovery Bay, Negril, and many other areas all around Jamaica. It is easily collected on rock-tables just below low water mark at Prospect, where algal growths on the shell provide effective camouflage.

CERITHIUM VARIABILE C. B. Adams 1845

Dwarf Cerith **pl. 7**

Southern Florida to Texas and the West Indies.

Description: Length ½ inch. Colour light to dark brown. Sculpture similar to *Batillaria minima*, but the inside of the aperture is much thicker and there is a broad channel where the outer lip joins the body whorl. The operculum is paucispiral.

Occurrence: Relatively common on algae in 1 to 3 feet at Priory and Runaway Bay. Dead shells are quite common on the beaches at Negril and just east of Montego Bay.

CERITHIUM EBURNEUM Bruguière 1792

Ivory Cerith **pl. 7**

The Bahamas, south-east Florida and the Greater Antilles.

Description: Length ¾ to 1 inch. Colour usually white, speckled with light brown; some specimens are all white. Sculpture of 4 to 6 beaded rows of 18 to 22 beads per whorl. There may or may not be former varices present.

Occurrence: As for *C. litteratum.* It is particularly common at Negril, where dead shells litter the beaches.

CERITHIUM ALGICOLA C. B. Adams 1848

Middle-spined Cerith **pl. 7**

Southern Florida and the West Indies.

Description: Length 1 to 1½ inches. Colour white with occasional brown splotches. Sculpture similar to *C. litteratum*, but it lacks the sharp nodules just below the suture, and has its most prominent nodules in the centre of the whorl. Former varices are present in about 50% of specimens.

Occurrence: As for *C. litteratum*, but it is not as common. Beach specimens at Negril lose their brown coloration after prolonged exposure to the sun.

CERITHIUM AURICOMA Schwengel 1940
Schwengel's Cerith **pl. 7**

Florida, Jamaica and Puerto Rico.

Description: Length about $1\frac{1}{2}$ inches. Colour white with light brown stripes. Sculpture of 8 axial ribs per whorl on the last 4 whorls, and numerous fine spiral threads over the entire shell. Former varices are usually present.

Occurrence: In about 30 feet off the Bogue Islands, at Negril and Rocky Point, at the mouth of the Salt River (Clarendon). It is a rather rare species, and we have found no beach specimens.

CERITHIUM MUSCARUM Say Fly-specked Cerith **pl. 7**

Southern Florida and the West Indies.

Description: Length $\frac{3}{4}$ to 1 inch. Colour grey, with spiral rows of red-brown flecks. Sculpture of 9 to 11 nodulous axial ribs per whorl; there is a single spiral cord at the base of the shell. All the whorls are markedly convex.

Occurrence: This is a rare species in Jamaica. It occurs at Negril in about 20 feet, and dead shells are occasionally found on the beaches there. It appears to be more common in the southern Caribbean.

SUPERFAMILY EPITONIACEA

FAMILY JANTHINIDAE

Members of this unusual family are modified for a pelagic existence, far out of sight of land. They produce a float of air bubbles, which is attached to the shell, and in which the eggs are placed. (See fig. 15.)

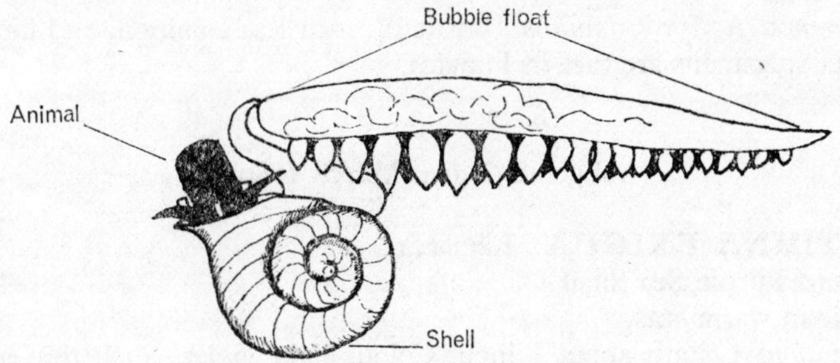

FIG. 15. A Purple Sea Snail (*Janthina globosa*) supported by its bubble float

The shells are graceful, delicate and purplish in colour. The animal is blind, has no operculum and is carnivorous. All species are hermaphroditic.

Genus *Janthina* Röding 1798

Subgenus *Janthina* Röding 1798

JANTHINA JANTHINA Linné 1758
Common Purple Sea Snail **pl. 7**
Pelagic in warm seas.
Description: Diameter 1 to $1\frac{1}{2}$ inches. Colour pale violet above, deep purple below. Thin and fragile with an expanded outer lip. This is a very beautiful and well-known shell, which lives on the surface of the sea far out in the Atlantic Ocean.
Occurrence: These shells are quite often blown ashore on the north coast, but we have few records from south coast beaches. Occasionally, perfect living specimens can be taken from the water just before they hit the beach. The unusual animal merits close examination. This species preys on the beautiful jellyfish *Velella,* better known as the 'By-the-Wind Sailor'.

Subgenus *Violetta* Iredale 1929

JANTHINA GLOBOSA Swainson 1822
Globe Purple Sea Snail **pl. 7**
Pelagic in warm seas.
Description: Diameter about $\frac{3}{4}$ inch. Colour uniform pale violet. Shell thin, fragile and globular. Quite as beautiful, but less well known than *J. janthina.*
Occurrence: As for *J. janthina,* but it is much less common, and large, perfect specimens are rare in Jamaica.

Subgenus *Jodina* Mörch 1860

JANTHINA EXIGUA Lamarck
Dwarf Purple Sea Snail **pl. 7**
Pelagic in warm seas.
Description: Length about $\frac{1}{3}$ inch. Colour light violet. Shell thin and fragile, distinguished by a prominent notch in the centre of the outer lip.

Occurrence: As for *J. janthina*, but this is a very uncommon species on Jamaican beaches. Our specimens were found on the beach at Robins Bay.

FAMILY EPITONIIDAE

This is the distinctive family of 'Staircase' shells which have been favourites with collectors everywhere for many years. There are about 200 species, many of which occur only in very deep water. They are found in nearly all seas, from the Arctic to the Antarctic.

All the Jamaican species are of delicate, and sometimes exquisite, construction. The animals are capable of excreting a purple dye which occasionally colours dead shells. *Epitoniidae* are carnivorous, and many species, in turn, are eaten by fish. Most Caribbean genera feed on anemones or corals, and are consequently found in association with these animals.

Genus *Cirsotrema* Mörch 1852

CIRSOTREMA DALLI Rehder 1945
Dall's Wentletrap **pl. 7**
South-east U.S., the West Indies to Brazil.
Description: Length 1 to $1\frac{1}{2}$ inches. Colour a rather dirty white. The whorls are strongly shouldered and the sculpture is elaborate, with numerous foliated costae. Surface pitted with small holes when the costae are close together. This is a very beautiful shell; beach-worn specimens have usually lost their delicate sculpture and much of their beauty.
Occurrence: In 30 to 60 feet, on mud or soft sand bottoms, in Ocho Rios harbour and at Negril. We have two beautiful specimens dredged from about 40 feet off the Bauxite pier at Ocho Rios, and one from the beach at Negril. This is a rare and much sought after species in Jamaica.

Genus *Opalia* H. and A. Adams 1853

Subgenus *Dentiscala* de Boury 1886

OPALIA CRENATA Linné 1758
Crenulated Wentletrap **pl. 7**
Lower Florida to Trinidad.
Description: Length $\frac{3}{4}$ to 1 inch. Colour greyish white; fresh specimens sometimes have a light purple cast, but this is a secretion from the animal and nothing to do with the actual shell colour. The whorls are strongly shouldered and the suture is crenulated. There are several very weak ribs and the surface is microscopically pitted. The basal ridge is well developed. Operculum thin and yellowish.
Occurrence: Uncommon under rocks in 1 to 15 feet at the Police Training School pool at Port Royal, at Robins Bay and in many areas along the north coast.

Subgenus *Nodiscala* de Boury 1889

OPALIA PUMILIO Mörch 1874 Pumilio Wentletrap
South-east U.S. and the West Indies.
Description: Length up to $\frac{5}{8}$ inch. Colour greyish white; the remarks on the purplish cast of *O. crenata* also apply to this species. Shell long and thin with numerous rounded costae. Outer lip round and thickened. Surface microscopically pitted.
Occurrence: As for *O. crenata*, with which it is sometimes found.
Forms: Opalia pumilio morchiana Dall is a form with between 9 and 15 costae, strongly angulated at the periphery of the whorls. We have two specimens of this form from about 30 feet in Ocho Rios harbour.

Genus *Epitonium* Röding 1798

Subgenus *Epitonium* Röding 1798

EPITONIUM OCCIDENTALE Nyst 1871
West Atlantic Wentletrap **pl. 7**
Southern Florida to Trinidad.
Description: Length $\frac{3}{4}$ to 1 inch. Colour white, with a glossy sheen. The shell has numerous thin, axial costae acutely angled at the shoulder of the whorls (there are 12-15 costae on the body whorl). Shell thin, but surprisingly strong. Spiral sculpture absent. Operculum thin and yellowish.

Occurrence: In 10 to 40 feet on sandy or muddy bottoms at Negril, off the Bogue Islands and in Ocho Rios harbour. It is not uncommon on the beach at Negril, but is usually found in a damaged condition.

EPITONIUM ALBIDUM Orbigny 1842
Bladed Wentletrap **pl. 7**
Southern Florida to Argentina.
Description: Length $\frac{1}{2}$ to $\frac{3}{4}$ inch. Colour white. Sculpture of numerous, low, blade-like costae. These are not angled at the shoulder of the whorls; there are 12 to 14 on the body whorl. Spiral sculpture absent.
Occurrence: As for *E. occidentale*, with which it is sometimes found. Beach specimens are not uncommon.

Subgenus *Gyroscala* de Boury 1887

EPITONIUM LAMELLOSUM Lamarck 1822
Lamellose Wentletrap **pl. 7**
Southern Florida, the Caribbean and Europe.
Description: Length 1 to $1\frac{1}{4}$ inches. Colour whitish with irregular brown markings; these may be almost entirely absent in beach specimens. Sculpture of numerous sharp, thin, white, blade-like axial costae. Operculum paper thin and yellow.
Occurrence: As for *Opalia crenata*. It is a more common species, however, and good specimens can be found on the beach at Holland Bay (St Thomas), and on many sandy beaches all around the island.

Subgenus *Asperiscala* de Boury 1909

EPITONIUM CANDEANUM Orbigny 1842
Candé's Wentletrap **pl. 7**
Florida to Barbados.
Description: Length about $\frac{1}{2}$ inch. Colour white to light brown. Sculpture of 18 to 25 costae on the body whorl. These are not angled at the shoulder.
Occurrence: Uncommon in sand under rocks, at 10 to 45 feet, in Ocho Rios harbour, at the Bogue Islands and at Negril. Dead shells are occasionally found in drift sand, and we have seen a few occupied by hermit crabs in shallow water at Priory.

EPITONIUM NOVANGLIAE Couthouy 1838

New England Wentletrap **pl. 7**

Massachusetts to Brazil.

Description: Length about ½ inch. Colour white; occasionally light brown. Sculpture of numerous axial costae, which usually have a characteristic hook-like projection at the shoulder of the whorl. There are many fine spiral threads between the costae, and these are crossed by microscopic axial lines.

Occurrence: Relatively common in sand under rocks, at about 25 feet, at Ocho Rios, Discovery Bay and the mouth of Kingston harbour. Dead shells are sometimes found in drift sand at Negril and Prospect. Like many other *Epitoniidae*, this species is also dredged from very deep water.

SUPERFAMILY HIPPONICACEA

FAMILY HIPPONICIDAE

These are cap-shaped, non-spiral shells with sedentary habits. Most secrete a shelly platform. Shells of the genus *Hipponix* are very thick and solid; those of *Cheilea*, delicate and finely sculptured. The family prefers relatively shallow water, and is widely distributed in temperate and tropical seas. They feed on detritus extracted from the water.

Genus *Cheilea* Modeer 1793

CHEILEA EQUESTRIS Linné 1758

False Cup and Saucer **pl. 7**

South-east Florida and the West Indies.

Description: Diameter ¾ to 1¼ inches. Colour white. Sculpture of irregular folds, crossed by many fine radiating threads. Shell is cup-shaped, with a 'half-cup' at the apex internally.

Occurrence: Reasonably common attached to rocks in 4 to 30 feet off the southern side of the Palisadoes, at Negril and in Ocho Rios harbour. Specimens are occasionally washed ashore, usually in a damaged condition.

Genus *Hipponix* Defrance 1819

HIPPONIX ANTIQUATUS Linné 1767

White Hoof Shell **pl. 7**

South-east Florida and the West Indies.

Description: Length about ½ inch. Colour dull white. Sculpture of many concentric overlapping folds, crossed by microscopic incised lines. The shell is solid and heavy. Beach worn specimens often have the concentric folds worn smooth.

Occurrence: In 2 to 10 feet, on rocks, at Prospect, Discovery Bay and in rocky areas all around Jamaica. Marine growths on the living shell often afford it excellent camouflage. It is a very common species.

HIPPONIX SUBRUFUS SUBRUFUS Lamarck 1822

Orange Hoof Shell **pl. 7**

South-east Florida and the West Indies.

Description: Very similar to *H. antiquatus*, but the shell is coloured orange, and there are numerous radiating cords and concentric ridges. These are absent in beach worn specimens.

Occurrence: As for *H. antiquatus*, but it is less common. Dead shells are often found on the beaches along the Palisadoes.

SUPERFAMILY CALYPTRAEACEA

FAMILY CALYPTRAEIDAE

This unusual family includes a number of non-spiral species whose shells possess internally either a thin, horizontal shelf, or a cup-like process descending from the apex, which help to support the soft parts of the mollusc.

In the genus *Crepidula*, not only can individuals change sex, but their more or less sedentary habit has given rise to a bivalve-type manner of feeding, in which particles of food are trapped by mucus in the mantle. Some small, shallow water species are exceedingly common in the West Indies.

Genus *Crucibulum* Schumacher 1817

CRUCIBULUM AURICULA Gmelin 1791

West Indian Cup and Saucer

West Florida and the West Indies.

Description: Diameter 1 inch. Colour dull white. Sculpture of coarse

Plate 5 TURBANS, TOP SHELLS, TURRETS, STAR SHELLS, WORM SHELLS, SUNDIALS, TURCICULAS

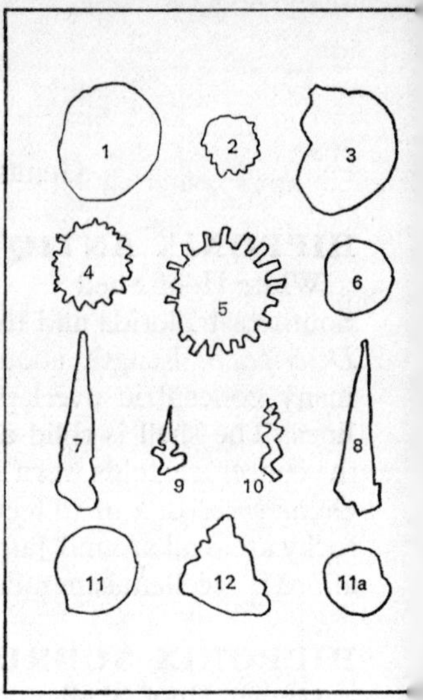

Scale: reduced by 35%

1. **Cittarium pica** Linné 57
West Indian Top Shell

2. **Astraea tecta** Solander 66
Imbricated Star Shell

3. **Turbo canaliculatus** 62
Hermann Channelled Turban

4. **Astraea caelata** Gmelin 63
Carved Star Shell

5. **Astraea phoebia** 63
Röding Long-spined Star Shell
This species may, or may not, have an umbilicus.

6. **Astraea tuber** Linné 63
Green Star Shell

7. **Turritella variegata** 77
Linné Variegated Turret Shell

8. **Turritella exoleta** 77
Linné Eastern Turret Shell

9. **Vermicularia spirata** 78
Philippi West Indian Worm Shell
The length of this species depends entirely on the tightness of its coils.

10. **Vermicularia knorri** 77
Deshayes Knorr's Worm Shell
The length of this species depends entirely on the tightness of its coils.

11. **Architectonica nobilis** 79
Röding Common Sundial
11a, ventral view. Circular, corneous operculum just visible.

12. **Turcicula imperialis** 60
Dall Imperial Turcicula
This species probably occurs off Jamaica and throughout the Greater and Lesser Antilles.

Plate 6 WORM SHELLS, SUNDIALS, HORN SHELLS, MODULUS, PLANAXIS SHELLS

Scale: reduced by 20%

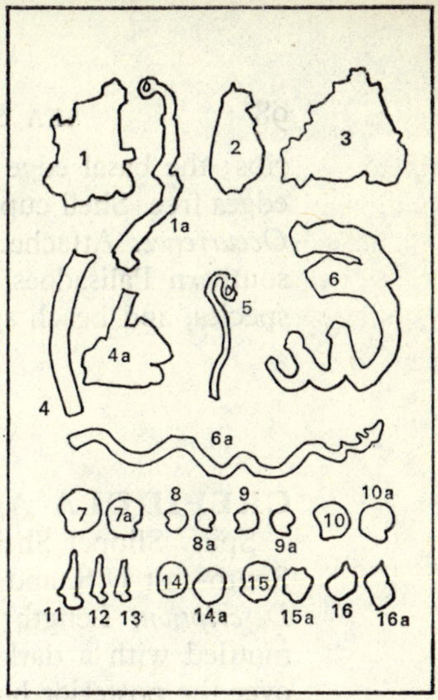

1. **Serpulorbis riisei** Mörch 83
Riise's Worm Shell
Loosely coiled shells may be longer. **1a,** solitary, loosely coiled example.

2. **Serpulorbis decussata** 83
Gmelin Decussate Worm Shell
Loosely coiled shells may be longer.

3. **Petaloconchus irregularis** 82
Orbigny Irregular Worm Shell

4. **Petaloconchus mcgintyi** 83
Olsson and Harbison
McGinty's Worm Shell
Section of terminal tube. **4a,** complete shell. Note straight terminal tube.

5. **Petaloconchus erectus** 82
Dall Erect Worm Shell
Solitary specimen; more often found in tightly coiled clusters.

6. **Siliquaria anguillae** 84
Mörch Slit Worm Shell
Characteristic longitudinal slit not visible. **6a,** smooth form. This may be a separate species. Note the longitudinal slit.

7. **Heliacus infundibuliformis** Gmelin 79
Channelled Sundial
Final whorl may be detached. **7a,** ventral view.

8. **Heliacus bisulcatus** 78
Orbigny Orbigny's Sundial
8a, ventral view.

9. **Philippia krebsi** Mörch 79
Kreb's Sundial
9a, ventral view.

10. **Heliacus cylindricus** 78
Gmelin Cylinder Sundial
10a, ventral view.

11. **Cerithidea pliculosa** 87
Menke Plicate Horn Shell

12. **Cerithidea costata** 86
da Costa Turret Horn Shell

13. **Batillaria minima** Gmelin 87
False Cerith
This is an exceedingly variable species in colour pattern; deformed shells are very common.

14. **Modulus carchedonius** 85
Lamarck Angled Modulus
14a, ventral view.

15. **Modulus modulus** Linné 85
Atlantic Modulus
15a, ventral view.

16. **Planaxis nucleus** 84
Bruguière Black Atlantic Planaxis
16a, side view.

ribs; the basal edge is crenulated. The internal apical cup has all its edges free. Shell cup-shaped, moderately heavy and strong.

Occurrence: Attached to rocks, or other shells, in 3 to 20 feet off the southern Palisadoes and in Ocho Rios harbour. It is not a common species, and beach specimens are usually damaged.

Genus *Crepidula* Lamarck 1799

CREPIDULA ACULEATA Gmelin 1791

Spiny Slipper Shell **pl. 7**

South-east U.S. and the West Indies.

Description: Length ¾ to 1 inch. Colour cream to light brown, often mottled with a darker brown. A distinctive horizontal shelf extends over the posterior half of the interior of the shell. Exterior rough and more or less spiny. The base of the shell is roughly oval.

Occurrence: Commonly found attached to hard surfaces in 1 to 6 feet at Discovery Bay and the old Coal Wharf, Port Royal. It is equally common in other relatively sheltered areas all around Jamaica, and often adheres to other shells.

Subgenus *Ianacus* Mörch 1852

CREPIDULA PLANA Say 1822

Eastern White Slipper Shell **pl. 7**

Eastern U.S. and the West Indies.

Description: Length about 1 inch. Colour white. The shell is thin and flat with sharp edges, and takes its shape from the surface to which it adheres (it may be either concave or convex). The interior shelf is less than half the length of the shell.

Occurrence: In 3 to 50 feet in Kingston harbour, at Discovery Bay and Negril. It often adheres to other shells, alive and dead, and is found on the operculums of *Fasciolaria* and *Murex* along the harbour side of the Palisadoes. Dead shells are common on the beaches there.

FAMILY CAPULIDAE

This is the rather uncommon family of Cap Shells, some of which superficially resemble, and are often mistaken for, the common Hoof Shells of the genus *Hipponix*.

There is only one known Jamaican species, which is easily separated from the Hoof Shells by its coiled apex and delicate construction. It is believed to obtain food by filtering algae-laden water through its gills, like a bivalve. Elsewhere, certain species permanently attach themselves to Scallops for the same purpose. All *Capulidae* lack operculums.

Genus *Capulus* Montfort 1810

Subgenus *Krebsia* Mörch 1877

CAPULUS INTORTUS Lamarck 1822
Incurved Cap Shell **pl. 7**
Florida and the West Indies.
Description: Length about ½ inch. Colour white. Sculpture of numerous spiral cords and irregular growth lines; the effect is delicate and attractive. The apex is spirally curved, and the aperture oval. There may be a light-brown periostracum.
Occurrence: Very uncommon on rocks in about 20 feet in Kingston harbour and at Ocho Rios. Dead shells are occasionally found on the beaches along the Palisadoes.

SUPERFAMILY STROMBACEA

FAMILY XENOPHORIDAE

This unique group of gastropods (known as Carrier Shells) is noted for its extraordinary habit of cementing fragments of coral, shells and other debris to its own shell. The resulting camouflage is very effective, and from above, the shells closely resemble small piles of rubble.

The family which includes about 20 species, is related to the Strombus conchs, and is mainly confined to tropical seas. There are three Atlantic species, only one of which occurs in shallow water. Like the conchs, members of this family are vegetarian.

Genus *Xenophora* G. Fischer 1807

Subgenus *Xenophora* G. Fischer 1807

XENOPHORA CONCHYLIOPHORA Born 1780
Atlantic Carrier Shell **pl. 8**
South-east U.S. and the West Indies.
Description: Diameter 2 to 2½ inches, excluding foreign attachments.

Colour: base and aperture brown, exterior brownish-cream. The broad periphery of the whorls has numerous fragments of shells, coral and stones firmly cemented to it. From above, the camouflage effect is excellent and the Carrier Shell looks like a small nondescript pile of rubble. Base slightly concave; juveniles with a narrow umbilicus which is generally closed in the adult shell. Operculum thin, dark brown and fan-shaped.

Occurrence: Relatively common in 6 to 30 feet on eel grass beds at Negril, Montego Bay and off the Port Royal Cays. Specimens which live on the site of the submerged town of Port Royal occasionally attach to themselves fragments of pottery which date back to the seventeenth century.

XENOPHORA CARIBAEA Petit 1856

Caribbean Carrier Shell

Florida Keys and the West Indies.

Description: Similar in most respects to *X. conchyliophora*, but it is lighter in structure, has fewer and smaller attachments and is umbilicated. In addition, the edges of the shell extend beyond the base to form a cape.

Occurrence: This is a very rare shell, but dead specimens have been taken by hermit crabs into fish pots set in 100 feet off Runaway Bay. It may occur in limited numbers in 100 to 400 feet all along the north coast.

FAMILY STROMBIDAE

This is the familiar family of Conch shells. There are about 80 known species, widely distributed in tropical seas where they prefer shallow water. Four of the five West Indian species are common and quite easily collected. The sexes are separate.

The animals are equipped with sharp, narrow operculums, and have colourful eyes set on prominent stalks. They are extremely active and appear to possess a certain degree of intelligence. In the Caribbean, and many parts of the world, the animals are used both as food and fish-bait. In spite of many 'expert' reports to the contrary, all Conchs are vegetarians.

Juvenile *Strombidae*, which do not possess the characteristic thickened lip, are commonly mistaken for Cone Shells (*Conidae*). Comparison with the distinctive spires of adult specimens usually resolves any doubt.

Genus *Strombus* Linné 1758

STROMBUS GIGAS Linné 1758 Queen Conch **pl. 12**

Bermuda, south-east Florida and the West Indies.

Description: Length 5 to 12 inches. Colour yellowish on the exterior, rich pink, yellow or peach within the aperture. Shell heavy and solid, with a row of blunt spines on the whorls below the suture. Periostracum thick and brown; it tends to become brittle and flakes off when the animal dies. Operculum large and claw-like. Live adult shells may weigh over 5 pounds. This is a beautiful and widely known species.

Occurrence: In 1 to 60 feet on eel grass beds at Negril, in Kingston harbour and in most areas around Jamaica. It is extensively used as food and bait by fishermen, and in areas near fishing villages it may now be quite rare. Elsewhere, it is still a common species. Juveniles up to 3 or 4 inches in length usually prefer very shallow water, and are nocturnal in habit.

Forms: The name *Strombus samba* Clench has erroneously been given to heavy, worn, coral encrusted specimens. It is not a true form.

STROMBUS PUGILIS Linné 1758

West Indian Fighting Conch **pl. 8**

South-east Florida and the West Indies.

Description: Length 3 to 4 inches. Colour deep orange, with the tip of the siphonal canal commonly purple. The early whorls bear thin longitudinal ribs, which develop into nodules and finally into spines on the last two whorls. Operculum brown and claw-shaped, with saw-like teeth on its outer margin. The animal is extremely active when removed from the water. This beautiful species is a favourite of all collectors.

Occurrence: Common in 4 to 50 feet in Kingston harbour, Montego Bay and on muddy bottoms all around Jamaica. At certain times of the year this species is abundant in very shallow water between the Gypsum Pier and the Palisadoes Airport.

Forms: There is a rare and beautiful all-purple form which has been found at Lucea.

STROMBUS RANINUS Gmelin 1791

Hawk-wing Conch **pl. 8**

South-east Florida and the West Indies.

Description: Length 2 to 4 inches. Colour variable, usually greyish with brown mottling, but sometimes tan or rust red with white speckling. There are a number of blunt spines, the largest being the two last. Operculum brown, smooth and claw-like.

Occurrence: Common in 1 to 20 feet on eel grass beds all around the Island, but particularly in Kingston harbour and at Braco, Trelawny. It is a widespread species and thrives in most environments. It is, perhaps, the most common *Strombus* in Jamaica.

Forms: There is a thick, wrinkled dwarf form, less than 1½ inches long, named *S. raninus nanus* (pl. 8). We have found a very few examples on the south side of the Palisadoes, and it is a rare shell in Jamaica.

STROMBUS COSTATUS Gmelin 1791 Milk Conch **pl. 8**

South-east Florida and the West Indies.

Description: Length 4 to 7 inches. Colour pale yellowish white. Older specimens sometimes have an aluminium-like glaze on the parietal area and outer lip. The shell is very heavy and solid, with short blunt spines. Operculum similar to *S. gigas*. Occasional specimens are brick red.

Occurrence: As for *S. gigas*, with which it is often found. *Costatus* is a much less common species, however, and juvenile shells do not appear to seek very shallow water as do those of *gigas*.

STROMBUS GALLUS Linné 1758
Rooster-tail Conch **pl. 8**

South-east Florida and the West Indies.

Description: Length 4 to 6½ inches. Colour similar to that of *S. raninus*, but commonly with additional purplish maculations on the spire and body whorl. The shape of the shell is also very like that of *raninus*, but there are long extensions at both ends of the outer lip. *Raninus* has only 2 long spines on the body whorl, while *gallus* has 3 to 5. Operculum very similar to that of *raninus*. It is a beautiful and graceful shell.

Occurrence: Very uncommon on eel grass beds in 1 to 40 feet at Priory, in Port Antonio, at Negril and all along the north coast. We have found specimens at Whitehouse (Westmoreland) on the south coast; between there and Morant Point, it is very rare. Our best specimens came from 2 feet of water off the 'Columbus Cottages' at Priory. It is perhaps the West Indian shell most in demand among collectors.

Forms: There is a very rare and beautiful all-pink colour form, which we have found in the Grenadines (pl. 8).

SUPERFAMILY CYPRAEACEA

FAMILY ERATOIDAE

SUBFAMILY TRIVIINAE

Members of this subfamily are small, and more or less globular, with the body whorl completely concealing the spire.

As in the Cowries – a closely related family – the mantle of *Triviinae* envelops the shell, the mantle edges meeting at the characteristic furrow in the centre of the body whorl. The animals themselves are very different from the Cowries, and are thought to be exclusively carnivorous.

There are some 100 species with a worldwide distribution.

Genus *Trivia* Broderip 1837

TRIVIA PEDICULUS Linné 1758

Coffee Bean Trivia **pl. 22**

Southern Florida and the West Indies.

Description: Length about $\frac{1}{2}$ inch. Colour brownish pink with three pairs of dark brown spots on the back. Sculpture of 16 to 19 ribs, with a straight longitudinal dorsal furrow. The ribs cross the outer lip.

Occurrence: Relatively common under rocks in 2 to 15 feet at Prospect, Negril, and many areas on the north and south coasts. It is commonly washed ashore, but living shells are not easily found.

TRIVIA SUFFUSA Gray 1882 Suffuse Trivia **pl. 22**

South-east Florida and the West Indies.

Description: Length $\frac{1}{3}$ inch. Colour bright pink with poorly defined brownish spots and speckling. Outer lip white (cf. *T. quadripunctata*). Sculpture of 18 to 23 beaded ribs which cross the outer lip. There is a weak brownish-pink spot on each side of the anterior furrow.

Occurrence: Reasonably common in 6 to 30 feet at Negril, Ocho Rios harbour and many other areas on the north and south coasts. It is quite often washed ashore, and dead shells can usually be found on the Long Bay beach, Negril.

TRIVIA QUADRIPUNCTATA Gray 1827
Four-spotted Trivia **pl. 22**
South-east Florida, Yucatan and the West Indies.
Description: Length about $\frac{1}{4}$ inch. Colour bright pink with four small, dark brown spots touching the central dorsal furrow. There are 19 to 24 riblets which cross the outer lip. This species can be separated from *T. suffusa* by the lack of brown speckling, and the fact that its riblets are not beaded. The bright pink colour tends to fade when the shell has been exposed to the sun for some days.
Occurrence: As for *T. suffusa*. This is a common shell on many Jamaican beaches, but is seldom found alive. It is widely used in the manufacture of shell jewellery throughout the West Indies.

TRIVIA ANTILLARUM Schilder 1922
Antillean Trivia **pl. 22**
South-east Florida and the West Indies.
Description: Length $\frac{1}{8}$ to $\frac{1}{4}$ inch. Colour deep purple. There are 18 to 22 riblets which cross the outer lip, and sometimes also cross the dorsal furrow. This is an exquisite little shell, with a colour seldom seen in gastropods.
Occurrence: In 30 to 60 feet at Orange Bay and in Ocho Rios harbour. It is an uncommon species which occurs down to at least 600 feet. We have never seen a beach specimen, but commercial dredgings at Ocho Rios indicate that it occurs in much shallower water than previously suspected.

TRIVIA LEUCOSPHAERA Schilder 1931
Little White Trivia **pl. 22**
South-east U.S. and the West Indies.
Description: Length $\frac{1}{4}$ inch. Colour white. There are 20 to 24 riblets which cross the outer lip. We have not seen a specimen without a dorsal furrow, but Abbott (*Caribbean Sea Shells*) indicates that it may be absent in some specimens. The anterior and posterior ends of the aperture are slightly more produced than in *T. nix*.
Occurrence: This is a rare species in Jamaica. The only specimens we have came from dredgings in Ocho Rios harbour in about 25 feet.

TRIVIA NIX Schilder 1922 White Globe Trivia **pl. 22**
South-east Florida and the West Indies.
Description: Length $\frac{3}{8}$ inch. Colour white. There are usually 22 to 26 riblets with a strong dorsal furrow. It can be separated from *T. leucosphaera* by its larger size. Immature specimens of this and other

Trivia are smooth and transparent, and there is invariably a trace of the apex which is concealed later by the final whorl.
Occurrence: As for *T. pediculus*, but it is much less common. Good specimens are often found on the beach at Negril.
Note: A live *T. nix* in an aquarium frequently floated upside down on the surface of the water with its mantle extended – 'launching' itself from the side of the tank. It is not known whether it does this in its natural environment.

FAMILY CYPRAEIDAE

This is the family of beautiful, polished, porcelainous Cowry shells that have delighted man for thousands of years. There are some 200 species.

Many of the living animals are themselves of great beauty, and in all species the extended mantle envelops the shell. The family is characterised by the semi-globular shape and the very large body whorl which conceals the spire. The sexes are separate.

Cowries occur in relatively shallow water in all tropical seas and recent research suggests that different species may be carnivorous, herbivorous or both. Previously, the family was generally considered exclusively carnivorous.

Genus *Cypraea* Linné 1758

Subgenus *Trona* Jousseaume 1884

CYPRAEA ZEBRA Linné 1758 Measled Cowry **pl. 9**
South-east Florida and the West Indies.
Description: Length $1\frac{1}{4}$ to 4 inches. Colour chocolate brown with many white spots. A few of the spots near the concealed apex generally have brown centres. There is a pale, narrow dorsal band crossing the shell from front to back (this is common in many cowries). Surface highly polished.

Juvenile cowries often look very unlike the adult forms. In the case of *C. zebra*, the young shells are strongly banded with grey and brown, and lack spots.
Occurrence: This is a common species in 4 to 30 feet at Discovery Bay, Cow Bay, off the southern side of the Palisadoes, and in rocky areas all around Jamaica. The animal lives in crevices in the rocks and under plates of coral; damaged specimens are not uncommon on some beaches.

Note: Cypraea cervus from southern Florida and Cuba, which is larger and more spotted than the average *C. zebra*, has long been classified as a separate species. *C. zebra*, however, shows considerable variation in size, shape and colour pattern, and it may be that *C. cervus* is merely one of its forms. Abbott (*Caribbean Sea Shells*) reports that *C. cervus* may be the female form. Most authorities, however, believe that they are in fact separate species in the same subgenus.

Subgenus *Luria* Jousseaume 1884

CYPRAEA CINEREA Gmelin 1791
Atlantic Grey Cowry **pl. 9**
South-east Florida and the West Indies.
Description: Length $\frac{3}{4}$ to $1\frac{1}{2}$ inches. Colour orange, grey or chocolate brown, usually with two faint lighter bands. Base ivory, with flecks of grey or brown sometimes on the outer surfaces of the base. There are, occasionally, purple interspaces between the teeth. Surface polished.
Occurrence: Common in 1 to 15 feet under rocks at Prospect, Discovery Bay and in rocky areas all around Jamaica. This is the most common West Indian cowry, and its dead shells are familiar to all West Indian beachcombers.

Subgenus *Erosaria* Troschel 1863

CYPRAEA SPURCA ACICULARIS Gmelin 1791
Atlantic Yellow Cowry **pl. 9**
Southern Florida, Yucatan and the West Indies.
Description: Length $\frac{1}{2}$ to $1\frac{1}{4}$ inches. Colour orange, flecked with white; base white. The base is flatter, and the shell less globular, than the other two local species. There are sometimes small round indentations on the exterior near the base. Shell polished. The beautiful animal of this species merits close examination.
Occurrence: Relatively common in 3 to 20 feet at Prospect, Negril, and in many areas all around Jamaica. It is perhaps less frequently found than the other two species, and lives under rocks and plates of coral in sandy areas.

CYPRAEA MUS Linné 1758 Mouse Cowry **pl. 9**
Southern Caribbean.
Description: Length about 2 inches. Colour grey and white (mouse coloured), with two irregular dark brown stripes on the back. These

may be solid or composed of small individual markings. There is a dark brown stain around the aperture.

Occurrence: Relatively uncommon in shallow water on and under rocks in the southern Caribbean. It does not occur in Jamaica, but is frequently found in West Indian collections, and is included here as a matter of interest.

FAMILY OVULIDAE

Members of this family are closely related to the Cowries, and the mantles envelop the shells in the same way. In the genus *Cyphoma*, the mantles of the various species are strikingly attractive.

The shells are smooth and polished; the aperture extends for the length of the shell, and the body whorl conceals the spire. West Indian *Ovulidae* live on Sea Whips or Sea Fans, and members of several different species are sometimes found on the same fan.

Genus *Neosimnia* Fischer 1884

NEOSIMNIA ACICULARIS Lamarck 1810

Common West Indian Simnia

South-eastern U.S. and the West Indies.

Description: Length ½ inch. Colour greenish yellow or light pink. Shell thin, fragile and transparent. Columella area flattened, and bordered by a long pale ridge.

Occurrence: Relatively common on sea fans in 1 to 30 feet at Negril, Prospect and wherever sea fans occur. The mollusc's colour blends perfectly with its background, and it is very easily overlooked. Beach specimens are almost invariably damaged.

NEOSIMNIA UNIPLICATA Sowerby 1848

Single-toothed Simnia **pl. 9**

Virginia to Florida, and the West Indies.

Description: Length ½ to ¾ inch. Colour yellowish, pink, or light purple. Very similar to *N. acicularis*, but with a twisted spiral plication at the posterior end of the columella. It tends to be proportionally slightly narrower than *acicularis*.

Occurrence: As for *N. acicularis*. Both species are found on sea fans in association with *Cyphoma*, but *uniplicata* seems to prefer slightly deeper water.

Genus *Cyphoma* Röding 1798

CYPHOMA GIBBOSUM Linné 1758

Flamingo Tongue **pl. 9**

South-east U.S. and the West Indies.

Description: Length ¾ to 1 inch. Colour cream on the base and in the centre of the shell; remainder bright orange, which lightens progressively towards the base. There is a prominent transverse dorsal ridge. Shell polished. The mantles of the animals of this genus cover most of the shell when extended, and are unusually beautiful; in this species the animal is orange with many black rings.

Occurrence: Very common in 1 to 20 feet on sea fans and sea whips at Runaway Bay, the Port Royal Cays and in rocky areas all around Jamaica. The shell is often cast ashore, where the sun soon fades the bright orange colour to a uniform white. On sea fans the living shell is poorly camouflaged, but on certain sea whips they are practically undetectable to the inexperienced eye.

CYPHOMA SIGNATUM Pilsbry and McGinty 1939

Fingerprint Cyphoma **pl. 9**

Lower Florida Keys to the West Indies.

Description: Length 1 to 1¼ inches. Colour light buff with a cream-buff tint deep inside the aperture. Very similar to *C. gibbosum* in shape, but it is proportionally longer, slimmer and lower. The animal has a pale yellow mantle with close set black, transverse lines, and is immediately distinguishable from *gibbosum*.

Occurrence: As for *C. gibbosum*, but it is a very uncommon species. Our specimens were taken in about 10 feet from a sea fan at Negril bearing both species, and it has also been taken live at Discovery Bay in about 35 feet.

SUPERFAMILY HETEROPODA

FAMILY ATLANTIDAE

These are small, spiral, transparent snails with a compressed and fragile planorboid shell which is kept upright by a sharp keel. The animals have retained a thin operculum.

All heteropods are pelagic and possess light, transparent bodies and very large eyes. The live molluscs are occasionally washed ashore, but their shells are far more easily collected by dredging in deep water on soft substrate.

Genus *Atlanta* Leseur 1817

ATLANTA PERONI Leseur 1817 Peron's Atlanta **pl. 22**
Pelagic in warm waters of the Atlantic and Pacific.
Description: Diameter about $\frac{1}{2}$ inch. Colour whitish. The shell is thin and fragile, and the keel is rarely found intact. It is quite distinct from any other West Indian mollusc.
Occurrence: This species is pelagic throughout the West Indies. Live specimens are very occasionally washed ashore, but dead shells are not uncommon on the bottom in depths over 50 feet. Our specimens are from Orange Bay and Rio Bueno.

SUPERFAMILY NATICACEA

FAMILY NATICIDAE

These are smooth, polished, often colourful shells which are found in all oceans, in both shallow and deep water.

The living animal appears too large for the shell, which in fact is partially covered by the extended foot. The operculum serves as an effective trap-door and completely seals the aperture. In *Polinices* and *Sigatica* the operculum is thin and corneous; in *Natica* thick and calcareous.

Naticidae spend most of their lives burrowing beneath the surface of mud or soft sand bottoms. They are voracious carnivores, and eat both bivalves and smaller members of their own class. The collar-shaped, sand-encrusted objects, often found in shallow water, are the egg cases of this family.

SUBFAMILY POLINICINAE

Genus *Polinices* Montfort 1810

POLINICES LACTEUS Guilding 1834
Milk Moon Shell **pl. 10**
South-eastern U.S. and the West Indies.
Description: Length $\frac{3}{4}$ to $1\frac{1}{2}$ inches. Colour uniform white. Shell solid and polished with a small, deep umbilicus partly covered by the callus. Operculum thin, and either red or yellow; periostracum thin and yellowish.
Occurrence: In 1 to 6 feet at Priory, the Port Royal Cays and in many areas all around Jamaica. This is a very common shell which lives in the

sand beneath small rocks, in shallow water. Dead shells are common on the beaches.

POLINICES HEPATICUS Röding 1798

Brown Moon Shell **pl. 10**

South-east Florida and the West Indies.

Description: Length 1 to 2 inches. Colour purplish-brown or orange-brown, with a white base and interior. Umbilicus large and deep; operculum thin and light brown. The shell is thick, heavy and very strong.

Occurrence: Relatively common in 10 to 40 feet in sand, at Negril, Whitehouse (Westmoreland) and in Kingston harbour. It is less common along the north coast, but beach specimens can be found at Orange Bay.

SUBFAMILY SININAE

Genus *Sinum* Röding 1798

Note: This genus lacks an operculum.

SINUM PERSPECTIVUM Say 1831

Common Baby's Ear **pl. 10**

South-eastern U.S., the Gulf States and the West Indies.

Description: Diameter 1 to 1½ inches. Colour uniform white. Shell flat with a large aperture. Sculpture of many fine, wavy spiral lines.

Occurrence: Uncommon in 6 to 30 feet at Negril, Montego Bay and in Ocho Rios and Kingston harbour. Dead shells are occasionally washed up on the Long Bay beach at Negril, and at Rocky Point (Clarendon).

SINUM MACULATUM Say 1831

Maculated Baby's Ear **pl. 10**

South-eastern U.S. to the West Indies.

Description: Diameter 1 to 1½ inches. Colour pale brown and white. Shell similar to *S. perspectivum*, but with a weaker sculpture. The unusual animal, which is three or four times the size of the shell and more or less enfolds it, merits close examination.

Occurrence: This is a common shell on eel grass beds in 1 to 6 feet along the harbour side of the Palisadoes, and dead specimens can be found there on the beach. The distribution of the species in Jamaica is very restricted; we have found it in few other areas, and never in quantity.

SUBFAMILY NATICINAE

Genus *Natica* Scopoli 1777

NATICA LIVIDA Pfeiffer 1840 Livid Natica **pl. 10**
North Carolina to the West Indies.
Description: Length ½ inch. Colour greyish-brown with poorly defined spiral bands of a darker colour. The callus is brown, and the aperture and columella light brown; the base is white. Umbilicus almost filled by the callus. The shell is light and smooth.
Occurrence: Relatively uncommon in sand at Priory, Runaway Bay and Montego Bay, in 1 to 20 feet. Good specimens are quite often found occupied by small hermit crabs in very shallow water. Dead shells are common on the beach near the parish border of Trelawny and St James.

NATICA MENKEANA Philippi 1852 Menke's Natica **pl. 10**
West Indies.
Description: Length about ½ inch. Colour tan to yellowish brown, sometimes with two faint dotted reddish brown spiral bands. Shell light and polished. Umbilicus narrow.
Occurrence: Uncommon in 30 feet in Ocho Rios harbour. We have not found this species in any other locality, but there are indications that it occurs on muddy bottoms in similar areas along the north coast, and it has also been reported from Bull Bay.

NATICA CANRENA Linné 1758
Colourful Atlantic Natica **pl. 10**
South-east U.S. and the West Indies.
Description: Length 1 to 2 inches. Colour variable, but usually a pattern of brown, wavy, zigzag lines superimposed on alternating brown and white spiral bands. Callus white and large; umbilicus narrow. Operculum thick and white with about 10 spiral grooves. This is a colourful and unusually attractive shell.
Occurrence: Common in 2 to 25 feet in Kingston harbour, Montego Bay, and in sandy or muddy areas all around the Island. Like other members of this family, *canrena* lives buried beneath the surface of the sea floor. The dead shells of the larger specimens are often carried into fish pots by hermit crabs.

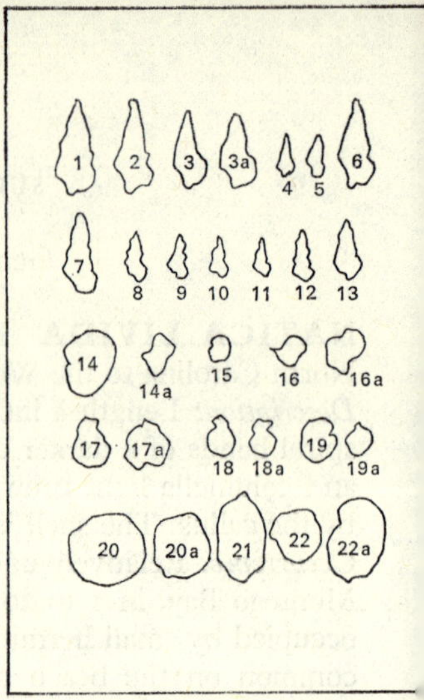

Plate 7 CERITHS, WENTLE-TRAPS, PURPLE SEA SNAILS, HOOF SHELLS, CAP SHELLS, SLIPPER SHELLS, CUP AND SAUCER SHELLS

Scale: natural size

1. Cerithium eburneum 88
Bruguière Ivory Cerith

2. Cerithium algicola 88
C. B. Adams Middle-spined Cerith This may be only a form of **1.** Occasional specimens are all-brown.

3. Cerithium litteratum 87
Born Stocky Cerith
Typical colour variety. **3a,** colour variety. Some shells are entirely orange.

4. Cerithium muscarum 89
Say Fly-specked Cerith

5. Cerithium variabile 88
C. B. Adams Dwarf Cerith

6. Cerithium auricoma 89
Schwengel Schwengel's Cerith

7. Epitonium lamellosum 93
Lamarck Lamellose Wentletrap

8. Epitonium occidentale 92
Nyst West Atlantic Wentletrap

9. Epitonium albidum 93
Orbigny Bladed Wentletrap

10. Epitonium candeanum 93
Orbigny Cande's Wentletrap

11. Epitonium novangliae 94
Couthouy New England Wentletrap

12. Cirsotrema dalli Rehder 91
Dall's Wentletrap

13. Opalia crenata Linné 92
Crenulated Wentletrap

14. Janthina janthina Linné 90
Common Purple Sea Snail
14a, side view.

15. Janthina exigua Lamarck 90
Dwarf Purple Sea Snail
May be a juvenile form of **14.**

16. Janthina globosa 90
Swainson Globe Purple Sea Snail
16a, dorsal view.

17. Hipponix antiquatus 95
Linné White Hoof Shell
17a, ventral view.

18. Capulus intortus 99
Lamarck Incurved Cap Shell
18a, ventral view.

19. Hipponix subrufus subrufus Lamarck Orange Hoof Shell 95
19a, ventral view.

20. Cheilea equestris Linné 94
False Cup and Saucer
20a, ventral view.

21. Crepidula plana Say 98
Eastern White Slipper Shell

22. Crepidula aculeata 98
Gmelin Spiny Slipper Shell.
22a, dorsal view.

Plate 8 CONCHS, CARRIER SHELLS

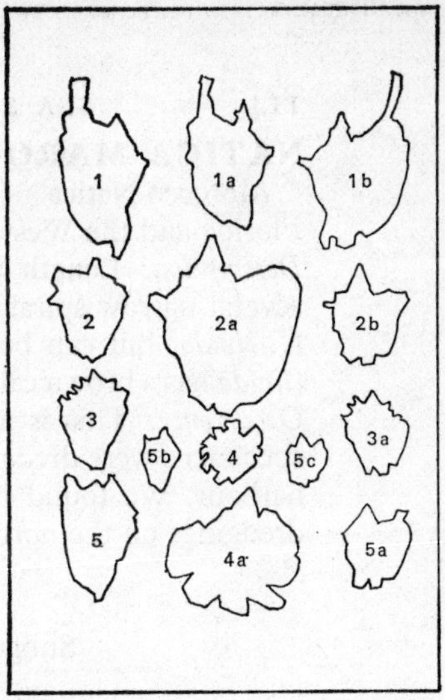

Scale: reduced by 50%

1. Strombus gallus 102
Linné Rooster Tail Conch
The typical colour variety. **1a,** the rare pink colour variety. **1b,** ventral view of **1.**

2. Strombus costatus 102
Gmelin Milk Conch
The rare reddish colour variety. **2a,** ventral view. The parietal area acquires an aluminium colour in old specimens. **2b,** rare dwarf example.

3. Strombus pugilis 101
Linné West Indian Fighting Conch
3a, ventral view. Occasional specimens are entirely purple.

4. Xenophora conchyliophora Born 99
Atlantic Carrier Shell
Specimen with shell attachments. **4a,** ventral view. Specimen with pebble attachments.

5. Strombus raninus 101
Gmelin Hawk-wing Conch
5a, ventral view.
5b, *S. r. nanus* Dwarf Hawk-wing Conch.
5c, ventral view.

NATICA MAROCHIENSIS Gmelin 1791
Morocco Natica **pl. 10**
Florida and the West Indies.
Description: Length ¾ to 1 inch. Colour dark or light brown, with several narrow spiral bands of zigzag brown markings. It is similar to *N. livida*, but can be separated from it by the white umbilical callus (*livida* has a brown callus).
Occurrence: This is a rare and coveted species in Jamaica. Our only specimens were dredged in about 20 feet in Montego Bay and Kingston harbour. We found no trace of it in the several massive commercial dredgings on the north coast.

Subgenus *Glyphepithema* Rehder 1943

NATICA FLORIDANA Rehder 1943 Florida Natica **pl. 10**
South-east Florida, Grand Cayman Island and Jamaica.
Description: Length about ¾ inch. Colour: there are two broad spiral bands of light brown, each bordered on either side by rectangular, small brown spots. The shell bears a superficial resemblance to *N. canrena*, but it has no axial brown markings.
Occurrence: Uncommon in about 25 feet off the Bogue Islands, in Ocho Rios harbour and at Bull Bay. We have found no live shells, and beach specimens are rare.

NATICA CAYENNENSIS Récluz 1850
Cayenne Natica **pl. 10**
West Indies.
Description: Length about 1 inch. Colour similar to *N. canrena* but not as vivid; the pattern is less well defined. There are prominent wrinkles near the suture. Live specimens are easily distinguished from *canrena* by the operculum, which is paucispiral, pustulose and has a large, arched ridge in the centre.
Occurrence: As for *N. canrena*, but it is much less common, and prefers deeper water. Our specimens were dredged in about 25 feet of water, 100 yards east of Fort Augusta in Kingston harbour. It is quite commonly dredged by shrimp trawlers in the southern Caribbean in about 150 feet.

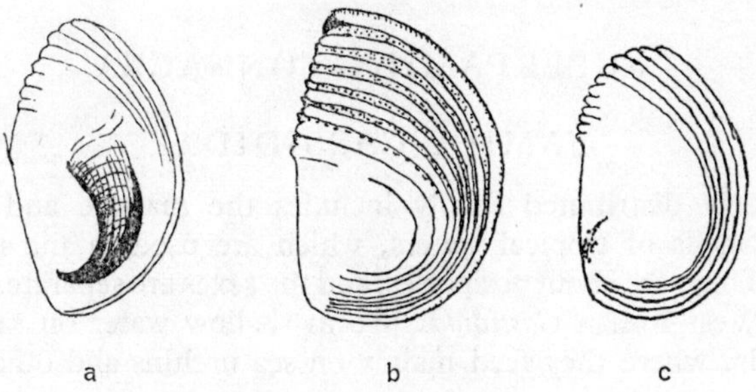

FIG. 16. Three *Naticidae* opercula. a, *Natica cayennensis*. b, *N. canrena*. c, *N. menkeana*

Genus *Stigmaulax* Mörch 1852

STIGMAULAX SULCATA Born 1778

Sulcate Natica **pl. 10**

West Indies.

Description: Length ½ to 1 inch. Colour cream, sometimes with irregular yellowish-brown markings. Sculpture of many strong axial ribs crossed by fine radial threads. Umbilicus large and deep.

Occurrence: Relatively uncommon in 20 to 50 feet off the Bogue Islands, at Discovery Bay, and in Ocho Rios harbour. Commercial dredgings at Ocho Rios produced them in fair numbers; we have found only one beach specimen.

Genus *Sigatica* Meyer and Aldrich 1886

SIGATICA SEMISULCATA Gray 1839

Scratched Moon Shell **pl. 10**

Southern Florida and the West Indies.

Description: Length about ½ inch. Colour white. This species resembles *Polinices lacteus*, but has a simple umbilicus without a callus, and a well impressed suture below which are 4 or 5 thin, incised lines. The shell is much lighter in structure than *P. lacteus*.

Occurrence: This is a very uncommon species. It occurs in sand off the Bogue Islands and in Ocho Rios harbour, in depths of 20 to 60 feet. We have not found beach specimens.

SUPERFAMILY TONNACEA

FAMILY CASSIDIDAE

This widely distributed family includes the massive and beautiful Helmet Shells of tropical waters, which are used in the making of cameos. There are about 70 species, and the sexes are separate.

Most West Indian *Cassididae* live in shallow water on sand or eel grass beds, where they feed mainly on sea urchins and other echinoderms.

In the southern Caribbean, the larger Helmets are commonly eaten; like octopus, they are palatable but tough. The very rare and valuable *Morum dennisoni* Reeve is a member of this family.

Genus *Morum* Röding 1798

MORUM ONISCUS Linné 1767 Atlantic Woodlouse **pl. 11**
South-east Florida and the West Indies.
Description: Length $\frac{3}{4}$ to 1 inch. Colour white with extensive and irregular brown speckling. Sculpture of 3 spiral rows of blunt tubercles. Parietal wall glazed, minutely pustulose, and occasionally bright pink. Nuclear tip generally red or pink. There is a thin, semi-transparent periostracum.
Occurrence: Relatively common in 1 to 12 feet under rocks at Folly (Port Antonio), Prospect, and in rocky areas all around Jamaica. Dead shells are commonly washed ashore.
Forms: A small, unusual, all black colour form occurs at Orange Bay.

MORUM DENNISONI Reeve 1842 **pl. 11**
The Bahamas, the West Indies and Central America.
Description: Length $1\frac{1}{2}$ to 2 inches. Colour whitish, with four spiral bands of light brown on the body whorl, and irregular brown patches on the spire and outer lip. The whorls are longitudinally compressed to form blade-like plications. The surface is cancellated, with numerous narrow ribs. Parietal shield broad, thick and minutely pustulose. In fresh specimens it is blood red.
Occurrence: This is one of the rarest and most valuable shells in this part of the world, and of the few known specimens less than half a dozen are believed to be in private collections. Our specimens were dredged in 150 feet of water, 200 miles east-south-east of Kingston and in 1,000 feet off Discovery Bay. It is included here partly as a matter of

interest, and partly because its depth range puts it within reach of amateur dredging and local fish pots.

Note: The Veliger, Vol. 10, No. 2, of October 1st, 1967, carries an interesting monographic account of the species by S. Peter Dance and William R. Emerson.

Genus *Sconsia* Gray 1847

SCONSIA STRIATA Lamarck 1822 Royal Bonnet **pl. 11**

Florida Keys and West Indies.

Description: Length 1½ to 2½ inches. Colour white, with spiral rows of rectangular brown spots. Sculpture of spiral striations. The shell is polished and strong; there may, or may not, be previous varices present. *S. striata* is considerably slimmer and more attenuated at both ends than either of the members of the following genus (*Phalium*).

Occurrence: This is a very rare species in Jamaican waters. The only specimen we have came from 120 feet, on a sandy bottom, about ½ mile off Salem, near Runaway Bay. Occasional shells are probably carried by hermit crabs into fish pots along the north coast, but we have yet to find one. Shrimp trawlers operating in the Gulf of Mexico have recently demonstrated that the species, in that area, is not as rare as was once thought.

Genus *Phalium* Link 1807

PHALIUM GRANULATUM Born 1778

Scotch Bonnet **pl. 11**

North Carolina to the West Indies.

Description: Length 1½ to 4 inches. Colour cream, with spiral rows of rectangular brown spots. Sculpture of many spiral and axial grooves. The depth of the spiral grooves, in particular, is very variable; some specimens may be almost smooth, while in others the grooves may be deeply etched. The white parietal area is pustulose, and the outer lip is strongly and regularly toothed. Operculum thin and light brown. Former varices sometimes present.

Occurrence: Relatively common in 6 to 50 feet at Cow Bay, in Kingston harbour, Montego Bay, and in sandy or muddy areas all around Jamaica. It also occurs in much deeper water, throughout the West Indies, and specimens from more than 100 feet tend to be thicker, rougher, and more likely to bear prominent former varices.

PHALIUM CICATRICOSUM Gmelin 1791
Smooth Scotch Bonnet **pl. 11**

Bermuda, south-east Florida and the Caribbean.

Description: Length 1 to 2½ inches. Colour very similar to *P. granulatum*. Surface of shell smooth and polished. Slightly more attenuated in shape than *P. granulatum*, and considerably lighter in structure.

Occurrence: Relatively common in 1 to 20 feet in sand, and occasionally on mossy rocks, at Priory, Ocho Rios harbour, Negril and all around Jamaica. Dead shells are washed ashore more frequently than those of *granulatum*.

Forms: There is an uncommon form with a single row of small pustules on the shoulder of each whorl. It occurs in Ocho Rios harbour in about 20 feet.

Genus *Casmaria* H. and A. Adams

CASMARIA ATLANTICA Clench
Atlantic Agate Helmet **pl. 11**

The Bahamas and south through the West Indies to Trinidad.

Description: Length 1 to 2 inches. Colour very similar to *Phalium cicatricosum*. In shape it differs from this species in being more attenuated and having the columella broad, twisted and sharply truncated below. There is no pustuled parietal shield as in *cicatricosum*.

Occurrence: This is a rare, deep water species. Our single specimen reportedly came from about 200 feet, 10 miles south of Kingston, but the accuracy of this information may be questioned. There is no doubt, however, that it occurs in this general area of the Caribbean.

Genus *Cassis* Scopoli 1777

CASSIS TUBEROSA Linné 1758 King Helmet **pl. 12**

South-east Florida and the West Indies.

Description: Length 4 to 10 inches. Colour cream, mottled with irregular brown crescent-shaped marks. Parietal shield with a large brown patch in the centre; outer lip with dark brown between the teeth. Shell heavy and strong, with a reticulated surface and triangular parietal shield. The purplish base colour of the parietal shield and outer lip in live specimens, tends to fade slowly when the animal dies. The three Jamaican members of this genus are exceptionally handsome shells, and the animals make a tough but palatable meal.

Occurrence: Common in 1 to 30 feet, on eel grass beds and sand, at

Negril, Prospect, Port Antonio, and in many similar localities around Jamaica. Live specimens are often found partially buried under the sand (this is particularly common with juvenile shells), and in some areas adult shells may be exposed by extreme low tides.

CASSIS FLAMMEA Linné 1758 Flame Helmet **pl. 12**

Bahamas and the Antilles.

Description: Length 3 to 5 inches. Colour and shape very similar to *C. tuberosa*, with the following three important differences – (i) the sculpture consists of axial growth lines only, (ii) the parietal shield has rounded corners, (iii) there are no brown markings between the teeth on the outer lip.

Occurrence: As for *C. tuberosa*, with which it is often found. We have seen it in quantity only at Negril, but its distribution is apparently as wide as that of *C. tuberosa*, and it is generally regarded as a common species.

CASSIS MADAGASCARIENSIS Lamarck 1822
Emperor Helmet **pl. 12**

The Bahamas, south-east Florida and the Greater Antilles.

Description: Length 4 to 12 inches. Colour: exterior cream, parietal shield salmon. (The exterior of juvenile shells is flecked with brown.) The inside edge of the parietal shield is stained dark brown, and there are patches of the same colour between the teeth of the outer lip. Sculpture of axial growth lines, and broad spiral cords. The parietal shield has rounded corners. This species is easily separated from *C. tuberosa* and *C. flammea* by the uniform exterior colour. The operculums of all three species are long, narrow and rounded at both ends.

Occurrence: As for *C. tuberosa*, but it prefers deeper water and is very much less common, particularly on the south coast.

Forms: The form *spinella* Clench is rarely found in Jamaica. It differs from typical *madagascariensis* by its larger size, and smaller, more regular, and more numerous tubercles. We are inclined to question its recognition as a genuine form because of the existence of many intergrades.

Genus *Cypraecassis* Stutchbury 1837

CYPRAECASSIS TESTICULUS Linné 1758
Reticulated Cowrie Helmet **pl. 11**

Bermuda, south-east Florida and the West Indies.

Description: Length 1 to 3 inches. Colour light orange-brown or grey,

generally with irregular spiral bands of dark brown markings. Parietal shield and outer lip cream-orange, with bright orange streaks. Sculpture reticulated. Adults lack both periostracum and operculum.
Occurrence: Very common in 1 to 10 feet at Priory, the southern side of the Palisadoes, Runaway Bay, and in many areas all around Jamaica. These shells live underneath rocks or coral slabs, usually buried in the sand. This accounts for the unblemished surface of live specimens, which are nearly always free of algal growths.
Forms: The form *crumena* Bruguière differs from typical *testiculus* in possessing a single or double row of ridges on the shoulder of the body whorl. Our specimens come from Priory, but it is found wherever *testiculus* occurs in quantity.

FAMILY CYMATIIDAE

These are strong, colourful, moderately elongate shells usually characterised by a thick, hairy periostracum and prominent former varices. The inside of the outer lip is generally toothed.

There are some 100 widely distributed species, most of which are found in shallow, tropical waters. The family is active and carnivorous, and most genera are believed to subsist on bivalves or echinoderms.

Genus *Charonia* Gistel 1848

CHARONIA VARIEGATA Lamarck 1816
Trumpet Triton **pl. 12**
South-east Florida and the West Indies.
Description: Length 10 to 18 inches. Colour buff, brown and purple-pink; the colour pattern is generally arranged in broad irregular spiral bands. Sculpture of axial and spiral cords on the early whorls; spiral cords only on the later whorls. In mature adults, the shoulder of the body whorl is angular. Very young specimens are bright coral pink with a reticulated surface. The adult shell is still used occasionally in Jamaica as a fisherman's trumpet, with a hole drilled high on the spire.
Occurrence: This beautiful species is common in Jamaica. It occurs from 1 to 30 feet, on a variety of surfaces, at Negril, Runaway Bay, Port Antonio, the Port Royal Cays, and in both rocky and sandy areas all around the Island. The best specimens come from eel grass beds; those from coral reefs are usually encrusted and poorly preserved. Juvenile shells seek shelter under rocks.
Forms: There is an uncommon and striking colour form with a bright orange cast over the entire shell. Our specimens come from Priory.

Genus *Cymatium* Röding 1798

Subgenus *Linatella* Gray 1857

CYMATIUM POULSENI Mörch 1877
Poulsen's Triton **pl. 13**
Florida south through the West Indies to Venezuela.
Description: Length 2 to 3 inches. Colour light brown, sometimes with darker spiral bands. Sculpture of 18 to 20 flattened spiral cords, and very fine axial growth lines. The outer lip is crenulated, and the aperture wide.
Occurrence: Reported from Whitehouse (Westmoreland) in 15 feet, but the record needs confirmation. We have not found this species in Jamaica, but there are indications that it occurs along the south-west coast in moderately deep water (*Johnsonia* records its depth range as between 12 and 209 fathoms).

Subgenus *Cabestana* Röding 1798

CYMATIUM LABIOSUM Wood 1828 Lip Triton **pl. 13**
Description: Length ¾ inch. Colour variable, but usually light orange, purplish, brown or white. Sculpture of coarse spiral cords, the most prominent at the shoulder of the whorls. These are crossed by strong axial cords which form nodules where the cords intersect. Siphonal canal short and slightly umbilicated; outer lip thickened.
Occurrence: Relatively common under rocks in 15 to 40 feet, in sandy or muddy areas at Negril, Montego Bay, Orange Bay, Ocho Rios harbour, and many other protected places all along the north coast. Occasional specimens are washed ashore. Recent commercial dredgings indicate that this species is far more common than previously suspected.

Subgenus *Ranularia* Schumacher 1817

CYMATIUM CARIBBAEUM Clench and Turner 1957
Dog-head Triton **pl. 13**
South-east Florida and the West Indies.
Description: Length 1½ to 3 inches. Colour dark or light orange, with white on the varices. Parietal area light orange, with a dark brown

patch crossed by whitish plications. Sculpture of noduled spiral cords, squarish at the shoulder. Siphonal canal extended, but quite strong. The outer lip usually has 7 large teeth. Operculum dark brown and semi-circular.

Occurrence: Relatively common on mud or sand bottoms, in 10 to 30 feet, at Negril, the Bogue Islands, in Ocho Rios and Kingston harbours, and off the Port Royal Cays. Damaged shells are occasionally washed ashore.

CYMATIUM NICOBARICUM Röding 1798

Gold-mouthed Triton **pl. 13**

Bermuda, south-east Florida and the West Indies.

Description: Length 1 to 2½ inches. Colour grey, mottled with purplish-brown. Aperture bright orange in adults; lined with white and brown spiral bands in juveniles. The outer lip has a series of single white teeth (cf. *C. pileare*, the following species). Operculum pale brown and very thin.

Occurrence: Common in 1 to 20 feet, on and under rocks, at Montego Bay, Priory, Port Royal, and in many rocky areas all around Jamaica. Dead shells are commonly cast ashore.

Subgenus *Septa* Perry 1810

CYMATIUM PILEARE MARTINIANUM Orbigny 1842

Atlantic Hairy Triton **pl. 13, 14**

South-east U.S. to the West Indies.

Description: Length 2 to 6 inches. Colour variable, but generally dark or light brown with occasional paler spiral bands; some specimens are light orange or occasionally dark red. The parietal area is dark brown. Sculpture of spiral cords, usually alternating thick and thin; these are irregularly noduled near the shoulder. Outer lip thickened, with small white teeth set in pairs on a reddish background (cf. *C. nicobaricum*, the preceding species). Operculum thin and brown.

Occurrence: As for *C. nicobaricum*, with which it is often found; *pileare*, however, generally prefers slightly deeper water. Giant specimens measuring as much as 6 inches, occur off Whitehouse (Westmoreland) in about 30 feet; our examples from this area are the dark red variety.

CYMATIUM RUBECULUM OCCIDENTALE

Mörch 1881 **pl. 13**

Florida, the Bahamas, Mexico and the Caribbean.

Description: Length 1 to 1¾ inches. Colour light orange or yellow, with a white spiral band on the body whorl, and two white patches on the varices. Sculpture and shape very similar to *C. pileare*, but with the important differences that *rubeculum* has finely beaded spiral cords, an obtuse point to its spire, and single teeth on the greatly thickened outer lip.

Occurrence: This is one of the rarer West Indian *Cymatiidae*. Our specimens came from commercial dredgings at the Bogue Islands and in Ocho Rios harbour, from depths of 30 to 50 feet. Apparently it has not been reported previously from Jamaica.

CYMATIUM KREBSI Mörch 1877 Kreb's Triton **pl. 13**

Florida to the West Indies.

Description: Length 1½ to 3 inches. Colour uniform white, with occasional thin, yellowish bands. Sculpture of 5 to 7 strong, nodulous ridges; these are crossed by 6 or 7 heavy spiral cords. Outer lip greatly thickened; siphonal canal rather short.

Occurrence: This is a rare Jamaican species. Our specimens were dredged from about 100 feet, 10 miles due south of Port Royal. We have seen no other examples from Jamaican waters.

CYMATIUM VESPACEUM Lamarck 1822
Dwarf Hairy Triton **pl. 13**

Southern Florida to the West Indies.

Description: Length 1 to 1½ inches. Colour white to light orange; the varix is brown and white. Sculpture of several thin spiral cords, the one at the shoulder with sharp, widely spaced nodules. There are numerous, widely spaced growth lines. Whorls squarish at the shoulder. Outer lip thickened, with many small, paired teeth; parietal area with several folds, the largest uppermost. Siphonal canal long, slender and delicate.

Occurrence: Relatively common in 10 to 30 feet at Negril, Discovery Bay, Ocho Rios harbour, Bull Bay and in many other areas all around Jamaica. It prefers sandy surfaces in the vicinity of coral reefs.

Forms: There is an uncommon dark brown colour form with a lighter spiral band. We have found it at Negril and Ocho Rios.

Subgenus *Gutturnium* Mörch 1852

CYMATIUM MURICINUM Röding 1798
Knobby Triton **pl. 13**

Bermuda, south-east Florida and the West Indies.

Description: Length 1 to 2 inches. Colour yellowish-white, maculated with brown; occasional specimens are all brown. There is often a lighter spiral band on the body whorl; interior of aperture usually dark orange. Sculpture of coarse spiral cords, nodulose at the shoulder. Parietal shield thick and white; outer lip with several large white teeth. Siphonal canal fairly short and recurved.

Occurrence: Common in 1 foot of water, on and under rocks at the pool at the Police Training School, Port Royal. Elsewhere, it occurs in deeper water, and it is relatively common at Negril, the Bogue Islands, in Ocho Rios harbour, and at the Port Royal Cays, in depths of 3 to 30 feet.

Subgenus *Monoplex* Perry 1811

CYMATIUM PARTHENOPEUM von Salis 1793

von Salis' Triton **pl. 13**

Bermuda, Florida, and the West Indies south to Brazil.

Description: Length 2½ to 6 inches. Colour light brownish-yellow with dark patches on the varices. Sculpture of 5 or 6 strong, low spiral cords, which are usually nodulose near the shoulder, and many finer threads. There are fine axial growth lines. Outer lip thickened with paired white teeth, interspersed with dark brown. Parietal area dark brown with many white plicae; siphonal canal short.

Occurrence: We have not yet found this species in Jamaica, but include it here because it is not uncommon in many areas close to the Island. Our specimens come from Barbados and Dominica, in 40 to 80 feet, and it may well occur in Jamaican waters within that depth range.

Subgenus *Cymatium* Röding 1798

CYMATIUM FEMORALE Linné 1758

Angular Triton **pl. 13**

South-east Florida and the West Indies.

Description: Length 3 to 7 inches. Colour dark orange or brown, with purplish-white nodules on the last two or three varices. Sculpture of several broad, flat, nodulous spiral cords with finer threads between them. Outer lip thickened and triangular in shape; siphonal canal short and recurved. Operculum thin, yellow and disproportionately small for the size of the shell. The periostracum is unusually thick and mat-like. This is a very beautiful species, and the shells of young adults have a delicate, porcelainous texture.

Occurrence: Not uncommon in 10 to 30 feet on eel grass beds at Port

Royal, Negril, Montego Bay, Ocho Rios and many other areas all around Jamaica. We have found live adult specimens in 2 feet of water, partially buried in the eel grass, at the Old Coal Wharf, Port Royal, but the species does not usually favour such shallow water.

Genus *Distorsio* Röding 1798

DISTORSIO CLATHRATA Lamarck 1816
Atlantic Distorsio **pl. 14**
South-east U.S., the Gulf States and the Caribbean.
Description: Length 1 to 3 inches. Colour white to tan, with parietal area and base of outer lip light brown. Sculpture of spiral and axial cords, which give a strongly reticulated surface. Siphonal canal short and curved to the right of the shell's axis. Aperture crowded with teeth of uneven shape and size. Periostracum thick, brown and mat-like.
Occurrence: We have not found this species in Jamaican waters. It is included here because its distribution in the Caribbean suggests that it may occur locally. It is normally a shallow water species, but our examples come from the southern Caribbean in about 80 feet. Shrimp trawlers in that area bring them up in large numbers.

DISTORSIO MCGINTYI Emerson and Puffer 1953
McGinty's Distorsio **pl. 14**
Florida and south through the West Indies.
Description: Length 1 to $1\frac{1}{2}$ inches. Colour white. It is similar to *D. clathrata*, but smaller, lacks the strongly reticulated sculpture, and has a shorter siphonal canal.
Occurrence: Like *clathrata*, this species has not been found in Jamaica. It is included here for the same reasons. It occurs in very much deeper water than *clathrata*; our examples were dredged off Barbados in about 200 feet. It is a very uncommon species.

FAMILY BURSIDAE

This is an unusually attractive family of some 60 species, which occur only in tropical seas. The shells are characterised by prominent former varices, a short siphonal canal and oval, horny operculums.

All members of the family are active nocturnal predators, and like the *Cymatiidae* they prefer small, thin-shelled bivalves.

Genus *Bursa* Röding 1798

BURSA THOMAE Orbigny 1842
St Thomas Frog Shell **pl. 14**
South-east Florida and the West Indies.
Description: Length ¾ to 1 inch. Colour white, with purplish-brown on the varices and on top of the nodules, and a tinge of lavender over much of the shell. The aperture is a striking shade of lavender, which tends to fade in dead specimens. There are two prominent varices per whorl, placed axially one below the other. This is perhaps the most attractive example of an unusually beautiful genus.
Occurrence: Uncommon from 1 to 45 feet at Priory, the Bogue Islands, Discovery Bay, Bull Bay, and in rocky areas all around the Island. Our live-collected specimens were taken at Priory both on and under rocks, in a few inches of water at low tide. Dead shells occasionally wash ashore.

Subgenus *Colubrellina* Fischer 1884

BURSA CUBANIANA Orbigny 1842
Granular Frog Shell **pl. 14**
South-east Florida and the West Indies.
Description: Length 1 to 2 inches. Colour orange-brown, with three white patches on the varices, and occasionally two or three white spiral bands. Aperture tinted orange-brown. Sculpture of several spiral rows of beads. The varices are placed axially one below the other.
Occurrence: This is the most common member of the family in Jamaica. It can be collected under rocks in 1 to 20 feet at Prospect, Priory, Runaway Bay and in rocky areas all around the Island. Dead shells are common on many beaches.

BURSA CORRUGATA Perry 1811 Gaudy Frog Shell **pl. 14**
South-east Florida and the Caribbean.
Description: Length 2 to 3 inches. Colour dark or light orange-brown. Sculpture of two or more spiral rows of nodules on each whorl. The edge of the outer lip is delicately frilled, and the frills are retained as part of the former varices. The parietal shield is tan coloured and covered with white pustules and plicae. The former varices do not line up below each other.
Occurrence: Uncommon on and under rocks in 10 to 90 feet at the Bogue Islands, Discovery Bay, Ocho Rios, Bull Bay, and in many

rocky areas around Jamaica. Beach specimens usually lack the delicate frills on the varices.

Subgenus *Bufonaria* Schumacher 1817

BURSA SPADICEA Montfort 1810
Chestnut Frog Shell **pl. 14**
South-east Florida and the Caribbean.
Description: Length 1 to 3 inches. Colour brownish-yellow, sometimes with darker spiral bands; inside of outer lip light orange. Sculpture of many spiral rows of small beads; surface smooth and polished. The former varices more or less line up under each other. Operculum thin, dark brown to black, and oval.
Occurrence: This is a relatively deep water species which we have not yet found in Jamaica. It is included here because its known distribution suggests that it occurs locally, probably in depths greater than 50 feet. Our specimens were dredged by shrimp trawlers off Barbados in about 100 feet; it is locally abundant.

Subgenus *Tutufa* Jousseaume 1881

BURSA TENUISCULPTA Dautzenberg and Fischer
Fine-sculptured Frog Shell
South-east Florida and the West Indies.
Description: Length 2 to 3 inches. Colour dull ash-grey. Sculpture of 5 to 7 spiral rows of numerous, small, evenly sized beads. The former varices do not line up under each other. The shell is proportionately longer and narrower than the other members of the family.
Occurrence: This is a very rare species. The only specimen we have seen was dredged dead in about 100 feet, ½ mile north of Salem, St Ann. We have heard of no other Jamaican record.

FAMILY TONNIDAE

These are large, rounded, thin-shelled gastropods, found mostly in tropical seas. The animal is usually larger than the shell itself, and has a very broad, flat foot.

Adult *Tonnidae* lack operculums, but most species have a thin periostracum. All members of the family are carnivorous, and many species subsist mainly on echinoderms.

Plate 9 COWRIES, CYPHOMAS, SIMNIAS

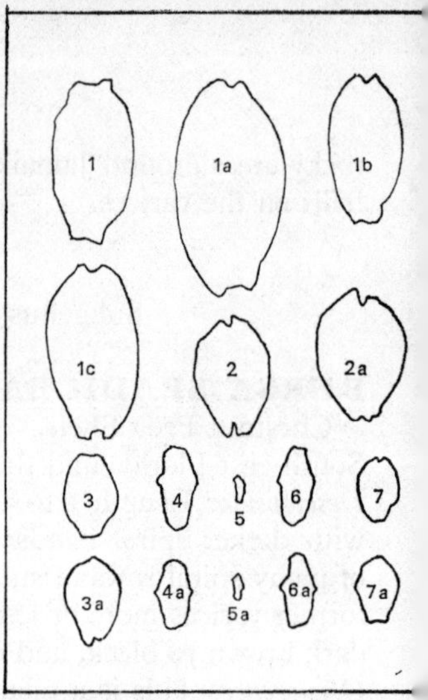

Scale: reduced by 25%

1. Cypraea zebra Linné 105
Measled Cowry
Immature shell without spots. **1a,** mature specimen. Note spire concealed by body whorl. **1b,** intermediate specimen. Apex remains visible, and colour pattern is not yet clearly defined. **1c,** ventral view of **1.**

2. Cypraea mus Linné 106
Mouse Cowry
This species does not occur in Jamaica. **2a,** dorsal view.

3. Cypraea cinerea 106
Gmelin Atlantic Grey Cowry
3a, ventral view.

4. Cyphoma gibbosum 108
Linné Flamingo Tongue
4a, ventral view.

5. Neosimnia uniplicata 107
Sowerby Single-toothed Simnia
The purple colour variety. **5a,** ventral view.

6. Cyphoma signatum 108
Pilsbry and McGinty
Fingerprint Cyphoma
6a, ventral view.

7. Cypraea spurca acicularis Gmelin 106
Atlantic Yellow Cowry
7a, ventral view.

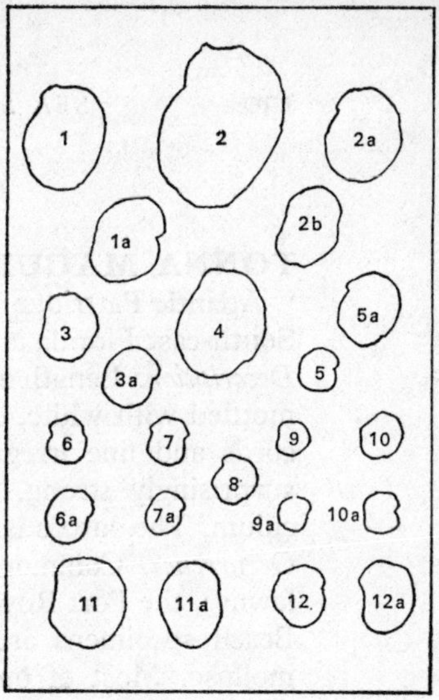

Plate 10 MOON SHELLS, BABY'S EARS

Scale: reduced by 35%

1. **Natica cayennensis** Récluz Cayenne Natica 114
1a, ventral view. The interior of the aperture is purple. See Fig. 16, page 115, for operculum.

2. **Natica canrena** Linné Colourful Atlantic Natica 111
Typical colour variety. **2a,** predominantly white colour variety. **2b,** ventral view of **2.** Note the characteristic shelly operculum.

3. **Polinices lacteus** Guilding Milk Moon Shell 109
3a, ventral view. The operculum is horny and may be red or yellow.

4. **Polinices hepaticus** Röding Brown Moon Shell 110

5. **Stigmaulax sulcata** Born Sulcate Natica 115
5a, dorsal view. Umbilicus is large and very deep.

6. **Natica floridana** Rehder Florida Natica 114
6a, ventral view.

7. **Natica livida** Pfeiffer Livid Natica 111
7a, ventral view.

8. **Natica marochiensis** Gmelin Morocco Natica 114

9. **Sigatica semisulcata** Gray Scratched Moon Shell 115
9a, ventral view.

10. **Natica menkeana** Philippi Menke's Natica 111
10a, ventral view. See Fig. 16, page 115, for operculum.

11. **Sinum perspectivum** Say Common Baby's Ear 110
11a, ventral view. This species has no operculum.

12. **Sinum maculatum** Say Maculated Baby's Ear 110
This species has no operculum. **12a,** dorsal view. The shells of this genus are more or less concealed by the very large animal which cannot retract within the shell.

Genus *Tonna* Brünnich 1772

TONNA MACULOSA Dillwyn 1817
Atlantic Partridge Tun **pl. 11**
South-east Florida and the West Indies.
Description: Length 2 to 5 inches. Colour variable, but generally brown mottled with white, or all brown. Sculpture of many broad flat spiral cords and fine irregular growth lines. Shell thin and polished, but surprisingly strong. Adults have a thin periostracum but no operculum. The suture is deep, but not channelled.
Occurrence: Common in about 12 feet at Negril, Florida beach (Trelawny), the Port Royal Cays and in sandy areas all around the Island. Beach specimens are common, but the living animal is an elusive mollusc. Most of our live-collected specimens were recovered from fish-pots set in shallow water.

TONNA GALEA Linné 1758 Giant Tun **pl. 11**
Circumtropical.
Description: Length 4 to 7 inches. Colour light to dark brown, either uniform or mottled. Sculpture of many broad, flat, spiral cords; shell thin but strong, with an almost flat spire. The nuclear whorls are blue-black and the suture is channelled.
Occurrence: Very uncommon in 30 to 100 feet at Robins Bay, Cow Bay, and Bull Bay. It usually lives on eel grass beds in sandy areas, and may occur in limited numbers all around the Island. Undamaged shells are occasionally taken into fish pots by large hermit crabs; shell fragments are found on some beaches after rough weather.

Order Neogastropoda

SUPERFAMILY MURICACEA

FAMILY MURICIDAE

This is a major marine family embracing a large number of species with a world-wide distribution. The *Muricidae* are characterised by long spines, and delicate frills on the varices and siphonal canal. They all have a brown, corneous operculum, which fits tightly into the aperture.

All species are voraciously carnivorous, and many are not above eating members of their own family. The Royal Tyrian Purple dye, which coloured the togas of the Roman emperors, was extracted from *Murex brandaris*, a Mediterranean species.

SUBFAMILY MURICINAE

Genus *Murex* Linné 1758

Subgenus *Phyllonotus* Swainson 1833

MUREX POMUM Gmelin 1791 Apple Murex **pl. 15**

South-east U.S. to the West Indies.

Description: Length 2 to $4\frac{1}{2}$ inches. Colour generally a combination of cream, tan and dark brown. Interior of aperture orange, yellow or ivory, usually with one black spot on the upper end of the parietal wall. and two or three on the outer lip. The shell is heavy and strong with a very rough surface.

Occurrence: Very common in 1 to 40 feet all along the harbour side of the Palisadoes, and at Negril, Runaway Bay, Port Antonio and in many other areas all around Jamaica. This hardy species flourishes in nearly all environments, and is the most common *Murex* in the Island.

Forms: There is a beautiful colour form with a rich pink aperture, and a faint pinkish flush over the exterior surface of the shell. This form is not uncommon in the southern Caribbean, and we have recently found what appear to be local examples at Bull Bay. It is very rare in Jamaica.

Subgenus *Chicoreus* Montfort 1810

MUREX SPECTRUM Reeve 1846 Pink Murex **pl. 15**

The Greater Antilles.

Description: Length 2 to 3 inches. Colour dark reddish-brown, with pink pearly whorls: occasional specimens may be pale yellow. Outer lip with 8 frondose spines; aperture small and nearly circular. There are three axial rows of small tubercles between the varices. In mature adults, the teeth of the outer lip, and part of the parietal area, are stained a yellowish-orange. This species is superficially similar to some juvenile specimens of *M. brevifrons*, but the latter are easily distinguished by their much wider apertures. *M. spectrum* is an exceptionally beautiful shell.

Occurrence: Very uncommon in about 35 feet at the Bogue Islands, Discovery Bay, Ocho Rios harbour, Orange Bay and Bull Bay (the last locality is the only south coast record). We have found only one beach specimen which was washed ashore after very heavy weather. Dead specimens from Orange Bay were found at the base of a coral reef, in about 60 feet, which suggests that the live shells in that area normally live within the reef.

Note: Johnsonia, No. 17 page 56, states that this species is not thought to occur in the Western Atlantic. This should be disregarded.

It is apparently closely related to *M. florifer*, which does not occur in Jamaica, and we have included it here in the same subgenus (*Chicoreus*), though it may well merit its own subgenerical status.

MUREX BREVIFRONS Lamarck 1822

West Indian Murex **pl. 15**

Lower Florida Keys and the West Indies.

Description: Length 3 to 6 inches. Colour variable, usually cream with narrow dark brown spiral bands; some specimens are entirely cream or dark brown. Interior of aperture white. The shell has three axial varices with many frond-like spines; the development of the spines varies greatly among individuals.

Occurrence: As for *M. pomum*, but it is far less common on the north coast. Beautiful specimens are easily collected in Kingston harbour.

Forms: There is an unusual pink and white colour form which occurs in about 15 feet off the Government Explosives Store at Port Henderson. We have not found it in any other locality.

Subgenus *Murex* Linné 1758

MUREX RECURVIROSTRIS RUBIDUS

F. C. Baker 1897 Rose Murex

Florida, the Bahamas, Jamaica and Puerto Rico.

Description: Length 1 to 2 inches. Colour cream, pink, orange or red. There are two or three rows of axial tubercles between the varices. The siphonal canal is thin and extended, and there are generally two or three short spines at the base. This species is very similar in shape to *M. woodringi*, but can be separated from it by differences in colour, and the latter's longer spire and smaller tubercles between the varices.

Occurrence: This is a rare species in Jamaica. Our only specimen came from about 30 feet off Cave (Westmoreland), on a sandy bottom. We know of no other Jamaican records. This specimen differs in several respects from the typical Florida *recurvirostris rubidus*, and may merit the status of a sub-species.

MUREX WOODRINGI Clench and Farfante 1945
Woodring's Murex **pl. 15**

Jamaica, Puerto Rico, and north-eastern South America.

Description: Length $1\frac{3}{4}$ to $2\frac{3}{4}$ inches. Colour cream with heavy maculations of light or dark grey; in beach specimens the grey fades to brownish-orange. There is generally a narrow, dark brown band at the shoulder. There are usually three or four axial rows of small nodules between the varices. Siphonal canal very long and fragile, with a few short spines at the base. The nodule at the shoulder of each varix may, or may not, be extended into a short spine. This species is easily confused with *M. recurvirostris rubidus*. *M. woodringi* is one of the most delicately beautiful of West Indian *Muricidae*.

Occurrence: Common in 1 to 20 feet in Kingston harbour. Perfect examples are easily collected in the 'lagoons' in the centre of the Palisadoes just east of Port Royal. Elsewhere in Jamaica, it is an uncommon species and prefers a soft, muddy substrate.

MUREX ANTILLARUM Hinds 1844
Antillean Murex **pl. 15**

Southern Florida through the Lesser Antilles.

Description: Length 2 to 4 inches. Colour cream to purple-brown, often with orange varices. There may be three indistinct darker bands on each whorl. The varices and base of the long siphonal canal are very spinose; the adult shell is solid and strong. There is an excellent illustration of this species on the current 4 cent Jamaican postage stamp (the specimen illustrated is preserved in the Institute of Jamaica).

Occurrence: This spectacular shell occurs in 40 to 150 feet at Cow Bay, the Bogue Islands, off Priory and in sandy or muddy areas all around the Island (it has not been found in Kingston harbour, however). It is rarely obtained in perfect condition, and is one of the choice West Indian *Muricidae*.

MUREX CIBONEY Clench and Aguayo 1945 **pl. 15**

Cuba south to Trinidad.

Description: Length 2 to 3 inches. Colour white, often with brown spiral lines; some specimens show more brown than white. Sculpture similar to that of *M. antillarum*, but the spines on the varices are poorly developed. The shell is solid and strong.

Occurrence: Uncommon from about 350 to 1,500 feet throughout the Antilles. A beautiful live specimen was dredged in about 700 feet just south of Black River in 1971, but it does not seem to be easily found anywhere along the south coast. Our examples were brought up by shrimp trawlers south of Trinidad.

Note: Abbott reports that this species is probably a form of *M. antillarum.* However, there are significant differences between juvenile examples of the two species, quite apart from the more obvious adult disparities, and we believe that *ciboney* is a valid species.

MUREX CHRYSOSTOMA Sowerby 1834
Gold-mouthed Murex **pl. 15**
The Caribbean and south to Brazil.
Description: Length $1\frac{3}{4}$ to 3 inches. Colour purplish-cream or buff, with scattered dark markings and thin, broken, brown spiral lines. There are two or three axial rows of nodules on each whorl; the nodule at the shoulder is very much larger than the others. Aperture white, usually with three spots of yellow on the inside of the outer lip. There are usually two or three small spines at the base of the siphonal canal, which is of medium length. It is a very attractive shell.
Occurrence: This is a very rare species in Jamaica, if it in fact occurs here at all. Our specimens reportedly came from a fish pot off Cave (Westmoreland). This was probably set in about 60 feet of water. We have no other Jamaican record, and its occurrence in these waters needs further confirmation. In the southern Caribbean it is often dredged by shrimp trawlers.

MUREX CAILLETI Petit 1856 Caillet's Murex **pl. 15**
The Bahamas, southern Florida and south through the Lesser Antilles.
Description: Length $1\frac{1}{2}$ to $2\frac{1}{2}$ inches. Colour white to dark cream, with several thin, brown, spiral lines. Some specimens have two indistinct light brown bands on the last whorl. Sculpture of three thick varices per whorl, with two or three nodulose axial ridges between them. Siphonal canal long and very slender; unlike *M. antillarum,* its width remains constant.
Occurrence: This is a rare deep water species in Jamaica. It has been dredged in 600 feet off Discovery Bay and Salem (St Ann), and in 1948 a fisherman's seine net at Yallahs (St Thomas) accidentally brought up 26 of them at one 'drawing', proving that the species can be found also in relatively shallow water. We have no other accurate locality records.
Forms: M. cailleti variation *kugleri* is a form with a more extended spire, and a number of short, sharp spines on the varices.

MUREX CABRITI Bernardi 1858 Cabrit's Murex **pl. 15**
South Florida, the Gulf of Mexico and the West Indies.
Description: Length 1 to 3 inches. Colour white or yellowish-pink. There are usually three or four sharp spines on each varix, and the extended siphonal canal may bear three rows of spines along much of

its length. There are usually three axial rows of small nodules between varices. In general shape, it is not unlike *M. woodringi*.
Occurrence: This is a rare Jamaican species which is found in relatively deep water. Our specimens were dredged off the Pedro Cays in about 140 feet. It may occur at depths up to 400 feet in several areas around Jamaica. In the southern Caribbean it is occasionally dredged by shrimp trawlers.

MUREX BEAUI Fischer and Bernardi 1857 Beau's Murex
South Florida, the Gulf of Mexico and the West Indies.
Description: Length 3 to 5 inches. Colour cream to pale brown. The spiny varices may, or may not, bear thin, wavy webs. There are 5 or 6 axial rows of small, evenly sized tubercles between the varices. This is a very beautiful species, particularly when the 'webbing' on the varices is present and extensive.
Occurrence: The only specimen we have seen came from a fish pot set in about 80 feet at Negril. The specimen contained a hermit crab, which probably brought up the shell from a greater depth. *M. beaui* is generally considered a deep water species.
Note: The presence of 'webbing' is apparently directly related to the type of substrate on which the shell occurs. Growth on a hard, rocky bottom is thought to inhibit 'webbing'.

MUREX TRYONI Hidalgo 1880 Tryon's Murex **pl. 15**
Florida and the West Indies south to Grenada.
Description: Length 1 to 2 inches. Colour uniform white or grey. This species is similar in shape to *M. cabriti*, but smaller and more delicate. The varices bear three sharp, evenly sized spines; the siphonal canal is long, slender and more or less spinose. There are usually three axial rows of small tubercles between the varices.
Occurrence: Rare in 150 feet on sand at Negril. We dredged two dead specimens in this locality and are aware of no other Jamaican records.

MUREX PULCHER A. Adams 1853
Beautiful West Indian Murex **pl. 15**
Caribbean to Brazil.
Description: Length $1\frac{1}{2}$ to $2\frac{1}{2}$ inches. Colour cream to light yellow, variegated with reddish-brown; there is a band of darker colour at the shoulder of the whorl, another at the base of the whorl, and a third at the narrowing point on the siphonal canal. Sculpture of moderately prominent varices which support delicate webbing, particularly towards the base; there is usually a short open spine at the shoulder of each whorl. There are 2 to 4 nodulous inter-varical ridges. Siphonal

canal extended and quite sharply recurved; occasionally, the previous canal is retained. The species is well named.

Occurrence: Rare in 100 to 200 feet in Discovery Bay and Orange Bay. This species is nocturnal in habit, and spends daylight hours lightly buried in the sand. In the eastern Caribbean, specimens are occasionally brought up by shrimp trawlers, and it has been found on the beach in Barbados.

MUREX TRISTICHUS Dall 1889 Tristichus Murex **pl. 22**
Florida, Cuba and Jamaica.

Description: Length about $\frac{5}{8}$ inches. Colour cream to pale salmon pink. Sculpture of well developed blade-like varices with denticulated edges. Shell surface smooth and polished. Siphonal canal moderately extended turned up and to the right. Aperture small and tear-drop shaped. Shell light and rather delicate.

Occurrence: This is one of the rarest West Indian *Muricidae*, and one of the smallest members of the family. Two live specimens were taken from a blade of *Agaricia* coral at 210 feet in Discovery Bay in 1970. Previous West Indian records were off Cuba at depths of about 2,400 feet. It is an exquisite and valuable species.

Subgenus *Favartia* Jousseaume 1880

MUREX CELLULOSUS Conrad 1846 Pitted Murex **pl. 16**
Bermuda, south-east U.S. and the Greater Antilles.

Description: Length $\frac{3}{4}$ to 1 inch. Colour yellowish or greyish-white; occasionally with patches of dark purple at the shoulder, the centre of the varices, and within the aperture. Axial sculpture of 5 to 7 fluted varices per whorl; spiral sculpture of many strong ridges which become frondose where they cross the varices. Siphonal canal sharply recurved.

Occurrence: Relatively common at Negril, the Bogue Islands, Ocho Rios, and Orange Bay, in about 25 feet. Dead shells are not uncommon on the beaches at Negril, but usually lack the tip of the siphonal canal.

Forms: There is a form which is smooth between the varices; we have difficulty in separating this species from *M. mcgintyi* (from the Bahamas), and there is some doubt that they are separate species.

MUREX CELLULOSUS NUCEUS Mörch 1850
Globose Pitted Murex **pl. 16**
Bermuda, south-east U.S. and the Greater Antilles.

Description: Length $\frac{3}{4}$ to 1 inch. Colour yellowish, with light brown

varices. This form is very similar to *M. cellulosus*, but it is shorter, globular, and far more solidly constructed; the aperture is rounded and the siphonal canal shorter.
Occurrence: This is a very uncommon species in Jamaica. Our specimens come from the Bogue Islands, Priory, and Bull Bay, in depths of 1 to 30 feet. Shallow water examples are usually heavily encrusted with calcareous growths, making immediate identification uncertain. The shell occurs both on sand and moss-covered rocks.

Genus *Muricopsis* Bucquoy, Dautzenberg and Dollfus 1882

MURICOPSIS OXYTATUS M. Smith 1938
Hexagonal Murex **pl. 16**
Florida Keys and the West Indies.
Description: Length 1 to 2 inches. Colour white or light brown with pink or dark brown spines. Occasionally there is a spiral brown band in the centre of the body whorl; aperture white. There are 5 to 7 axial ribs on each whorl; these are covered with a number of short spines, which in juvenile specimens are thin and sharp. The spire is extended.
Occurrence: Relatively common at the Bogue Islands, Discovery Bay, Orange Bay and many other areas all around Jamaica. Depth range 10 to 30 feet; specimens have been found both in open muddy or sandy areas, and on coral reefs.
Note: Beach specimens, which are uncommon, usually lack spines on the varices.

MURICOPSIS species Little Pink Murex **pl. 16**
Jamaica.
Description: Length ¾ to 1 inch. Colour white, with extensive brown or black patches on the whorls. These are crossed by two white spiral lines which are extended into short spines on the varices. Aperture a striking shade of pink (specimens from Ocho Rios and Runaway Bay); occasionally white (specimens from the Bogue Islands). In general shape, this species is very similar to *M. oxytatus*, but with fewer, shorter, and more irregular spines on the varices.
Occurrence: This beautiful little shell is uncommon in 10 to 30 feet at the Bogue Islands, Runaway Bay and in the Ocho Rios area. It has been collected live on the reef opposite the Silver Seas Hotel (Ocho Rios) in about 15 feet. Specimens from this locality are heavily encrusted with a calcareous growth, which conceals the colour pattern. We have found no beach specimens.
Note: This may be a new species. If so, it will be described jointly by Dr R. T. Abbott and the author.

MURICOPSIS PHILIPPIANA Dall 1889
Orange-banded Murex **pl. 16**
The Greater Antilles.
Description: Length $\frac{3}{4}$ to $1\frac{1}{2}$ inches. Colour cream, generally with a broad brown or orange spiral band in the centre of the body whorl, and sometimes a thin broken line of the same colour towards the base of the shell. Sculpture of about six smooth, raised axial ribs per whorl: in occasional specimens these bear short spines on the early whorls. There are usually microscopic spiral threads within the coloured band. Inside the outer lip there is a series of thin, rather long teeth. The siphonal canal is short and slightly recurved; the operculum is wine red. The shell is solid and strong.
Occurrence: This is a rare species in Jamaica, but it is widely distributed. Live specimens have been collected on algae-covered rocks in 1 to 10 feet at Priory and the Port Royal Cays, and we have also dredged dead shells from about 35 feet off the Bogue Islands and in Ocho Rios harbour.

MURICOPSIS PUDICUS Reeve Little Spiny Murex **pl. 16**
The Greater Antilles.
Description: Length $\frac{1}{2}$ to $\frac{3}{4}$ inch. Colour dirty white, usually with two or three thin, broken, orange spiral bands. The shell is very similar to the preceding species (*M. philippiana*), but there is a short, sharp spine at the shoulder of each axial rib, and several smaller ones extending farther down. The long, thin teeth within the outer lip are set back some distance from its edge.
Occurrence: This is a very rare species in Jamaica. We have found dead shells in about 40 feet off the Bogue Islands, on a muddy substrate. We have no other Jamaican records, and the shells we have recovered were all long dead and buried in the mud. This has led to speculation that the species is now locally extinct, but this must be regarded as unlikely, and further exploration of the Bogue Islands area will probably turn up a live specimen and end the speculation.

SUBFAMILY TYPHINAE

Genus *Typhis* Montfort 1810

TYPHIS FORDI Pilsbry Ford's Typhis **pl. 16**
The Bahamas and Jamaica.
Description: Length $\frac{3}{4}$ to $1\frac{1}{4}$ inches. Colour uniform white. Sculpture of numerous fine axial lines and finer growth lines. The varices are

delicately sculptured, extended and blade-like. The shell is unmistakably characterised by the presence of one tubular projection at the shoulder of each whorl, just below the suture. The siphonal canal is recurved, the apex extended, and the shell is light and delicate. It is difficult to imagine a more delicately beautiful shell than this species. *Occurrence:* Rare on sand or mud bottoms, at about 30 feet, in Ocho Rios harbour, off the Bogue Islands, and at Negril (where one beach specimen was found). We can trace no previous local record. The clean exterior surface of the shell suggests that it normally lives buried in the substrate.
Note: The Jamaican form of this species differs in minor respects from typical *fordi* from the Bahamas. It may merit sub-specific status.

SUBFAMILY PURPURINA

Genus *Drupa* Röding 1798

Subgenus *Morula* Schumacher 1817

DRUPA NODULOSA C. B. Adams 1845
Blackberry Drupe **pl. 16**
Southern Florida and the West Indies.
Description: Length $\frac{1}{2}$ to $\frac{3}{4}$ inch. Colour black, occasionally reddish-brown. Sculpture of about 10 beaded axial cords. Aperture with 4 or 5 white teeth on the thickened outer lip.
Occurrence: This is one of the most common shells of its size in Jamaica. It occurs in 1 to 12 feet on, and under, rocks at Runaway Bay, Port Henderson and in rocky areas all around the Island. Many specimens are heavily encrusted with coralline growths.

Genus *Risomurex* Olsson and McGinty 1958

RISOMUREX ROSEUS Reeve 1856 Pink Drupe **pl. 16**
West Indies.
Description: Length about $\frac{1}{2}$ inch. Colour bright reddish-pink, with black or brown spiral bands; aperture pink. Sculpture of nodulose axial ribs; outer lip thickened and toothed within. Mature adults have a small, glazed parietal shield.
Occurrence: This beautiful little shell is very uncommon in Jamaica. Our specimens come from 20 to 60 feet at the Bogue Islands, Discovery Bay, Ocho Rios harbour and in the reefs at Orange Bay.

RISOMUREX MURICOIDES C. B. Adams
Beaded Drupe **pl. 16**
Florida and the West Indies.
Description: Length ½ to ¾ inch. Colour black or brown, with two or three spiral rows of white nodules. Aperture with white teeth. Sculpture of several nodulose spiral ribs. Siphonal canal short and recurved. The shell is similar to *R. roseus*, but larger and more elongate.
Occurrence: Relatively uncommon in 1 to 10 feet, under rocks, at Robins Bay, Prospect, and in several other rocky areas around the Island. Beach specimens are rare.

Genus *Purpura* Bruguière 1789

PURPURA PATULA Linné 1758
Wide-mouthed Purpura **pl. 17**
South-east Florida and the West Indies.
Description: Length 2 to 3½ inches. Colour dull greyish-brown; columella salmon pink. Sculpture of 6 or 7 spiral rows of sharp nodules. The shell gets progressively smoother with age. Aperture very wide; inside of outer lip crenulated and marked with brown. The animal emits a greyish fluid when disturbed, and this is in fact a strong purple dye; it also has a most noxious smell which is extremely difficult to remove from the fingers (this can be avoided by picking up the shell by the back, so that the fluid drains away from the hand).
Occurrence: This is a very common species at Port Henderson, Runaway Bay, and in the inter-tidal zone of rocky, exposed areas all around the Island. It clings tenaciously to the surface on which it is resting, and is difficult to remove without damaging the lip. It often seeks shelter in cavities in the rocks, and two or three shells will sometimes share the same quarters.

Genus *Thais* Röding 1798

Subgenus *Stramonita* Schumacher 1817

THAIS RUSTICA Lamarck 1822 Rustic Rock Shell **pl. 17**
Bermuda, south-east Florida and the West Indies.
Description: Length 1½ inches. Colour variable; generally cream or grey, banded with dark brown or black. The nodules are often rust-red. Interior of aperture white, with purple stains on the inside of the outer lip, and on the inside lower corner of the columella. Sculpture of one or more spiral rows of blunt nodules at the shoulder and mid-area of the

whorls. There are also many fine growth lines and spiral threads. Outer lip strongly crenulated, with the teeth occurring in groups of 2 or 3 (cf. *T. haemastoma floridana*, the following species).
Occurrence: Relatively common in the lower inter-tidal area, under rocks, along the southern side of the Palisadoes, and in similar exposed areas all around the Island. Beach specimens are often found at Bull Bay and Orange Bay.

THAIS HAEMASTOMA FLORIDANA Conrad 1837
Florida Rock Shell **pl. 17**
South-eastern U.S. and the Caribbean.
Description: Length 2 to 3 inches. Colour variable, usually light grey mottled with blue-black and orange. Occasional specimens are a uniform shade of light orange. Interior of aperture and columella salmon-pink in fresh specimens, with dark markings on the inside of the outer lip. Sculpture of many fine spiral lines, sometimes with 1 or 2 spiral rows of small nodules; fine fixial growth lines are also present. There is a strong ridge just above the columella which runs back into the aperture. Unlike the previous species (*T. rustica*), the teeth on the outer lip are not in groups.
Occurrence: Common on and under rocks, in the inter-tidal area, on both sides of the Palisadoes, and in rocky areas all around Jamaica. We have collected excellent examples at the old Coal Wharf, Port Royal. Dead shells are commonly adopted by hermit crabs.

Subgenus *Mancinella* Link 1807

THAIS DELTOIDEA Lamarck 1822
Deltoid Rock Shell **pl. 17**
Bermuda, Florida and the West Indies.
Description: Length 1 to 2 inches. Colour greyish-cream, usually mottled with broad, broken spiral bands of dark brown. Interior white; columella bright purple in fresh specimens. Sculpture of 1 or 2 spiral rows of large, blunt nodules, and fine spiral threads. There are occasional growth lines. The shell is heavy and strong and often covered with coralline deposits.
Occurrence: Very common in the lower inter-tidal zone, on and under rocks, at Prospect, Port Royal, and in exposed areas all around the Island. Dead shells are often washed ashore.

Genus *Aspella* Mörch 1877

Subgenus *Aspella* Mörch 1877

ASPELLA PAUPERCULA C. B. Adams 1850
Little Aspella **pl. 16**
West Indies.
Description: Length $\frac{3}{4}$ to 1 inch. Colour a uniform dull white, sometimes with a yellowish tinge. There are 3 principal axial varices, with a smaller one between each. There are a few spiral ridges and many spiral threads. Suture deep and pitted; aperture small, and outer lip feebly toothed within. In fresh specimens there are usually 2 small yellowish-purple patches on the inside of the outer lip. Siphonal canal short and recurved.
Occurrence: Uncommon under rocks in 6 to 20 feet at Orange Bay, the Bogue Islands, Ocho Rios and Negril. Dead shells are sometimes found on the beach at Bull Bay.

Genus *Ocenebra* Gray 1847

OCENEBRA INTERMEDIA C. B. Adams 1850
Frilly Dwarf Triton **pl. 16**
West Indies.
Description: Length $\frac{3}{4}$ to 1 inch. Colour greyish-white, sometimes with faint brown spiral bands. Sculpture of 6 or 7 frilled axial varices, and several spiral ridges. There is a deep suture and a short siphonal canal. The aperture is oval and the outer lip delicately frilled. The sculpture is not unlike that of *Murex cellulosus nuceus*.
Occurrence: This is a relatively uncommon species except at Port Henderson, where it occurs under rocks, just below the low tide mark, in quite large numbers (next to the pier at the Government Explosives Store). Elsewhere, it occurs at Montego Bay, Discovery Bay, and along the north coast in 4 to 20 feet on rocky bottoms.

FAMILY MAGILIDAE

Members of this family are closely associated with corals and sea fans (*Gorgonia*). The shells are strong and coarsely sculptured; many bear short spines or prominent ribs.

Most species live in shallow water in tropical seas, and are relatively common. The operculums of the West Indian species are thin, and either wine red or pale yellow in colour; the shells, externally, are all white, and often encrusted with coralline algae. Shells of the genus *Coralliophila* lack a radula, and feed suctorially on coral polyps.

Genus *Coralliophila* H. and A. Adams 1853

CORALLIOPHILA ABBREVIATA Lamarck 1816
Short Coral Shell **pl. 16**
South-east Florida and the West Indies.
Description: Length 1 to 1¼ inches. Colour white with the aperture tinted pale orange. Sculpture of numerous spiral cords and rounded axial ridges; there is a wide umbilicus. Operculum thin and light orange.
Occurrence: Common on or near living coral, in 1 to 35 feet, at Negril, Port Antonio, the Port Royal Cays and all around Jamaica. The shells are often wedged into crevices in the coral, and are well camouflaged.
Note: Research now in progress suggests that this species subsists by tapping the body fluids of coral polyps without actually killing them.

CORALLIOPHILA CARIBAEA Abbott 1958
Caribbean Coral Shell **pl. 16**
South-east Florida and the West Indies.
Description: Length ½ to 1 inch. Colour greyish-white; aperture usually tinted purple. Characterised by its triangular shape. Sculpture of about 12 rounded axial ridges and numerous spiral cords, which form nodules at the periphery. There is a slit-like umbilicus; the operculum is corneous, opaque and wine red. The body whorl of old specimens may be somewhat irregular in shape.
Occurrence: Very common in the base of sea fans at Negril, Runaway Bay, the Port Royal Cays and wherever sea fans occur. It also occurs on coral, from 1 to 200 feet, and dead shells are often washed ashore.

CORALLIOPHILA ABERRANS C. B. Adams 1850
Globular Coral Shell **pl. 16**
West Indies.
Description: Length ½ to ¾ inch. Colour, dirty white. Sculpture of many close set, rounded cords and numerous axial ridges. Whorls are well rounded, giving the shell a globular shape. There is a relatively deep suture. Operculum wine red.
Occurrence: This is a very uncommon species. Our live-collected specimens came from the vicinity of the pool at the Police Training

Plate 11 BONNETS, TUNS, SMALL HELMETS

Scale: reduced by 35%

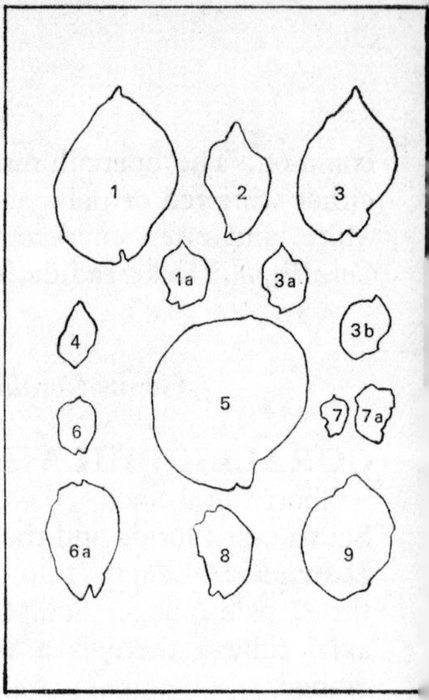

1. **Phalium granulatum** 117
Born Scotch Bonnet
Occasional specimens retain former varices.
1a, ventral view of small example.

2. **Sconsia striata** 117
Lamarck Royal Bonnet

3. **Phalium cicatricosum** 118
Gmelin Smooth Scotch Bonnet
Occasional specimens retain former varices.
3a, ventral view of small example. **3b,** form with pustules on the shoulder of each whorl.

4. **Casmaria atlantica** 118
Clench Atlantic Agate Helmet

5. **Tonna galea** Linné 130
Giant Tun

6. **Cypraecassis testiculus** Linné 119
Reticulated Cowrie Helmet
This is the form *crumena* Bruguière, with a single or double row of ridges on the shoulder of the body whorl. **6a,** typical variety. Adult shells lack both periostracum and operculum.

7. **Morum oniscus** Linné 116
Atlantic Woodlouse
7a, dorsal view of large example.

8. **Morum dennisoni** 116
Reeve
Specimen illustrated is possibly the largest known.

9. **Tonna maculosa** 130
Dillwyn Atlantic Partridge Tun
This species has a periostracum but no operculum.

Plate 12 HELMETS, CONCHS, TULIP SHELLS, TRITONS, SPINDLES

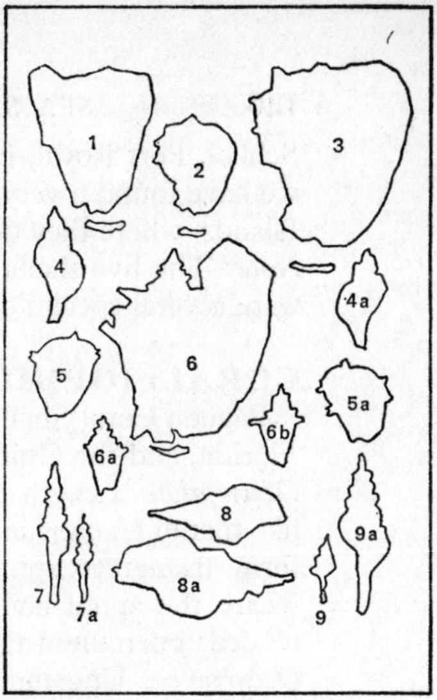

Scale: reduced by 75%

1. Cassis tuberosa Linné 118
King Helmet
Note operculum below the illustration.

2. Cassis flammea Linné 119
Flame Helmet
Note operculum below the illustration.

3. Cassis madagascariensis 119
Lamarck Emperor Helmet
Note operculum below the illustration.

4. Fasciolaria tulipa 156
Linné True Tulip
Typical colour variety. **4a,** dorsal view of the uncommon red and cream colour variety.

5. Melongena melongena 153
Linné West Indian Crown Conch
Variety with a single row of spines. **5a,** ventral view. Typical variety with three rows of spines.

6. Strombus gigas Linné 101
Queen Conch
Note operculum below the illustration. **6a,** ventral view of juvenile example. Note absence of flared lip. **6b,** dorsal view of **6a.**

7. Fusinus eucosmius 155
Dall Ornamented Spindle
7a, dorsal view.

8. Charonia variegata 120
Lamarck Trumpet Triton
This is the rare colour variety with an orange cast over the shell. **8a,** dorsal view of the typical colour variety.

9. Fusinus closter 155
Philippi Closter Spindle
This species has not been found in Jamaican waters. **9a,** ventral view of a large specimen. Note that the brown coloration disappears in the later whorls.

School, Port Royal, in about 1 foot of water under a rock. Elsewhere, we have found a very few dead shells at Discovery Bay and the Bogue Islands, where they occur in about 20 feet.

Note: The live shells collected were not found on coral, though there were several species of coral in the vicinity.

CORALLIOPHILA SCALARIFORMIS Lamarck 1822

Pagoda Coral Shell **pl. 16**

Florida, and the Gulf of Mexico east to the Virgin Islands.

Description: Length $\frac{3}{4}$ to 1 inch. Colour white. Sculpture essentially like that of *C. aberrans*, but the spiral cords are thin and blade-like, and form triangular spines, particularly at the shoulder of the whorls where the spiral and axial sculpture intersects. Spire acute and extended; operculum thin and yellow.

Occurrence: Uncommon under rocks in 1 to 40 feet at Robins Bay, Runaway Bay, Prospect and several other areas around Jamaica. We have found only two live specimens; the remainder have contained hermit crabs.

Note: The note on *C. aberrans* also applies to this species.

SUPERFAMILY BUCCINACEA

FAMILY COLUMBELLIDAE

These are the small, beautifully coloured 'Dove Shells' which are widely distributed, and generally abundant in shallow water throughout the West Indies. There are some 400 species, all of which are vegetarian. The family is characterised by the thickening of the centre of the outer lip, and the high gloss on the surface of the cleaned shell. The sexes are separate.

Genus *Columbella* Lamarck 1799

COLUMBELLA MERCATORIA Linné 1758

Common Dove Shell **pl. 17**

South-east Florida and the West Indies.

Description: Length $\frac{1}{2}$ to $\frac{3}{4}$ inch. Colour pattern extremely variable – generally a combination of white, brown, black, orange or pink, commonly arranged in spiral bars or zigzags. The inside of the outer lip is usually yellow or orange. Sculpture of many fine spiral ridges.

The aperture is strongly toothed on both sides, with the outer lip incurved in the centre. There is a moderately thin periostracum in live specimens. The shell is solid and strong.
Occurrence: This is one of the most common shells on Jamaican beaches. It lives on and under rocks in 1 to 6 feet, in large numbers, along the southern side of the Palisadoes and in rocky areas all around the Island. The shells are usually well camouflaged by algal growths.

Genus *Pyrene* Röding 1798

PYRENE OVULATA Lamarck 1822 Ovate Dove Shell **pl. 17**
West Indies.
Description: Length $\frac{1}{2}$ to $\frac{3}{4}$ inch. Colour reddish-brown, with irregular white markings. In fresh specimens the ground colour is almost black. Sculpture of many faint spiral grooves. Aperture long and narrow; outer lip weakly toothed within, and slightly incurved in the centre. There is a thin periostracum.
Occurrence: As for *Columbella mercatoria* (the preceding species), but it is far less common in Jamaica.

Genus *Anachis* H. and A. Adams 1853

ANACHIS PULCHELLA Sowerby 1844
Beautiful Dove Shell **pl. 17**
Florida Keys and the West Indies.
Description: Length $\frac{1}{4}$ to $\frac{1}{2}$ inch. Colour yellowish-white with extensive brown maculations which are often orange in beach specimens. Sculpture of numerous axial ribs crossed by fine spiral threads. The columella callus is raised into a small lip at its edge.
Occurrence: This is not a common species in Jamaica. It occurs under rocks from 2 to 30 feet at Prospect, Robins Bay, Ocho Rios and Montego Bay. Dead shells are occasionally found in beach drift at Negril and along the southern side of the Palisadoes.

Genus *Nitidella* Swainson 1840

NITIDELLA OCELLATA Gmelin 1791
White-spotted Dove Shell **pl. 17**
Bermuda, lower Florida Keys and the West Indies.
Description: Length $\frac{1}{4}$ to $\frac{1}{2}$ inch. Colour dark brown, covered with white

spots. The shell is smooth except for a few spiral striae at the base. Outer lip toothed within; apex usually eroded.

Occurrence: Very common under rocks in about 1 foot of water near the pier at the Government Explosives Store, Port Henderson. Elsewhere, it is locally common, but its distribution is not as widespread as the other members of this family.

NITIDELLA NITIDA Lamarck 1822

Glossy Dove Shell **pl. 17**

South-east Florida and the West Indies.

Description: Length about ½ inch. Colour purplish-brown, maculated with white. The shell is smooth and polished. Aperture long and narrow; outer lip thickened in the centre and toothed within. The columella has two small ridges just below the centre.

Occurrence: Common under rocks in 1 to 6 feet along the southern side of the Palisadoes, and in rocky areas all around the Island. The shells are nearly always found in colonies of several individuals closely packed together.

NITIDELLA LAEVIGATA Linné 1758

Smooth Dove Shell **pl. 17**

Florida Keys and the West Indies.

Description: Length ½ to ¾ inch. Colour pattern variable, usually white with zigzag axial orange or brown markings; tip of spire generally light yellow. There may be a spiral row of dark spots above the suture and in the centre of the body whorl. Aperture oval, outer lip thin. There are two folds at the base of the columella; the periostracum is thin and straw-coloured.

Occurrence: Abundant on algae and under rocks in 1 to 3 feet along the southern Palisadoes, and in rocky areas all around Jamaica. This is perhaps the most common of the smaller gastropods on Jamaican beaches.

Genus *Mitrella* Risso 1826

Subgenus *Collumbellopsis* Bucquoy, Dautzenberg and Dollfus 1882

MITRELLA NITENS C. B. Adams 1850

Shiny Dove Shell **pl. 17**

West Indies.

Description: Length about ½ inch. Colour white, with a spiral series of brown flame-like markings. The shell is smooth and polished, with about 14 striae near the base. The aperture is nearly half the length of

the shell, and the outer lip is thickened and delicately toothed within. Sun-bleached specimens tend to lose their brown coloration.
Occurrence: This is a very uncommon species in Jamaica. It occurs at about 40 feet in Kingston harbour and at Ocho Rios, on a muddy substrate. Beach specimens are rare, but occasionally found along the eastern side of the causeway in Kingston harbour. It has also been dredged in 600 feet on a mud substrate off Discovery Bay.

FAMILY BUCCINIDAE

This is a huge, diversified family of over 2,000 species which includes the large, predatory 'Whelks', and extends from polar to tropical seas.

None of the larger species is found in Jamaican waters, and the shell known locally and throughout the West Indies as the 'Wilk' (Whelk), is in fact *Cittarium pica* of the family *Trochidae*. All *Buccinidae* are carnivorous and show a marked partiality for dead fish.

Genus *Bailya* M. Smith 1944

BAILYA PARVA C. B. Adams 1850
West Indian Baily Shell **pl. 17**
The Bahamas and the West Indies.
Description: Length about $\frac{3}{4}$ inch. Colour cream, with several irregular spiral brown and white bands. Sculpture of 10 or 12 thin axial ribs (which are sometimes prominent), crossed by finer spiral lines. There are small beads where the axial and spiral sculpture intersects. The whorls are rounded, and there is a deep suture. The shell is characterised by the greatly thickened lip. Within limits, this is a very variable species.
Occurrence: Common under rocks in 2 to 60 feet at Robins Bay, the Bogue Islands, Orange Bay, Runaway Bay, and in many rocky areas all around the Island. The shells are often found in pairs. Beach specimens are quite common.

BAILYA INTRICATA Dall 1883 Intricate Baily Shell
Southern Florida and the West Indies.
Description: Length about $\frac{3}{4}$ inch. Colour dirty white. Sculpture of 10 to 14 narrow axial ribs crossed by alternating thick and thin spiral cords, giving a reticulated surface. The whorls are shouldered. This species is similar to *B. parva* (the preceding species), but the slight difference in

colour, and the more elongate *intricata*'s strongly reticulated surface and shouldered whorls clearly separate them.

Occurrence: Uncommon under rocks in 1 to 30 feet at Robins Bay, Discovery Bay, Negril, and in many rocky areas on both the north and south coasts. Dead shells are washed up on the beach along the southern side of the Palisadoes. Like *B. parva* this species is often found in pairs.

Genus *Antillophos* Woodring 1928

ANTILLOPHOS CANDEI Orbigny 1853 Beaded Phos **pl. 17**

South-east U.S. and the West Indies.

Description: Length ¾ to 1¼ inches. Colour yellowish-white, often with extensive brown, spiral bands. Sculpture of about 16 axial ribs on the body whorl and many spiral threads. There are sharp pustules where the axial and spiral sculpture intersects, and several sharp ridges within the aperture, running back into the shell. The columella has two modest folds near the base.

Occurrence: Uncommon from 20 to 60 feet off the Bogue Islands, in Ocho Rios harbour and at Cow Bay. The only beach specimen we have seen came from Rio Bueno (Trelawny). The live shell is usually found lightly buried in mud or soft sand during the day.

ANTILLOPHOS OXYGLYPTUS Dall and Simpson 1901
Puerto Rican Phos

Jamaica and Puerto Rico.

Description: Length about ⅝ inch. Colour dirty white, sometimes with light brown maculations. Sculpture of about 16 axial ribs and many spiral threads.

Occurrence: This is a rare Jamaican species. The only specimen we have found came from the beach near Bluefields (Westmoreland), and had evidently been dead for a very long time. We cannot positively say that the species still lives in Jamaican waters.

Genus *Engoniophos* Woodring 1928

ENGONIOPHOS GUADELUPENSIS Petit 1852
Guadeloupe Phos

West Indies.

Description: Length ¾ to 1 inch. Colour white or any shade of brown. Sculpture of about 10 axial ribs crossed by numerous very fine spiral threads. Columella with one fold near the base.

Occurrence: Relatively common in about 20 feet on mud or sand substrate, at Negril, the Bogue Islands and in Ocho Rios harbour. Beach specimens are uncommon.
Forms: There is a form with slightly shouldered whorls and a narrow dark spiral band below the shoulder of the body whorl. It has been given the specific name of *Engoniophos unicinctus*, but is probably the same species as *guadelupensis*.

Genus *Engina* Gray 1839

ENGINA TURBINELLA Kiener 1836
White-spotted Engina **pl. 17**
Lower Florida Keys and the West Indies.
Description: Length about ½ inch. Colour black or reddish-brown (in dead specimens), with a spiral row of white nodules on the periphery of each whorl. Sculpture of numerous, fine spiral threads. Aperture narrow, and outer lip toothed within; parietal area glazed, with a number of folds and nodules.
Occurrence: Common on and under rocks in 1 to 12 feet at Negril, Runaway Bay, Port Henderson and in rocky areas all around the Island. It is commonly washed ashore.

Genus *Colubraria* Schumacher 1817

COLUBRARIA LANCEOLATA Menke 1828
Arrow Dwarf Triton **pl. 17**
South-east U.S. and the West Indies.
Description: Length 1½ to 2 inches. Colour cream with brown or purplish maculations; the tip of the spire is light purple in some specimens. Surface of the shell is finely reticulate, and the former varices are quite distinct. The border of the aperture is elevated to form a 'collar'. This is a delicate and attractive species.
Occurrence: Uncommon under rocks in 10 to 60 feet at Negril, the Bogue Islands, Ocho Rios harbour and Cow Bay. Dead specimens are occasionally found on the Long Bay beach at Negril.

COLUBRARIA OBSCURA Reeve 1844
Leaning Dwarf Triton **pl. 17**
Tortugas and the West Indies.
Description: Length 1½ to 2½ inches. Colour light brown with darker maculations, usually in the form of irregular spiral bands. Surface

reticulate, with spiral rows of small beads; former varices prominent. Outer lip thickened and toothed within. Easily separated from *C. lanceolata* by its fatter profile, and more polished surface.

Occurrence: As for *C. lanceolata.* It is equally uncommon, but because of its robust construction, beach specimens are more often found intact. We found three live specimens under the same rock at Negril in about 8 feet, but the species normally prefers deeper water.

Genus *Pisania* Bivona-Bernardi 1832

PISANIA PUSIO Linné 1758

Miniature Triton-Trumpet **pl. 17**

South-east Florida and the West Indies.

Description: Length 1½ to 2 inches. Colour purplish-black, generally with rather indistinct spiral bands of grey and brown 'arrow-heads'. There is usually a thin, broken, white band in the centre of the body whorl. Surface of shell smooth and polished; parietal area highly glazed. The outer lip has a number of small white teeth within, and the edge is crenulated. This is a beautiful representative of an unusually attractive family.

Occurrence: Relatively common on and under rocks, in 1 to 15 feet, at Montego Bay, Runaway Bay, Port Henderson and in rocky areas all around Jamaica. This species is almost invariably found in pairs, with the individuals seldom more than a foot apart.

Genus *Cantharus* Röding 1798

Subgenus *Pollia* Sowerby 1834

CANTHARUS TINCTUS Conrad 1846

Tinted Cantharus **pl. 17**

South-east U.S. and the West Indies.

Description: Length about 1 inch. Colour usually dark brown, with irregular patches of grey or white. Sculpture of 10 to 14 low axial ribs, and many spiral cords. Outer lip ridged within; edge crenulated.

Occurrence: Very uncommon under rocks, in 1 to 15 feet, at Priory, Port Royal and Bull Bay. Dead shells are occasionally found on beaches in rocky areas.

CANTHARUS AURITULUS Link 1807

Common Cantharus **pl. 17**

South-east Florida and south through the West Indies to Brazil.

Description: Length $\frac{3}{4}$ to $1\frac{1}{4}$ inches. Colour brown or orange, clouded with irregular white or grey markings. Sculpture of 9 to 11 low, axial ribs; otherwise very similar to *C. tinctus*, but broader and heavier.

Occurrence: Common in 1 to 10 feet under rocks at Folly (Port Antonio), Negril, Bull Bay, and in rocky areas all around the Island. Dead shells are often washed ashore.

CANTHARUS LAUTUS Reeve 1846

Gaudy Cantharus **pl. 17**

The West Indies.

Description: Length about $\frac{3}{4}$ inch. Colour reddish-brown, with one or more white spiral bands. Sculpture of 8 to 10 strong axial ribs crossed by a number of thin spiral cords; there are sharp nodules at the intersection of the axial and spiral sculpture.

Occurrence: This is a very uncommon species in Jamaica. Our specimens were dredged (dead) in about 30 feet off the Bogue Islands and in Ocho Rios harbour. There is also a record of a beach specimen from Bull Bay.

FAMILY MELONGENIIDAE

This is an attractive group of large, solid, generally colourful shells, distributed in tropical and temperate seas. They occur in, or just below, the inter-tidal zone and are easily collected. The animals are ferociously carnivorous, and the single West Indian species attacks both gastropods and bivalves. The sexes are separate.

The Australian Trumpet Shell (*Syrinx aruanus*), one of the Pacific members of the family, is the largest gastropod in the world, and sometimes measures over two feet long.

Genus *Melongena* Schumacher 1817

MELONGENA MELONGENA Linné 1881

West Indian Crown Conch **pl. 12**

Florida Keys (?) and the West Indies.

Description: Length 3 to 7 inches. Colour variable: usually cream with broad spiral bands of purplish-brown or blue; all-white 'albinos' are uncommon. There are generally 1 to 4 rows of broad, sharp spines on the body whorl, but occasional specimens lack spines altogether. There is a distinct channel at the suture. The operculum is large, dark brown and rounded.

Occurrence: Common on eel grass beds in 1 to 30 feet in Kingston harbour. Juvenile shells, which are beautifully marked, are abundant in the mangrove swamps adjacent to Gunboat Beach. The species is widely distributed along the south coast, and can be found in most sheltered, muddy areas, often partially buried in the substrate. It is an uncommon shell on the north coast, where the smooth form predominates.

FAMILY NASSARIIDAE

These are mainly small, solid, ovate shells with about 100 species distributed worldwide. Most genera live on mud bottoms in shallow, protected areas, and some species occur locally in great numbers. A few deep water species occur in the West Indies. The sexes are separate.

The family is omnivorous, and although they are active scavengers, they will also attack and eat bivalves. The classification of the Caribbean species is now under review.

Genus *Nassarius* Duméril 1806

NASSARIUS ALBUS Say 1826 Variable Nassa **pl. 17**

South-eastern U.S. to the West Indies.

Description: Length about ½ inch. Colour white or yellowish, generally with one or more spiral bands of brown dots. Sculpture of about 10 axial ribs and numerous spiral threads. There is a deep suture, and the shell has a broad, chunky appearance.

Occurrence: Common in 1 to 60 feet at Priory, Montego Bay and Ocho Rios. This species is found partially buried in mud or sand substrate, and is easily overlooked. Dead shells, which may be bleached white, are quite common on many beaches.

Note: Nassarius nanus Usticke 1959 closely resembles *N. albus*, but is much smaller and more elongate. *N. consensus* Ravenel, which is smooth and polished, with microscopic spiral threads, may in fact be a form of *albus* (see Abbott – *American Seashells*), though we doubt this.

NASSARIUS VIBEX Say 1822
Common Eastern Nassa **pl. 17**

Cape Cod, the Gulf States and the West Indies.

Description: Length $\frac{1}{2}$ to $\frac{3}{4}$ inch. Colour dark brown, with a light spiral band at the shoulder of each whorl. Sculpture of about 12 axial ribs and numerous spiral threads. Parietal shield well developed, and glazed white or yellowish.

Occurrence: Abundant on mud flats in brackish water at the old Coal Wharf, Port Royal, and in similar areas all around the Island. The unusually active animals merit study in their natural habitat. This species is easily distinguished from other Jamaican *Nassariidae* by the large parietal shield.

FAMILY FASCIOLARIIDAE

Members of this family are active, predatory gastropods which are found in shallow water in most warm seas. The group is characterised by the spindle shape, the extended spire and siphonal canal, the ridged interior of the outer lip, and the relatively large operculum, which is sharply pointed at the apex. The sexes are separate.

SUBFAMILY FUSININAE

Genus *Fusinus* Rafinesque 1815

FUSINUS CLOSTER Philippi Closter Spindle **pl. 12**

Lower Caribbean.

Description: Length about 7 inches. Colour white, with axial streaks of brown between the nodules; the colour is more pronounced in young specimens. Sculpture of sharp spiral cords, and axial ribs which are prominent in the earlier whorls.

Occurrence: Relatively common in about 200 feet around Barbados and Trinidad, and also off the north-eastern coast of Brazil. Shrimp trawlers operating in those areas frequently bring the shells up in their nets. It is unlikely to be found in Jamaican waters, but is included here as a matter of interest.

FUSINUS EUCOSMIUS Dall Ornamented Spindle **pl. 12**

Gulf of Mexico to Brazil.

Description: Length up to 6 inches. Colour white. Sculpture of 8 rounded axial ribs crossed by about 10 sharp spiral threads. Aperture

small, ovate, and located at the middle of the shell. This is an exceptionally beautiful species.
Occurrence: As for *F. closter* (the preceding species), with which it is dredged in about equal numbers in the southern Caribbean. It is also found in the Gulf of Mexico, and may well occur off Jamaica.
Note: There is evidence that specimens from Brazil grow much larger than those from its northern range.

SUBFAMILY FASCIOLARIINAE

Genus *Fasciolaria* Lamarck 1799

FASCIOLARIA TULIPA Linné 1758 True Tulip **pl. 12**
South-eastern U.S. and the West Indies.
Description: Length 3 to 10 inches. Colour very variable, but frequently brown clouded with white, yellow or tan, and overlaid with many brown spiral lines. Shell surface smooth except for axial wrinkles, and 2 or 3 spiral grooves just below the suture. The shell has an extended spire and siphonal canal; there are numerous thin ridges inside the outer lip.
Occurrence: Common in 1 to 15 feet on eel grass beds at Discovery Bay, in Kingston harbour, and also on coral reefs all around Jamaica. Good specimens are easily collected in shallow water close inshore, along the northern side of the Palisadoes. It will often burrow under the substrate in search of prey.
Forms: There is a beautiful red colour form which occurs in very limited numbers wherever the species is found.
Note: This is a voraciously carnivorous species and will decimate the molluscan population of an aquarium in a surprisingly short time.

Genus *Latirus* Montfort 1810

LATIRUS INFUNDIBULUM Gmelin 1791
Brown-lined Latirus **pl. 18**
Florida Keys and the West Indies.
Description: Length 2 to 3 inches. Colour brown with darker spiral cords. Sculpture of several rounded axial ridges crossed by narrow spiral cords. Aperture long and ovate; there are 3 weak columella folds. The shell is imperfectly umbilicate.
Occurrence: Uncommon in 10 to 30 feet off the Bogue Islands, in Discovery Bay, Cow Bay and on sandy and rocky bottoms all around the Island. Beach specimens are quite rare.

LATIRUS BREVICAUDATUS Reeve 1847
Short-tailed Latirus **pl. 18**

Lower Florida Keys and the West Indies.

Description: Length $1\frac{1}{2}$ to $2\frac{1}{2}$ inches. Colour reddish-brown with darker spiral threads. Sculpture of 8 or 9 broad, rounded, axial ribs, crossed by numerous spiral threads. The shell is similar to *L. infundibulum*, but the siphonal canal is shorter and the shell is proportionately broader.

Occurrence: Relatively common under rocks in 1 to 20 feet at Robins Bay, Prospect, and rocky areas all around the Island.

LATIRUS MCGINTYI Pilsbry McGinty's Latirus **pl. 18**

South-east Florida and Jamaica.

Description: Length $1\frac{1}{2}$ to $2\frac{1}{2}$ inches. Colour yellow to dark orange. Sculpture of 7 or 8 strong axial ribs crossed by several thin, sharp spiral cords. There are prominent nodules where the axial and spiral sculpture intersects. Mature shells are umbilicated. In general shape this species is similar to *L. brevicaudatus*, but the sculpture is sharper and less coarse.

Occurrence: As for *L. brevicaudatus*, but it is somewhat less common.

Note: This species differs from the typical *mcgintyi* of south-east Florida in being yellow or orange rather than cream. In other respects the shells are apparently identical.

Genus *Fusilatirus* McGinty 1955

FUSILATIRUS CAYOHUESONICUS Sowerby 1878
Key West Latirus **pl. 18**

Key West to the West Indies.

Description: Length about $\frac{3}{4}$ inch. Colour dark brown. Sculpture of 6 or 8 rounded axial ribs crossed by fairly strong spiral cords. In general shape, this species closely resembles a miniature *Latirus infundibulum*.

Occurrence: Uncommon under rocks in 3 to 15 feet at the Port Royal Cays, off the Bogue Islands, at Runaway Bay, and in several other areas around Jamaica. We have found it more common at Southeast Cay (off Port Royal) than any other locality.

Genus *Leucozonia* Gray 1847

Subgenus *Leucozonia* Gray 1847

LEUCOZONIA NASSA Gmelin 1791
Chestnut Latirus **pl. 18**
Florida to Texas, and the West Indies.
Description: Length 1½ to 2 inches. Colour dark brown with irregular axial cream streaks. There is a narrow spiral band of white on the lower third of the body whorl, and the tops of the nodules are crossed by thin white threads. Sculpture of about 10 low axial ribs crossed by numerous spiral cords, resulting in sharp nodules at the shoulder of the whorls. Spire sharp and extended; outer lip sharply angled in two places. Siphonal canal short.
Occurrence: Relatively common on and under rocks, in 1 to 6 feet along the southern side of the Palisadoes, and all along the north coast. It favours exposed areas, as does the following sub-species.

LEUCOZONIA NASSA LEUCOZONALIS
Lamarck 1822 **pl. 18**
Florida to Texas and the West Indies.
Description: Length 1½ to 2½ inches. Colour dark brown, with a narrow white spiral band on the lower third of the body whorl. Sculpture of very fine axial growth lines and spiral threads. Whorls rounded. The shell is heavy and solid, with a very short siphonal canal. Aperture yellowish, with the outer lip slightly angled where it is intersected by the spiral white band. The smooth sculpture of *nassa leucozonalis* contrasts strongly with that of the preceding species. Close comparison of the two shells, however, indicates why the former is a subspecies of true *nassa*.
Occurrence: Very common on weed-covered rocks in 1 to 6 feet, particularly in areas exposed to the open sea. It can be collected in large numbers at Bull Bay, Discovery Bay and in rocky areas all around the Island.

LEUCOZONIA OCELLATA Gmelin 1791
White-spotted Latirus **pl. 18**
Florida and the West Indies.
Description: Length about 1 inch. Colour brown or black, with white nodules, and several broken white bands near the base. Sculpture of about 9 large, blunt nodules at the periphery of the whorls. Siphonal canal short. The shell is less extended and more squat than the other Jamaican species of *Latirus* and *Leucozonia*.

Occurrence: Relatively common on rocks in and just below the inter-tidal area at Robins Bay, Holland Bay (St Thomas), and Old Pera (St Thomas). Its distribution, however, is limited, and we have found it in few other places.

FAMILY XANCIDAE

This family includes both the Vase Shells (*Vasinae*) and the Chanks (*Xancinae*). Most species are relatively common and occur in shallow water; all belong to tropical seas. There is only one representative of the group in Jamaican waters, though one of the Chanks (*Xancus angulatus*) occurs as close as the Pedro Cays.

The family is characterised by the massive construction and thick, persistent periostracums of the shells. They are all predatory molluscs and fear few adversaries.

SUBFAMILY XANCINAE

Genus *Xancus* Röding 1798

XANCUS ANGULATUS Solander 1786
West Indian Chank
Bermuda, the Bahamas, Key West, Cuba, Yucatan and the Pedro Cays.
Description: Length 7 to 14 inches. Colour cream or white; the interior is often tinged with light pink or dark orange. Periostracum light brown and thick. Sculpture of prominent, blunt tubercles on the shoulder, and undeveloped axial ridges extending downwards from the tubercles. The columella bears three prominent plications, and there is a deep umbilicus. The shell is thick and heavy.
Occurrence: This species is reported from the Pedro Cays, which are Jamaican territory, by V. A. Zans ('Natural History Notes of the Natural History Society of Jamaica', No. 74 of Sept. 1955). It does not occur in Jamaica itself. Dead shells are reported quite common at the Pedro Cays, and the live mollusc is found on sand and eel grass beds from the inter-tidal zone to about 60 feet.
Note: The geographical range of this species is most interesting, and the fact that it is known to occur both to the north and south of Jamaica suggests that one day it may well establish itself here.

Plate 13 HAIRY TRITONS

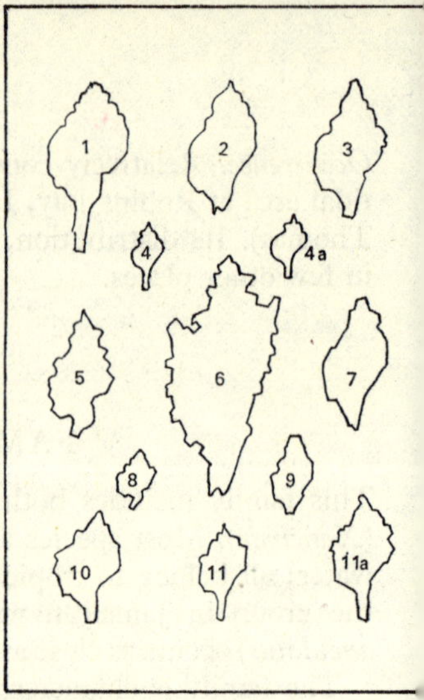

Scale: reduced by 30%

1. **Cymatium caribbaeum** Clench and Turner Dog-head Triton 121

2. **Cymatium pileare martinianum** Orbigny Atlantic Hairy Triton 122
Typical colour variety.

3. **Cymatium parthenopeum** von Salis von Salis' Triton 124

4. **Cymatium vespaceum** Lamarck Dwarf Hairy Triton 123
Typical colour variety.
4a, uncommon brown colour variety.

5. **Cymatium nicobaricum** Röding Gold-mouthed Triton 122

6. **Cymatium femorale** Linné Angular Triton 124
The operculum of this species is very small in relation to the size of the shell.

7. **Cymatium krebsi** Mörch Kreb's Triton 123

8. **Cymatium labiosum** Wood Lip Triton 121
Typical colour variety.

9. **Cymatium rubeculum occidentale** Mörch 122

10. **Cymatium poulseni** Mörch Poulsen's Triton 121

11. **Cymatium muricinum** Röding Knobby Triton 123
Dark colour variety.
11a, light colour variety.

Plate 14 HAIRY TRITONS, DISTORSIOS, FROG SHELLS

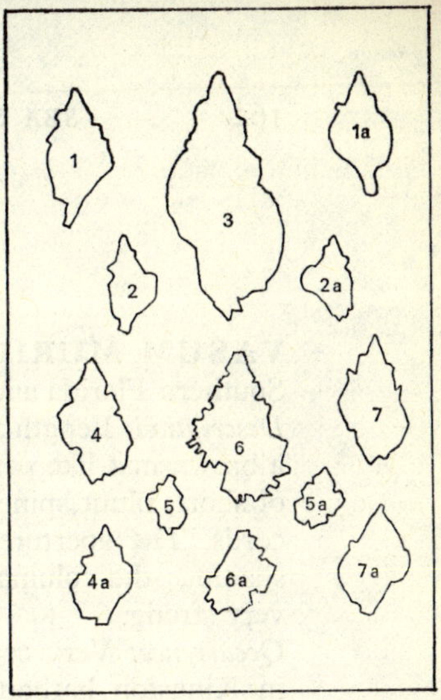

Scale: reduced by 30%

1. **Distorsio clathrata** Lamarck Atlantic Distorsio 125
   1a, ventral view. Note characteristic constricted aperture.

2. **Distorsio mcgintyi** Emerson and Puffer McGinty's Distorsio 125
   2a, ventral view. Note characteristic constricted aperture.

3. **Cymatium pileare martinianum** Orbigny Atlantic Hairy Triton 122
   This is the reddish colour variety. Compare the other variety on Plate 13.

4. **Bursa cubaniana** Orbigny Granular Frog shell 126
   4a, dorsal view.

5. **Bursa thomae** Orbigny St Thomas Frog Shell 126
   Note the characteristic lavender aperture. **5a,** dorsal view.

6. **Bursa corrugata** Perry Gaudy Frog Shell 126
   6a, dorsal view.

7. **Bursa spadicea** Montfort Chestnut Frog Shell 127
   7a, dorsal view.

SUBFAMILY VASINAE

Genus *Vasum* Röding 1798

VASUM MURICATUM Born 1778 Caribbean Vase **pl. 18**
Southern Florida and the West Indies.
Description: Length 2 to 4 inches. Colour white or cream, covered with a brown mat-like periostracum which is difficult to remove. Sculpture of short, blunt spines at the shoulder of the whorls, and coarse spiral cords. The aperture is glazed and white, usually with some purple shading; the columella bears five folds. Mature shells are heavy and very strong.
Occurrence: Very common on sand or eel grass beds in 1 to 15 feet in Kingston harbour, Runaway Bay, Negril and in many areas all around Jamaica. The shells prefer relatively quiet and protected water. In many areas specimens are so heavily encrusted with coral growths that the shape of the shell is unrecognisable from above.
Note: Vasum capitellus (pl. 18), which is smaller, more elongate and bears prominent spines at the shoulder of the whorls, has not yet been reported from Jamaica. However, it occurs in the Eastern Caribbean and in Puerto Rico, and may one day be found in Jamaican waters.

SUPERFAMILY VOLUTACEA

FAMILY OLIVIDAE

These are the well known Olive Shells, characterised by their graceful shape, high polish and striking variation in colour pattern.

There are about 300 species which live just below the surface of relatively shallow sand or mud bottoms in tropical seas. They are all carnivorous, and in the genus *Oliva*, there is no operculum. Olives usually show themselves only at night, when the mantle of the animal can be seen covering the entire shell.

Genus *Oliva* Bruguière 1798

OLIVA RETICULARIS Lamarck 1811 Netted Olive **pl. 19**
South-east Florida and the West Indies.
Description: Length $1\frac{1}{4}$ to $1\frac{3}{4}$ inches. Colour cream, with a variable

overlaid pattern of purplish-brown markings. The pattern is often in the form of small irregular triangles. The shell is highly polished and the spire moderately extended. Aperture purplish in fresh specimens.
Occurrence: Common in sand at depths of 2 to 40 feet at Negril, in Kingston harbour, and in muddy or sandy areas all around the Island. Members of this family tend to remain in the substrate during daylight, and live specimens are usually collected at night in shallow water with the aid of an electric torch.
Forms: There is an all-cream or white form, commonly found on the beaches just east of Montego Bay. It is named *O. reticularis olorinella* (pl. 19), and most examples have a slightly more extended spire than typical *reticularis*. Intergrades are numerous.
Note: O. fusiformis (pl. 19) is very similar but proportionately much fatter. It occurs in the Southern Caribbean.

OLIVA RETICULARIS GREENWAYI
Caribbean Golden Olive **pl. 19**
South-east Florida and the West Indies.
Description: Length $1\frac{1}{2}$ to 2 inches. Colour, partially or completely black or golden brown. Otherwise very similar to typical *reticularis*. This is an unusually beautiful shell, and an entirely golden brown specimen is a highly prized collector's item.
Occurrence: We have found the golden variety only off the Bogue Islands, in about 30 feet of water. Examples may occur in limited numbers wherever typical *reticularis* is found. Several old Jamaican collections indicate that it was relatively common about 100 years ago. The black variety is not uncommon in sand, in 3 feet, at Discovery Bay.

OLIVA CARIBAEENSIS Dall and Simpson 1901
Caribbean Olive **pl. 19**
Caribbean.
Description: Length 1 to 2 inches. Colour similar to *O. reticularis*, but the colour pattern is slightly blurred, and there may be a number of irregular, dark spiral lines. The species also differs from *reticularis* in having a much deeper and wider suture.
Occurrence: Very uncommon in about 25 feet at Discovery Bay, the Bogue Islands and in Ocho Rios harbour.

Genus *Olivella* Swainson 1831

Subgenus *Olivella* Swainson 1831

OLIVELLA NIVEA Gmelin 1791
West Indian Dwarf Olive **pl. 19**
South-east Florida, Bermuda and the West Indies.
Description: Length ½ to 1 inch. Colour white, with irregular white or purplish markings. These are usually strongest in two spiral bands at the base and shoulder of the body whorl. The suture is deep and quite wide.
Occurrence: Uncommon in about 20 feet in sand at Priory and Montego Bay. It probably occurs in limited numbers along both the north and south coasts. Beach specimens are rare.

OLIVELLA PETIOLITA Duclos 1835
Caribbean Dwarf Olive **pl. 19**
West Indies to Venezuela.
Description: Length ½ to ¾ inch. Colour white or yellowish, with purplish-brown arrow-shaped marks on the body whorl. The suture is grooved and the parietal callus well developed.
Occurrence: As for *O. nivea*, but it is a rare species in Jamaica.

Subgenus *Niteoliva* Olsson 1956

OLIVELLA MINUTA Link 1807 Minute Dwarf Olive **pl. 19**
The Caribbean.
Description: Length ¼ to ½ inch. Colour variable, usually grey or purple-brown in spiral bands. Most specimens have a narrow dark spiral band just below the suture, and also near the base of the body whorl. Suture indented into sides of the preceding whorls.
Occurrence: Uncommon in sand from 10 to 40 feet at Montego Bay and Ocho Rios. Beach specimens are occasionally found at Negril.

OLIVELLA VERREAUXI Ducros 1857
Verreaux's Dwarf Olive **pl. 19**
The West Indies.
Description: Length about ⅓ inch. Very similar to the preceding species (*O. minuta*), but usually more slender and with a more clearly defined pattern of zigzag lines. Our specimens have a yellowish cast and lack the narrow dark bands.

Occurrence: Relatively common in sand from 3 to 15 feet at Priory and Robins Bay. Dead shells occupied by small hermit crabs are sometimes found in very shallow water at Priory.

Subgenus *Mcgintiella* Olsson 1956

OLIVELLA ROTUNDA Dall 1889 Stout Dwarf Olive **pl. 19**
Jamaica, Puerto Rico and Barbados.
Description: Length about $\frac{3}{4}$ inch. Colour cream or pale yellow with faint irregular brownish markings. Suture deeply grooved and indented into the sides of the preceding whorls. Parietal callus pronounced; there are numerous columella plications.
Occurrence: Very uncommon in sand, in about 35 feet, at Negril and Ocho Rios. We have found few beach specimens. This species has not previously been recorded from Jamaica.

Genus *Jaspidella* Olsson 1956

JASPIDELLA JASPIDEA Gmelin 1791
Jasper Dwarf Olive **pl. 19**
South-east Florida to Barbados.
Description: Length about $\frac{3}{4}$ inch. Colour variable, usually white overlaid with a network of faint purplish-brown markings. The suture is without an indentation on the preceding whorls. There are 4 plaits at the base of the columella.
Occurrence: As for the preceding species (*Olivella rotunda*). It is an uncommon species in Jamaica.

FAMILY MITRIDAE

This is a most attractive family of nearly 600 species, widely distributed in warm, shallow seas.

The shells are strong and fusiform in shape; most genera are characterised by narrow apertures and between 3 and 5 plications on the columella. The animals are equipped with long retractable snouts which they use to prey on worms and bivalves.

Genus *Mitra* Lamarck 1799

MITRA NODULOSA Gmelin 1791 Beaded Mitre **pl. 19**
South-east U.S. and the West Indies.
Description: Length $\frac{3}{4}$ to $1\frac{1}{4}$ inches. Colour dark brown to orange. Sculpture of 15 to 17 rows of small neat beads. Columella with 4 white folds. Beach specimens may be worn smooth.
Occurrence: Very common in 1 to 12 feet under rocks at Runaway Bay, Prospect and in rocky areas all around the Island. Dead shells are often washed ashore.

MITRA BARBADENSIS Gmelin 1791
Barbados Mitre **pl. 19**
South-east Florida and the West Indies.
Description: Length 1 to 2 inches. Colour brown or orange, usually with a few flecks of greyish-white. Sculpture of many fine spiral cords. Columella with 4 or 5 white folds.
Occurrence: Relatively common in 1 to 20 feet, under rocks at Robins Bay, the Port Royal Cays, and in rocky areas all around Jamaica. Dead shells are not uncommon on the beaches.

MITRA FLUVIIMARIS Pilsbry and McGinty 1949
Gulf Stream Mitre **pl. 19**
Florida, Jamaica and Puerto Rico.
Description: Length about 1 inch. Colour white with occasional small brown patches. Sculpture of 14 to 18 thin spiral cords. Length of aperture a little more than half the shell.
Occurrence: This is a rare deep water species. Our single specimen was dredged in about 150 feet ten miles south of Kingston. It has been found on the beach in Puerto Rico.

Genus *Pusia* Swainson 1840

PUSIA CUBANA Aguayo and Rehder 1936
Cuban Mitre **pl. 19**
Northern Caribbean.
Description: Length about $\frac{1}{2}$ inch. Colour: top third of whorls white; lower two-thirds brownish-pink. There is a narrow darker dividing band. Sculpture of about 12 sharp axial ribs. Columella has 4 white folds, and the spire is sharp and extended.
Occurrence: Uncommon in 30 to 70 feet off the Bogue Islands, in

Ocho Rios harbour, and at Cow Bay. This species lives on sand or mud bottoms. Beach specimens are rare.

PUSIA GEMMATA Sowerby 1874 Little Gem Mitre **pl. 19**
Florida and the West Indies.
Description: Length $\frac{1}{4}$ to $\frac{1}{2}$ inch. Colour black-brown, with a white band on the periphery. There are 11 to 13 axial ribs on each whorl. The columella has 4 black folds.
Occurrence: Relatively uncommon under rocks in 6 to 20 feet in Ocho Rios harbour, and off the Bogue Islands. Beach specimens are found at Negril.
Note: Pusia hanleyi is a very similar species often confused with *gemmata. Hanleyi* is smaller, lighter in colour and has riblets below the suture. It occurs at Runaway Bay, under rocks, in about 4 feet of water.

PUSIA PUELLA Reeve 1845 Girl Mitre **pl. 19**
South-eastern U.S. to the Lesser Antilles.
Description: Length $\frac{1}{4}$ to $\frac{1}{2}$ inch. Colour black or dark brown, with white patches just below the suture. Sculpture of very fine spiral and axial threads; columella with 4 folds. Shell globular in shape.
Occurrence: Relatively uncommon under rocks in 1 to 20 feet at Robins Bay, Negril and Prospect. Beach specimens are rarely found.

PUSIA STAMINEA A. Adams 1851 **pl. 19**
Puerto Rico and Jamaica.
Description: Length $\frac{1}{2}$ to $\frac{5}{8}$ inch. Colour white, maculated with yellow-orange. Sculpture of many spiral lines of minute holes and axial threads; columella with 4 folds. Shell long and narrow with an extended spire; aperture about half length of shell.
Occurrence: This is a relatively rare species. We have found it only off the Bogue Islands and in Ocho Rios harbour, in about 30 feet of water.

PUSIA PULCHELLA Reeve 1845 Beautiful Mitre **pl. 19**
West Indies.
Description: Length $\frac{1}{2}$ to $\frac{3}{4}$ inch. Colour very variable: it may be red and orange, black and orange, or red and black – invariably with a narrow white spiral band on the lower third of the body whorl. Sculpture of about 20 axial ribs, with fine spiral grooves between them. There are 4 folds on the columella.
Occurrence: Not uncommon in holes in the reef and under rocks in 1 to

200 feet at Runaway Bay, Montego Bay, Prospect and in rocky areas all around the Island.

Forms: There are at least 3 quite distinct forms of this beautiful shell, all of which may well merit classification as sub-species. The shape and colour pattern varies considerably between these forms, but the sculpture remains constant. There is evidence that each form occurs in a different depth range.

PUSIA ALBOCINCTA C. B. Adams 1845

White-lined Mitre **pl. 19**

Florida and the West Indies.

Description: Length about 1 inch. Colour dark brown or black, with a white spiral band more or less in the centre of each whorl. Sculpture of 10 to 14 sharp axial ribs; columella with 4 folds. The shell is narrower than *P. pulchella*, the preceding species, and the spirc is sharp and extended.

Occurrence: Relatively uncommon in 6 to 30 feet off the Bogue Islands, at Discovery Bay and in Ocho Rios harbour. Beach specimens are occasionally found at Bull Bay.

FAMILY VOLUTIDAE

This aristocratic family of large and beautiful shells has been a favourite with collectors everywhere for hundreds of years.

There are some 200 species which live at various depths from the equator to the polar seas. Most genera lack operculums and all are carnivorous. The family is characterised by bulbous nuclear whorls and prominent plications on the columella.

SUBFAMILY VOLUTINAE

Genus *Voluta* Linné 1758

VOLUTA MUSICA Linné 1758 Music Volute **pl. 20**

The Caribbean.

Description: Length 2 to 3 inches. Colour pinkish-cream, with 2 or 3 spiral bands of fine, dark lines. There are a number of small blue-black spots scattered over the surface of the shell. Sculpture of several large rounded axial ridges, and a few broad spiral cords at the base. This is a beautiful member of a well known and widely distributed family of shells.

Occurrence: We have not found this species in Jamaica. However it is relatively common in sandy areas in the Southern Caribbean, and occurs as far north as Puerto Rico. It may yet turn up in Jamaican waters.

FAMILY CANCELLARIIDAE

These are the strongly sculptured, robust Nutmeg Shells, which are characterised mainly by the prominent spiral teeth present on the inside of the outer lip, and the lack of a true operculum.

They occur in relatively shallow water in both tropical and polar seas. Only one member of the family has been found in Jamaican waters, but there are two other Caribbean species which may eventually be discovered here. *Cancellariidae* are exclusively carnivorous.

Genus *Cancellaria* Lamarck 1799

CANCELLARIA RETICULATA Linné 1767
Common Nutmeg **pl. 20**
South-east U.S. and the West Indies.
Description: Length 1 to 2 inches. Colour cream, with indistinct brownish-orange spiral bands. Sculpture of many spiral cords and fine axial growth lines, giving a reticulated surface. There are 2 or 3 prominent folds on the columella; the inside of the outer lip is strongly ridged, the ridges running back into the aperture.
Occurrence: This is a very uncommon Jamaican species. It occurs between 15 and 50 feet at the Bogue Islands, and in Kingston and Ocho Rios harbours. There is also a record from deeper water off Old Harbour Bay. Our best specimens were dredged about 50 yards east of Fort Augusta, in Kingston harbour.
Forms: There is a form with a strongly reticulated surface sometimes classified as a separate species (*C. conradiana*) (pl. 19), which occurs off the south coast in about 150 feet.

FAMILY MARGINELLIDAE

These are small, colourful, highly polished shells with short spires and narrow apertures. They all lack an operculum.

There are about 600 species, with a very wide distribution in tropical and temperate seas. All the West Indian species are less than one inch long and live in shallow, sandy areas. Like the *Olivellidae*, members of this family conceal themselves during daylight, and envelop their shells with their mantles.

Genus *Prunum* Herrmannsen 1852

Subgenus *Leptegouana* Woodring 1928

PRUNUM APICINUM Menke 1828

Common Atlantic Marginella **pl. 20**

Florida and the West Indies.

Description: Length about ½ inch. Colour light golden-brown, with three spiral darker bands, sometimes overlaid with small white flecks. Nuclear whorls usually dark red. Similar to *P. guttatum* in shape, but the spire is more elevated. Shell smooth and polished.

Occurrence: Common in sand or mud in Ocho Rios harbour and at the Bogue Islands, in about 30 feet. Beach specimens are found at Negril after stormy weather.

PRUNUM GUTTATUM Dillwyn 1817

White-spotted Marginella **pl. 20**

South-east Florida and the West Indies.

Description: Length ½ to ¾ inch. Colour purplish-orange to cream; the body whorl is irregularly spotted with small white dots. There are 4 columella teeth, and the shell is highly polished.

Occurrence: Common in sand under rocks, in 1 to 15 feet, at Priory, Runaway Bay, Prospect, and in many areas all around the Island. Dead specimens of this beautiful little species are often washed ashore.

Genus *Hyalina* Schumacher 1817

Subgenus *Volvarina* Hinds 1844

HYALINA AVENA Kiener 1834

Orange-banded Marginella **pl. 20**

Description: Length ½ inch. Colour light tan, usually with 3 broad purplish-orange spiral bands. Outer lip thickened; columella with 4 teeth.

Occurrence: As for *Prunum guttatum*, with which it is often found living. This is a very common and attractive little shell.

HYALINA TENUILABRA Tomlin 1917
Pale Marginella **pl. 20**
Florida to the West Indies.
Description: Length ½ to ¾ inch. Colour white, occasionally light yellow. The apex is nearly flat and the aperture wide. The outer lip is thin and sharp, and there are 3 or 4 folds on the columella. Shell smooth and polished.
Occurrence: Common in sand in 15 to 35 feet (and probably deeper) at Montego Bay and in Ocho Rios harbour. Beach specimens are sometimes found at Negril.

FAMILY CONIDAE

This is a beautiful and specialised family of about 400 species, which have been collectors' favourites for hundreds of years.

Cone Shells are confined to warm tropical waters; they are characterised by the invariable conical shape and possession of a venomous stinging apparatus, which in some of the larger species can be dangerous to man. They all possess a tough periostracum and have narrow, horny operculums. The sexes are separate.

Genus *Conus* Linné 1758

CONUS REGIUS Gmelin 1791 Crown Cone **pl. 21**
The Bahamas, Florida, southern Mexico, and south to Brazil.
Description: Length 2 to 3 inches. Colour very variable, but generally white with irregular brown maculations. Sculpture of spiral threads, usually minutely pustulose. The spire may be moderately high or quite flat; it bears smooth, irregular tubercles.
Occurrence: Common in 1 to 40 feet, often partially buried in sand, all around Jamaica. Beautiful specimens can be collected at low tide on the mossy rock tables at Robins Bay and Prospect. Dead shells are common on beaches all around the Island.
Forms: There is a relatively uncommon orange-yellow form, sometimes flecked with brown and purple, known as *C. regius citrinus.* It occurs in shallow water off the Queen's Highway (near Discovery Bay), and at Prospect. It is an unusually attractive shell and is often mistaken for a separate species.

CONUS JASPIDEUS Gmelin 1791 Jasper Cone **pl. 21**

Southern Florida and the West Indies.

Description: Length ½ to 1 inch. Colour white with irregular brown or reddish maculations. Sculpture of shallow spiral grooves, and minutely beaded cords, which may reach as high as the shoulder. Spire smooth and moderately extended. Occasional specimens bear several spiral rows of small pustules over the entire body whorl, and these have been classified as *C. verrucosus*, a separate species. There is also an uncommon salmon pink colour variety known as form *vanhyningi*.

Occurrence: Moderately common in 3 to 20 feet in loose sand at Negril, Braco (Trelawny), Oracabessa (St Mary), and the Port Royal Cays. Dead shells are common on the beaches in certain areas, particularly just east of Montego Bay.

Note: The endless variations in size, colour and sculpture of this species have resulted in the unnecessary creation of many subspecies of *jaspideus*. We think that only *C. jaspideus havanensis* (the following species), of the Jamaican examples, merits sub-specific status. (For convincing treatment of this subject, see *Marine Mollusks of Grand Cayman Island* by R. T. Abbott.)

CONUS JASPIDEUS HAVANENSIS Aguayo and Farfante 1947 Havana Cone **pl. 21**

Cuba and Jamaica.

Description: Length ¾ inch. Colour white with irregular flames of reddish-brown or orange, usually separated by an irregular white spiral band near the centre of the body whorl. Sculpture of faint, minutely beaded spiral threads. Spire depressed and finely nodulose. There is a row of low nodules at the shoulder of the body whorl (very much as in *C. regius*). The shell is solid and strong.

Occurrence: This species apparently has not previously been reported from Jamaica. It is relatively common in about 40 feet on sand or mud off the Bogue Islands; we have also collected beach specimens at a point about 4 miles west of Lucea. These are the only known Jamaican records.

CONUS SPURIUS Gmelin 1791
West Indian Alphabet Cone **pl. 21**

The Bahamas and south through the Greater Antilles.

Description: Length 1½ to 3 inches. Colour white, with narrow, broken orange or brown bands, and irregular patches of solid colour. Sculpture of microscopic growth lines. Spire depressed and concave.

Occurrence: Relatively common in sand at about 25 feet in Ocho Rios harbour, Old Harbour Bay, Folly (Port Antonio), and Montego Bay.

Specimens are occasionally dredged in Kingston harbour. It is also found on beds of eel grass at Rocky Point (Clarendon). Beach specimens are uncommon, except near the Electric Power Station at Old Harbour Bay.
Forms: There is a very uncommon albino colour form; we have also seen an entirely black specimen with a lighter narrow spiral band on the body whorl (from Cow Bay).

CONUS DAUCUS Hwass 1792 Carrot Cone **pl. 21**
Florida and the West Indies.
Description: Length 1½ to 2 inches. Colour carrot-red, orange or yellow, often with spiral bands of minute brown dots. In fresh specimens, the aperture is purplish-pink within. There is generally a lighter spiral band in the centre of the body whorl. Sculpture of microscopic growth lines.
Occurrence: This beautiful species occurs all around the Island, but it is relatively uncommon. Live specimens have been collected from weed-covered rocks in 2 or 3 feet just west of Lucea and also at Priory. It generally prefers deeper water, however, and occurs in about 30 feet off the Bogue Islands, in Ocho Rios harbour and along the southern side of the Palisadoes. We have found all the colour variations living on the same rock, suggesting that the colour factor, in this species, is related neither to diet nor environment.

CONUS CENTURIO Born 1778 Centurion Cone **pl. 21**
Dominican Republic, Puerto Rico, the Virgin Islands, Jamaica, and the southern Caribbean.
Description: Length 2 to 2½ inches. Colour white, with 3 spiral bands of irregular brown markings on a yellow background. These markings occasionally intrude into the white areas between the bands. Some specimens are sprinkled with small, irregular brown dots. The area below the lowest spiral band is pale yellow. The tops of the whorls are marked with alternating brown and white areas. Sculpture of very fine growth lines and several spiral cords at the base of the shell. The spire is elevated and concave.
Occurrence: This beautiful species is perhaps the rarest Jamaican Cone. One live specimen was taken by the Sub-Aqua Club in about 90 feet at Cow Bay, on a soft mud bottom. There is also a record from off the Dunn's River falls (depth unknown). Live examples are occasionally dredged by shrimp trawlers in the southern Caribbean at depths of about 150 feet, though it has not been reported previously from this area.
Note: The specimens we have collected in the Caribbean differ in

several respects from typical *centurio* from the Dominican Republic. Possibly the Jamaican/Southern Caribbean variety – which is close to *C. sozoni* – merits recognition as a separate sub-species. (Abbott believes that *centurio* may be a subspecies of *sozoni*.)

CONUS JULIAE Clench 1942 Julia's Cone **pl. 21**

South-east Florida, Puerto Rico and Jamaica.

Description: Length 1½ to 2 inches. Colour variable; generally pale yellow, orange, brown or red. There is a spiral band of white in the centre of the body whorl, with small brown patches within the band; occasionally this is repeated at the shoulder of the whorl. Some specimens are overlaid with a series of fine spiral, broken lines. Sculpture of spiral rows of minute pustules; some specimens are quite smooth. Shoulder of the whorls more or less pustulose. Spire moderately extended and flat sided.

Occurrence: Like the preceding species, *juliae* has not been reported previously from Jamaica. It occurs in about 30 feet at Discovery Bay, in Ocho Rios harbour, and at the mouth of the Salt River (Clarendon), where it is uncommon in coral. It normally lives deep within the reefs, well protected and relatively inaccessible. A perfect adult specimen remains a choice collector's item. It is an unusually attractive shell.

CONUS RANUNCULUS Hwass 1792
Atlantic Agate Cone **pl. 21**

The West Indies.

Description: Length 2 to 4 inches. Colour extremely variable; usually greyish-white with irregular markings of chocolate brown, sometimes with incomplete, narrow spiral bands of brown dashes. Occasional specimens are entirely white; others are maculated with bright orange. Generally there is an indefinite white spiral band just below the centre of the body whorl. Sculpture of very fine axial growth lines, and a few rather weak spiral cords near the base of the shell. There are 4 or 5 spiral threads on the top of each whorl.

Occurrence: This is a very uncommon species in Jamaica. It occurs in about 40 feet at Priory, Prospect and Bull Bay. Shrimp trawlers in the southern Caribbean bring them up in large numbers from flat, muddy bottoms at depths of about 150 feet; our Jamaican examples, however. apparently prefer to live on coral reefs, though this requires confirmation.

CONUS GRANULATUS Linné 1758
Glory of the Atlantic Cone **pl. 21**
South-east Florida and the West Indies.
Description: Length $1\frac{1}{4}$ to 2 inches. Colour, bright red and brown; orange and brown; or pink and brown; all on a white background. There is usually a white spiral band just below the centre of the body whorl, with brown flecks on its upper edge. Sculpture of numerous coarse spiral cords and fine axial threads. The shell is relatively long and narrow, with a moderately extended spire. The bright red colour variety of this surpassingly beautiful species more than justifies its colloquial name.
Occurrence: Rare in 10 to 100 feet at Discovery Bay, Robins Bay, Orange Bay and off Bull Bay. This species lives in coral reefs, and as a result, is rarely dredged. In 1964 six specimens were found together in a small cave at Orange Bay at a depth of about 60 feet. Beach specimens are occasionally collected after rough weather.

CONUS PATAE Abbott 1971 **pl. 21**
Florida and Jamaica
Description: Length about $1\frac{1}{2}$ inches. Colour variable; generally a uniform greyish-white or occasionally pink, sometimes with irregular brown or orange spiral bands or solid patches of the same colour. There is usually an overlay of fine, dark, broken spiral lines. Sculpture of many very fine spiral cords and axial growth lines, giving a slightly reticulated surface. Spire moderately to slightly extended; there are 5 or 6 spiral threads on the top of each whorl. About 70% of the specimens we have seen bear prominent axial wrinkles on the body whorl.
Occurrence: This is a new species which has been described by Dr R. T. Abbott in *The Nautilus* of October 1971. Our specimens are paratypes. In Jamaica the species occurs in 30 to 80 feet at Orange Bay, Oracabessa, Ocho Rios, Discovery Bay and just east of Montego Bay. We have found no live shells, but the position of the dead shells found at Orange Bay suggests that the species there occurs on coral reefs. A fresh example from Oracabessa was found in loose coral rubble; the periostracum was brown and relatively thick, and looked much like that of *C. daucus*. This is not an exceptionally rare species in the localities where we have found it, and it is almost incredible that it has not previously been described.

CONUS AUSTINI Rehder and Abbott 1951 Austin's Cone
Tortugas to Yucatan and the West Indies. **pl. 21**
Description: Length $1\frac{1}{2}$ to $2\frac{1}{2}$ inches. Colour pure white; a few specimens may have a brown apex and scattered brown marks on the body

Plate 15 MURRAY SHELLS

Scale: reduced by 50%

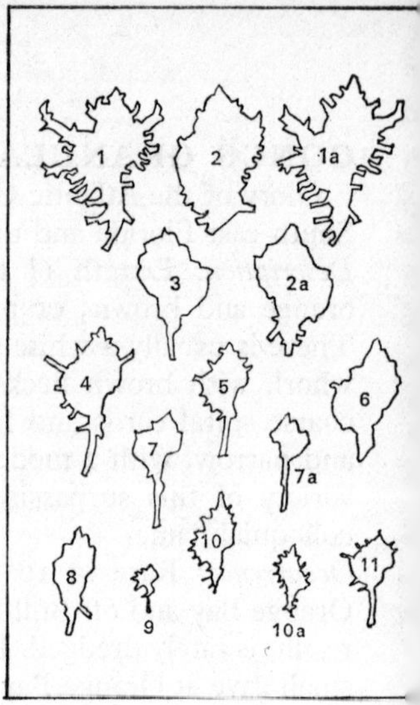

1. **Murex brevifrons** 132
Lamarck West Indian Murex
Typical colour variety.
1a, uncommon pink and white colour variety.

2. **Murex pomum** 131
Gmelin Apple Murex
2a, ventral view. This is the form with a distinctive pink aperture.

3. **Murex chrysostoma** 134
Sowerby Gold-mouthed Murex

4. **Murex antillarum** 133
Hinds Antillean Murex

5. **Murex cailleti** Petit 134
Petit's Murex
This is the variation *kugleri;* typical *cailleti* lacks sharp spines.

6. **Murex ciboney** 133
Clench and Aguayo
Specimen illustrated has lost the tip of its siphonal canal.

7. **Murex woodringi** 133
Clench and Farfante
Woodring's Murex
7a, ventral view.

8. **Murex pulcher** 135
A. Adams Beautiful West Indian Murex
Note the penultimate siphonal canal still in place.

9. **Murex tryoni** Hidalgo 135
Tryon's Murex

10. **Murex spectrum** 131
Reeve Pink Murex
10a, ventral view.

11. **Murex cabriti** 134
Bernardi Cabrit's Murex
Most specimens have a more spinose siphonal canal than the shell illustrated.

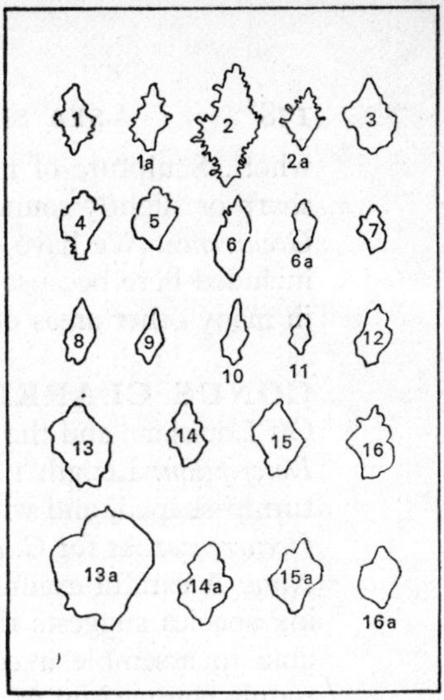

Plate 16 MUREX SHELLS, DRUPES, ASPELLAS, DWARF TRITONS, CORAL SHELLS

Scale: natural size

1. **Muricopsis** species 137
Little Pink Murex
This appears to be a new species. **1a,** dorsal view. Most specimens are heavily encrusted with coralline growths.

2. **Muricopsis oxytatus** 137
M. Smith Hexagonal Murex
2a, young specimen. The spines become thicker and blunter as the shell matures.

3. **Muricopsis philippiana** Dall 138
Orange-banded Murex

4. **Murex cellulosus** 136
Conrad Pitted Murex

5. **Murex cellulosus nuceus** Mörch 136
Globose Pitted Murex
Most examples of this species are heavily encrusted with coralline growths.

6. **Typhis fordi** Pilsbry 138
Ford's Typhis
6a, ventral view. Note the characteristic tubular projections at the shoulder of each whorl.

7. **Muricopsis pudicus** 138
Reeve Little Spiny Murex

8. **Drupa nodulosa** C. B. 139
Adams Blackberry Drupe

9. **Risomurex muricoides** C. B. 140
Adams Beaded Drupe

10. **Aspella paupercula** 142
C. B. Adams Little Aspella

11. **Risomurex roseus** 139
Reeve Pink Drupe

12. **Ocenebra intermedia** 142
C. B. Adams Frilly Dwarf Triton

13. **Coralliophila abbreviata** Lamarck 143
Short Coral Shell
13a, dorsal view.

14. **Coralliophila scalariformis** Lamarck 146
Pagoda Coral Shell
14a, dorsal view.

15. **Coralliophila caribaea** Abbott 143
Caribbean Coral Shell
Note purplish aperture.
15a, dorsal view.

16. **Coralliophila aberrans** C. B. Adams 143
Globular Coral Shell
16a, dorsal view.

whorl. Sculpture of numerous spiral cords of unequal size. Shoulder sharp or slightly rounded; apex extended and mildly concave.
Occurrence: We have not found this species in Jamaican waters. It is included here because it is quite commonly dredged in about 150 feet in many other areas of the Caribbean, and may yet turn up locally.

CONUS CLARKI Rehder and Abbott 1951 Clark's Cone
Off Louisiana and the lower Caribbean. **pl. 21**
Description: Length 1 to 2 inches. Very similar to *C. austini* but rather turnip-shaped, and with 27 to 30 strong spiral cords.
Occurrence: As for *C. austini*.
Note: A careful examination of 100 specimens of this and the preceding species suggests that they are in fact the same species. We were able to assemble a convincing series of over 50 intergrades, but a radula comparison is needed to substantiate this.

CONUS MUS Hwass 1792 Mouse Cone **pl. 21**
Southern Florida and the West Indies.
Description: Length 1 to 1½ inches. Colour grey or white, with extensive olive and brown mottling. The interior of the outer lip is purplish-brown, divided by a central spiral white band. There are irregular white knobs at the shoulder of the body whorl. Sculpture of numerous flat spiral threads. Periostracum dark brown, thick and persistent.
Occurrence: This is the most common *Conus* in Jamaica, and possibly throughout the West Indies. It occurs on and under algae-covered rocks, from the inter-tidal zone down to about 5 feet, all around the Island. It is particularly abundant at Prospect and Robins Bay (where it seems to flourish at points where fresh water springs emerge into the sea). Dead shells are common on most beaches where there are rocks or coral just off-shore.

CONUS MAZEI Deshayes 1874 Maze's Cone **pl. 21**
South-east Florida and the West Indies.
Description: Length 1½ to 2 inches. Colour white with numerous small, rectangular, yellow markings arranged in spiral rows. Sculpture of incised spiral lines, which extend over the entire body whorl. There are rows of delicate beads on the spire. The shell is thin, rather fragile, and unusually beautiful. Rare specimens have a smooth surface.
Occurrence: This is a deep water species usually found below 600 feet. Our specimens were dredged off Discovery Bay, in about 1,800 feet, on muddy substrate. The species is not uncommon at that depth in that particular locality, but dredging in similar depths along the south coast has so far failed to produce it.

CONUS DOMINICANUS Hwass 1792
Antillean Cone **pl. 21**
The lower Caribbean.
Description: Length $1\frac{1}{2}$ to $2\frac{1}{2}$ inches. Colour variable, usually white with irregular brownish-yellow patches, often arranged in two spiral bands. Sculpture of spiral rows of minute raised beads. The shoulder of the body whorl is usually nodulous. This species closely resembles *C. regius* in many respects, and for this reason is often incorrectly identified even in reputable museum collections.
Occurrence: This species does not occur in Jamaica, and is apparently confined to the lower Caribbean (the type specimen came from Dominica). Records from the Bahamas are almost certainly in error. It is found at depths of 20 to 60 feet, on and under rocks, and dead shells are occasionally washed ashore. It is included here as a matter of interest.
Note: This is the once much coveted '*cedo-nulli*', which fetched high prices at European shell auctions in the eighteenth and nineteenth centuries.

FAMILY TEREBRIDAE

Members of this family are long, narrow, brightly coloured shells with many whorls and sharp apexes. There are some 300 species, most of which live in shallow, sandy areas of tropical seas.

Like the *Conidae*, these molluscs are equipped with a venomous stinging apparatus. All of them are carnivorous. They should not be confused with the *Turritellidae*, some of which share a similar shape but have circular operculums. *Terebridae* bear claw-like operculums.

Genus *Terebra* Bruguière 1789

TEREBRA TAURINUS Solander 1786 Flame Auger **pl. 20**
South-east Florida and the West Indies.
Description: Length 4 to 6 inches. Colour cream overlaid with axial brown markings. Sculpture of axial ribs on the upper whorls; later whorls usually show only rather coarse growth lines. The shell is narrow and the apex sharp.
Occurrence: This beautiful species is very rare in Jamaica. Our specimens came from a depth of 40 feet off the Bogue Islands, and this is the only local record we know of. Shrimp trawlers operating south of Trinidad occasionally find them in quite large concentrations. They

prefer soft mud or sand substrate, and usually bury themselves just below its surface.

TEREBRA HASTATA Gmelin 1791
Shiny Atlantic Auger **pl. 20**
South-east Florida and the West Indies.
Description: Length 1 to $1\frac{1}{2}$ inches. Colour light brown with a white band below the suture. Sculpture of axial ribs; there is no spiral sculpture. The shell is polished.
Occurrence: Uncommon in 30 to 70 feet in Lucea harbour, Montego Bay, and at Yallahs (St Thomas). The mollusc lives buried just below the surface of mud or sand bottoms. Beach specimens are locally common after rough weather.

TEREBRA DISLOCATA Say 1822 Atlantic Auger **pl. 20**
South-east U.S., Texas, and the West Indies.
Description: Length $1\frac{1}{2}$ to 2 inches. Colour greyish, or yellowish-orange. Sculpture of about 25 axial ribs per whorl, divided by a deep spiral line. There are spiral cords between the ribs.
Occurrence: This is a very uncommon species in Jamaica, and is occasionally found on the beaches at Negril and Port Henderson. We have not found any living examples.

TEREBRA CINEREA Born 1778 Gray Auger **pl. 20**
South-east Florida and the West Indies.
Description: Length 1 to 2 inches. Colour grey, bluish-brown or orange-cream. Sculpture of numerous small axial riblets, which do not extend below the centre of each whorl. The surface is pitted with minute punctuations. The apex is sharp and white.
Occurrence: This species lives in the sand of soft beaches at the point where the waves break. As the wave recedes, the shells are often momentarily exposed above the sand, and this is the time to collect them. The species is common at Green Bay (St Catherine) and at Negril. It preys on the little bivalve *Donax denticulatus*, and is usually found in association with it.

TEREBRA PROTEXTA Conrad 1845
Fine-ribbed Auger **pl. 20**
North Carolina to Texas, and the West Indies.
Description: Length $\frac{3}{4}$ to 1 inch. Colour cream to brown. Sculpture of 16 to 22 fine axial ribs, crossed by 7 to 9 weak spiral lines.
Occurrence: Uncommon in about 10 feet on mud or sand bottoms, at Port Henderson and Negril. It has also been dredged off Rio Bueno and Discovery Bay in about 800 feet, and may occur in still deeper water.

FAMILY TURRIDAE

Members of this family are a highly advanced group of marine gastropods, which occur at various depths in both cold and tropical seas. There are numerous small abyssal species.

The family is characterised by the spindle shape and the presence of a distinct notch on the upper part of the outer lip. In many species the operculum is degenerate or absent, and some Turrids lack a periostracum. A few possess a neurotoxic poison gland. All Turrids are believed to be carnivorous.

SUBFAMILY TURRINAE

Genus *Polystira* Woodring 1928

POLYSTIRA ALBIDA Perry 1811
White Giant Turret **pl. 20**
South Florida, the Gulf of Mexico and the West Indies.
Description: Length 3 to 4 inches. Colour white. The shell is spindle-shaped, and there are 3 spiral ridges on each whorl. The spire is high and sharply pointed, and the siphonal canal long and open. There is a distinct notch on the upper part of the outer lip.
Occurrence: Uncommon on mud bottoms at Cow Bay and at Yallahs, in about 50 feet. This species has also been reported from the north coast, but we have never seen a beach specimen. The live shell is generally found more or less buried in the substrate.

SUBFAMILY COCHLESPIRINAE

Genus *Ancistrosyrinx* Dall 1881

ANCISTROSYRINX RADIATA Dall 1889
Common Star Turret **pl. 22**
South Florida, the Gulf of Mexico and the West Indies.
Description: Length $\frac{1}{2}$ to $\frac{3}{4}$ inch. Translucent and delicate. There is a row of sharp, triangular spines on the shoulder of each whorl. The spire is high and sharp, and the siphonal canal open and extended.
Occurrence: We have one record of a single specimen dredged at Negril in about 80 feet. It is more common in much deeper water, and was recently taken off Discovery Bay in 1500 feet on a soft mud substrate.

ANCISTROSYRINX ELEGANS Dall Elegant Star Turret
Key West and the Greater Antilles.

Description: Length 2 to 2½ inches. Colour white. Very similar to *A. radiata*, but larger, more robust and with more numerous and duller spines on the shoulders of the whorls.

Occurrence: This is an exceedingly rare deep water species. Two specimens were recently dredged in about 250 feet off Black River (St Elizabeth); this is a new Jamaican record.

SUBFAMILY CLAVINAE

Genus *Crassispira* Swainson 1840

Subgenus *Crassispirella* Bartsch and Rehder 1939

CRASSISPIRA FUSCESCENS Reeve 1843
Ebony Turret **pl. 22**
Florida Keys to the West Indies.

Description: Length about ¾ inch. Colour dark or reddish brown; some specimens are almost black. Sculpture of 17 to 20 oblique, axial ribs; there are fine spiral threads between the ribs. Just below the suture there is a strong spiral cord; the upper part of the outer lip is deeply notched.

Occurrence: Relatively common under rocks in 2 to 20 feet at Port Henderson, Runaway Bay, and many other areas around the Island. Dead shells are often washed ashore, where they are quickly bleached to a pale orange colour.

CRASSISPIRA species **pl. 22**
Jamaica.

Description: Length about ¾ inch. Colour white with lengthened brown spots between the axial ribs. These form three narrow revolving bands on the body whorl, and meet over the ribs towards the outer lip. The shape of the aperture is very similar to that of *C. fuscescens*, but the short siphonal canal is strongly recurved.

Occurrence: Very uncommon under rocks in about 45 feet at Orange Bay and in Ocho Rios harbour. We have found no beach specimens.

Note: We have been unable to identify this species, but it appears closest to the several Jamaican *Crassispira*, and we tentatively list it in this genus.

CRASSISPIRA species **pl. 22**
The Greater Antilles.
Description: Length about ½ inch. Colour black or brown, with flesh-coloured axial ribs and a narrow spiral band immediately below the suture. On the body whorl there are 3 or 4 spiral bands of the same colour near the base. Sculpture of numerous thin axial ribs, which do not extend to the suture or the base of the shell. In most respects (except colour), this species closely resembles a small version of *C. fuscescens.*
Occurrence: Relatively common in about 45 feet at Ocho Rios and Discovery Bay. We have also found it at Prospect in about 10 feet, under rocks among the eel grass beds, but it is not common in that area. Beach specimens are occasionally found at Robins Bay.
Note: We have been unable to identify this species with any certainty. It is very similar to the other small Jamaican *Crassispira*, but occurs in deeper water. There is no doubt that it is a valid species.

Subgenus *Compsodrillia* Woodring 1928

CRASSISPIRA LEUCOCYMA Dall 1883
White-knobbed Turret **pl. 22**
Southern Florida and the West Indies.
Description: Length ½ to ¾ inches. Colour brown or grey, with many white or yellowish nodules divided by one or two spiral lines. Apex sharp.
Occurrence: Very common in 1 to 6 feet on eel grass beds all along the harbour side of the Palisadoes. Elsewhere it is uncommon, but it is known to occur at Discovery Bay in about 25 feet.

Subgenus *Neodrillia* Bartsch 1943

DRILLIA CYDIA Bartsch 1943 Cydia Drillia **pl. 22**
Florida and the West Indies.
Description: Length ½ to ¾ inch. Colour white, occasionally with spiral bands of faint, minute brown spots. Sculpture of about 9 rounded axial ribs on the body whorl, crossed by many fine spiral threads. There are also numerous, very fine axial threads, giving the surface a microscopic reticulation. The aperture is channelled at both ends.
Occurrence: This is an uncommon species which occurs on and under rocks in about 25 feet at Montego Bay, Discovery Bay, Port Antonio and off the southern side of the Palisadoes. Dead shells are occasionally washed ashore, but the delicate lip is almost invariably damaged.

DRILLIA ALBINODATA Reeve 1846 **pl. 22**
White-noded Turret
South-east Florida and the West Indies.
Description: Length about ½ inch. Colour dark brown, with a spiral white band near the shoulder of the whorl. There are usually two white bands at the base of the shell. Sculpture of many rounded pustules within the white band. The shell is short and fat.
Occurrence: Very common under rocks in 1 to 6 feet at Robins Bay, Runaway Bay and in rocky areas all around Jamaica. Dead shells are frequently cast ashore.

DRILLIA INTERPLEURA Dall and Simpson 1901 **pl. 22**
Caribbean.
Description: Very similar in most respects to the preceding species (*D. albinodata*). However, it is not as fat, the nodules do not coalesce, and the aperture is wider.
Occurrence: As for *albinodata*, and it is equally common.

SUBFAMILY MANGELIINAE

Genus *Ithycythara* Woodring 1928

ITHYCYTHARA LANCEOLATA C. B. Adams 1850
Spear Turret **pl. 22**
Florida and the West Indies.
Description: Length ½ to ¾ inch. Colour white, tan, or occasionally light brown. There is a broken brown spiral band on the centre of the body whorl. Sculpture of 6 or 7 sharp, axial ribs on each whorl, which are barely interrupted by the suture. The narrow aperture is finely toothed within. The shell is thin and moderately fragile.
Occurrence: Relatively common in Ocho Rios harbour in about 25 feet; uncommon off the Bogue Islands at the same depth. We have no other confirmed records, but there are indications that it occurs between 20 and 60 feet on a muddy substrate in other areas around the Island.

Genus *Fenimorea* Bartsch 1934

FENIMOREA JANETAE Bartsch 1934 Janet's Turret **pl. 22**
Florida and the West Indies.
Description: Length 1 to 1½ inches. Colour: top quarter of whorls white; lower three-quarters chestnut brown at the top, bright pink at

the bottom. There is a narrow, broken brown spiral band between the white and chestnut sections. Sculpture of 10 to 14 strong, rounded axial ribs per whorl. Outer lip with a deep notch near the top. Operculum sickle-shaped and wine red.
Occurrence: This is one of the most beautiful little shells imaginable, and though it occurs elsewhere, it is apparently most common in Jamaica. It exhibits a considerable depth range, and we have found it exposed on algae at extreme low tide at Priory and also under rocks in 50 feet at Orange Bay. It is occasionally found in about 4 feet in sandy depressions in the eel grass beds at Prospect and Runaway Bay; dead shells have been collected on the beaches at Bull Bay and at several points just west of Lucea.

Genus *Monilispira* Bartsch and Rehder 1939

MONILISPIRA ALBOCINCTA C. B. Adams 1845

White-banded Turret

South-east Florida and the West Indies.
Description: Length about ½ inch. Colour dark brown or black, with a spiral white band near the shoulder of the whorl. There are usually two white bands on the base of the shell. Sculpture of 12 or 13 rounded pustules within the white band. The shell is relatively short and fat. This species is very similar to *Drillia albinodata.*
Occurrence: Very common under rocks in 1 to 6 feet at Robins Bay, Runaway Bay, and in rocky areas all around Jamaica. Dead shells are frequently cast ashore.

Genus *Kurtziella* Dall 1918

KURTZIELLA LIMONITELLA Dall 1918

Punctate Mangelia **pl. 22**

North Carolina, Florida and Jamaica.
Description: Length about ⅝ inch. Colour yellowish-white. Sculpture of strong, rounded ribs and rows of microscopic punctuations between them. Sinus in the form of a wide V.
Occurrence: This is a rare species in Jamaica. It occurs in Ocho Rios harbour at about 40 feet. We have found it nowhere else, and apparently it has not previously been reported from the West Indies.

SUBFAMILY DAPHNELLINAE

Genus *Daphnella* Hinds 1844

DAPHNELLA LYMNEIFORMIS Kiener 1840
Volute Turret **pl. 22**
South-east Florida and the West Indies.
Description: Length $\frac{1}{2}$ to $\frac{3}{4}$ inch. Colour cream with orange-brown maculations. Sculpture of strong, beaded axial ribs on the early whorls; the last two whorls have a finely reticulated surface. The body whorl is about two-thirds the length of the shell; the aperture is expanded and there is a distinct sinus.
Occurrence: Relatively uncommon in 1 to 30 feet at Negril, Ocho Rios harbour, Prospect, and in many areas all around the Island. In shallow water, these shells are usually found lightly buried in loose sand under rocks.

DAPHNELLA STEGERI McGinty 1955
Steger's Turret **pl. 22**
Florida, Puerto Rico and Jamaica.
Description: Length $\frac{1}{2}$ to $\frac{3}{4}$ inch. Colour greyish-white. Very similar to *D. lymneiformis* in general shape but there is a relatively strong spiral cord at the shoulder of each whorl, and a general sculpture of many spiral lines. The axial sculpture is conspicuous only on the first 3 or 4 whorls.
Occurrence: Uncommon in about 30 feet in Ocho Rios harbour. We have not found it anywhere else, but it probably occurs on mud bottoms in many areas along the north coast.

SUBCLASS OPISTHOBRANCHIA

Order Tectibranchia

Members of this order, known as the Bubble Shells, are an advanced group of molluscs with external, plume-like gills. Some have internal shells; all are carnivorous; and most live in tropical seas. Both sexes are present in the same individual.

FAMILY ACTEONIDAE

These are small – sometimes minute – molluscs with thin, delicate shells, which live in sandy areas in moderately shallow water.

The family is characterised by the short spire, few whorls and the presence of a single twisted fold on the columella.

Genus *Acteon* Montfort 1810

ACTEON PUNCTOSTRIATUS C. B. Adams 1840
Adams Baby-bubble **pl. 22**
Cape Cod to the Gulf of Mexico, and the West Indies.
Description: Length $\frac{1}{4}$ to $\frac{1}{3}$ inch. Colour white or pale purple. Sculpture of incised spiral lines, which extend from the base to cover the entire body whorl. There is a single twisted fold on the columella. The shell is light and highly polished.
Occurrence: Rare in 40 feet in Ocho Rios harbour on a soft, sandy bottom. It may also occur at Negril, but the record requires confirmation.
Note: One of the specimens we collected at Ocho Rios measures 14 mm. The species has been recorded previously with a maximum measurement of 7 mm. It is probable that the large specimen represents a subspecies, which is characterised by its larger size, the presence of spiral lines which reach to the suture of the whorls, and a delicate purple overlay.

FAMILY HYDATINIDAE

These are thin, oval, brightly coloured shells with depressed spires and simple outer lips. The animals themselves are surpassingly beautiful and partially envelop the shell.

They live in shallow, sandy areas in tropical seas all over the world. There are only two known Jamaican representatives of the family.

Genus *Micromelo* Pilsbry 1894

MICROMELO UNDATA Bruguière 1792
Miniature Melo **pl. 22**
Lower Florida Keys and the West Indies.
Description: Length about ½ inch. Colour white, with an overlaid network of fine reddish lines, usually in the form of three spiral threads inter-connected by a number of wavy axial threads. The shell is oval and moderately fragile. The living animal is delicately frilled, and brightly coloured in combinations of yellow, green and blue. It is certainly one of the most beautiful West Indian molluscs.
Occurrence: Very uncommon on algae or under rocks at, and just below, the low tide mark, in areas exposed to the open sea. Our live specimens came from Robins Bay and Prospect, and damaged shells are occasionally found on other beaches all around the Island.

Genus *Hydatina* Schumacher 1817

HYDATINA VESICARIA Solander 1786
Brown-lined Paper-bubble **pl. 22**
Southern Florida and the West Indies.
Description: Length 1 to 1½ inches. Colour white to tan, overlaid by many close set, wavy brown spiral lines. The shell is globose, moderately fragile and highly polished. Animal red with azure border.
Occurrence: Very uncommon in about 25 feet on sandy bottoms at Robins Bay and Orange Bay. It is probably quite widely distributed, but we have so far found it nowhere else. Occasional specimens are washed ashore at Robins Bay.

FAMILY BULLIDAE

Shells of this family are thin but strong, and oval in shape with a simple lip. They have an aperture longer than the shell itself.

The family is characterised by the depressed spire, and the short, concave columella. All species live in soft sandy or muddy areas, where they remain buried in the substrate during daylight. They are most easily collected at night, when they emerge to feed.

Genus *Bulla* Linné 1753

BULLA STRIATA Bruguière 1792 Striated Bubble **pl. 22**
West Florida to Texas and the West Indies.
Description: Length 1 to 1½ inches. Colour brown with darker mottling. Most specimens are smooth and polished; occasional shells have spiral grooves at the base and within the apical perforation. Examples from sandy, rather than muddy, areas tend to be very much lighter in colour.
Occurrence: Abundant in many shallow water areas all around the Island. The living animals shun daylight and hide just below the surface of the substrate. At night they can be seen – with the aid of an electric torch – crawling in 6 inches of water in great numbers, all along the harbour side of the Palisadoes. Their dead shells litter the beaches all around Jamaica.

FAMILY ATYIDAE

These are small, thin, delicate shells which are confined to shallow, muddy or sandy areas, often near the mouths of rivers.

The family is characterised by the disproportionately large animal in relation to the size of the shell, which is in fact more or less internal. Most of the Jamaican species have attractive, translucent shells.

Genus *Haminoea* Turton and Kingston 1830

HAMINOEA ELEGANS Gray 1825
Elegant Paper-bubble **pl. 22**
South-east Florida and the West Indies.
Description: Length ½ to ¾ inch. Colour brown or greenish yellow,

occasionally translucent white. Sculpture of numerous, fine, spiral grooves. The shell is oval, thin, and fragile.
Occurrence: Moderately common in about 20 feet at Negril, in Ocho Rios harbour, and in many sandy and muddy areas all around the Island. Dead shells are often found in drift sand at Negril.

HAMINOEA ANTILLARUM Orbigny 1841
Antillean Paper-bubble **pl. 22**
West Florida to the West Indies.
Description: Length $\frac{1}{2}$ to $\frac{3}{4}$ inch. Colour greenish-yellow. Sculpture of irregular growth wrinkles and microscopic wavy spiral threads. The columella is concave, and has a very narrow white callus. The shell is thin, but relatively strong.
Occurrence: Uncommon in about 6 feet at the Old Coal Wharf at Port Royal, and in many other muddy areas around the Island. Dead shells are occasionally found in drift sand at Negril.

SUPERFAMILY APLYSIACEA

FAMILY PYRAMIDELLIDAE

These are smooth, polished, elongate shells with sharp apexes. Most species live in sandy areas of warm, shallow waters. They are closely related to the externally shell-less 'Sea Hares' (*Aplysia*) and other Tectibranchs, but they lack a radula.

All members of the family are reported to be ectoparasites, with each species feeding on a particular host (usually a polychaete worm or bivalve mollusc). The exact classification of the group has been the subject of considerable disagreement among marine biologists.

Genus *Pyramidella* Lamarck 1799

PYRAMIDELLA DOLABRATA Linné 1758
Giant Atlantic Pyram **pl. 22**
The Bahamas and the West Indies.
Description: Length 1 to $1\frac{1}{2}$ inches. Colour white, with 2 or 3 brown spiral lines. The shell is highly polished and quite strong. The outer lip may, or may not, be toothed inside. Columella with 2 or 3 strong folds.
Occurrence: Common in about 6 inches of water on sand flats near Rose Hall (St James). At extreme low tide the flats are exposed, and the

shells then thrust themselves up from below the surface of the sand, where they normally live. At Montego Bay, Ocho Rios, and many other areas around the Island, they prefer deeper water. Dead shells are often washed ashore, and they are common on the beaches at Negril.

Order Sacoglossa

Members of this order are small, rather delicate opistobranchs, characterised by their unique feeding habits. They are herbivorous molluscs which live on the cell sap of red and green algae.

The order includes seven families, some of which lack shells. None of the *Sacoglossa* is more than ¾ inch long (probably as a result of their specialised diet), and many of them are very small indeed. The shelled species are not common in the West Indies. *Lobiger souverbii* Fischer 1856 is found as far north as Puerto Rico, and may occur in Jamaica though we have not yet found it.

Order Pteropoda

Pteropods are light, diaphanous, pelagic molluscs which are abundant in the open sea in nearly every part of the world. The order comprises two distinct suborders – *Thecosomata* with shells, and *Gymnosomata* without. Both are hermaphroditic.

The molluscs swim by using the membranous extensions of the molluscan foot to propel themselves through the water. Dead pteropods are occasionally found on the beach in drift sand, but the shells are far more easily recovered by dredging on muddy substrate in deep water.

Suborder Thecosomata

FAMILY CAVOLINIDAE

Genus *Cavolina* Abildgaard 1791

CAVOLINA TRIDENTATA Forskäl
Three-toothed Cavolina **pl. 22**
World-wide (40° N. to 40° S.), pelagic.
Description: Length ½ to ¾ inch. Colour pale brown. The shell is equally

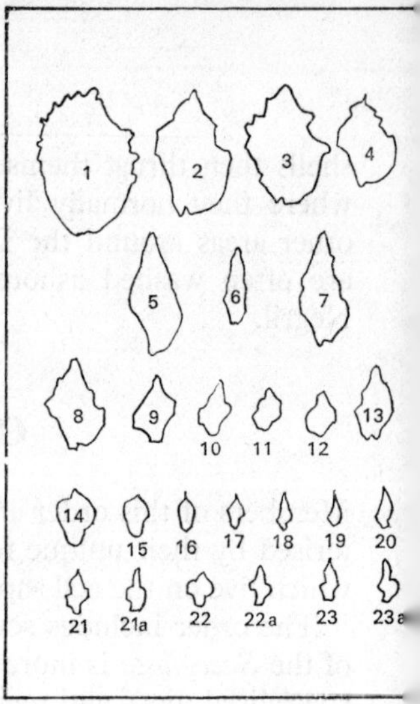

Plate 17 ROCK SHELLS, DWARF TRITONS, CANTHARUS SHELLS, NASSAS, PHOS SHELLS, ENGINAS, DOVE SHELLS, BAILY SHELLS

Scale: reduced by 20%

1. **Purpura patula** Linné 140
Wide-mouthed Purpura

2. **Thais haemastoma floridana** Conrad 141
Florida Rock Shell

3. **Thais deltoidea** Lamarck 141
Deltoid Rock Shell

4. **Thais rustica** Lamarck 140
Rustic Rock Shell

5. **Pisania pusio** Linné 152
Miniature Triton-Trumpet

6. **Colubraria lanceolata** Menke 151
Arrow Dwarf Triton

7. **Colubraria obscura** Reeve 151
Leaning Dwarf Triton

8. **Cantharus auritulus** Link 153
Common Cantharus

9. **Cantharus tinctus** Conrad 152
Tinted Cantharus

10. **Cantharus lautus** Reeve 153
Gaudy Cantharus

11. **Nassarius albus** Say 154
Variable Nassa

12. **Nassarius vibex** Say 155
Common Eastern Nassa

13. **Antillophos candei** Orbigny 150
Beaded Phos

14. **Columbella mercatoria** Linné 146
Common Dove Shell

15. **Pyrene ovulata** Lamarck 147
Ovate Dove Shell

16. **Nitidella laevigata** Linné 148
Smooth Dove Shell

17. **Nitidella nitida** Lamarck 148
Glossy Dove Shell

18. **Anachis pulchella** Sowerby 147
Beautiful Dove Shell

19. **Nitidella ocellata** Gmelin 147
White-spotted Dove Shell
Most adult shells have lost the tip of the spire, and have a truncated appearance.

20. **Mitrella nitens** C. B. Adams 148
Shiny Dove Shell

21. **Bailya parva** C. B. Adams 149
West Indian Baily Shell
21a, ventral view. This is the deeper water variety.

22. **Engina turbinella** Kiener 151
White-spotted Engina
22a, ventral view.

23. **Bailya parva** C. B. Adams 149
West Indian Baily Shell
23a, ventral view. This is the shallow water variety.

Plate 18 LATIRUS, VASE SHELLS

Scale: reduced by 15%

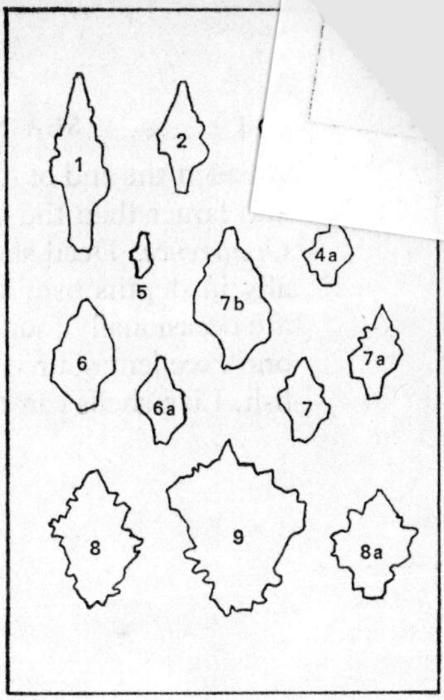

1. **Latirus infundibulum** 156
   Gmelin Brown-lined Latirus
   Live shells are usually darker than the specimen illustrated.

2. **Latirus mcgintyi** 157
   Pilsbry McGinty's Latirus

3. **Latirus brevicaudatus** 157
   Reeve Short-tailed Latirus

4. **Leucozonia ocellata** 158
   Gmelin White-spotted Latirus
   4a, dorsal view.

5. **Fusilatirus cayohuesonicus** 157
   Sowerby Key West Latirus
   Truncated spire of specimen illustrated is not typical.

6. **Leucozonia nassa leucozonalis** Lamarck 158
   Chestnut Latirus
   6a, ventral view.

7. **Leucozonia nassa** 158
   Gmelin Chestnut Latirus
   7a, dorsal view. **7b,** this variety is intermediate between **6** and true *L. nassa*.

8. **Vasum capitellus** 162
   Linné Spiny Vase
   8a, dorsal view. This species has not been found in Jamaica.

9. **Vasum muricatum** 162
   Born Caribbean Vase
   Columella is often stained purple.

broad at the end of the lips and at the anterior end. It is thin and fragile, and larger than the other West Indian *Cavolinidae*.

Occurrence: Dead shells are common on mud or sand substrate, generally in depths over 100 feet, all around Jamaica. Damaged specimens are occasionally found in drift sand at Negril. As with other pteropods, one excellent source of this species is the stomachs of plankton-eating fish. Live shells can be trawled in open water.

SUBCLASS PULMONATA

Order Basommatophora

Members of this order provide the link between land and marine molluscs. They are equipped with primitive lungs, and extract oxygen directly from the air; yet they are dependent upon plenty of moisture, and live in marshy areas close to the sea.

Some genera still require submersion for a certain period every day, but the more advanced groups have apparently abandoned all direct contact with the water.

Suborder Actophila

FAMILY ELLOBIIDAE

These are small, smooth shells with long, narrow apertures. They have strong folds on the columella, and the inside of the outer lip is often toothed or grooved.

They live in damp, marshy areas in close proximity to the sea, and are often found in large numbers under pieces of dead wood, or clinging to the underside of small plants.

Genus *Tralia* Gray 1840

TRALIA OVULA Bruguière 1789 Egg Tralia **pl. 22**
Bermuda, Florida and the West Indies.
Description: Length about ½ inch. Colour tan to dark brown. There is a white glaze on the parietal wall, which has 2 folds; on the columella there is a single fold. There is a slight inflection of the outer lip, and a single spiral ridge inside. The shape of the shell varies between oval and elongate.
Occurrence: Relatively common above the high tide line at Runaway Bay, and on the harbour side of the Palisadoes. Dead shells are found on many beaches around the Island.

Genus *Melampus* Montfort 1810

MELAMPUS COFFEUS Linné 1758
Coffee Melampus **pl. 22**
Florida to the West Indies.
Description: Length ½ to ¾ inches. Colour brown, generally with 2 or 3 light spiral bands. The columella has a single fold, and higher up on the parietal wall there is a double fold. Outer lip with many very small teeth. Shell cone-shaped and polished.
Occurrence: Very common under debris in the mangrove swamps and salt marshes all around Jamaica. The species occurs in large numbers in the swamps of the Palisadoes and the marshes around Falmouth.

Subgenus *Pira* H. and A. Adams 1855

MELAMPUS MONILE Bruguière 1789
Yellowish Melampus **pl. 22**
Florida to the West Indies.
Description: Length ½ to ¾ inch. Colour yellowish. This shell is similar to *M. coffeus*, but there is usually only one fold on the columella and one on the parietal wall.
Occurrence: As for *M. coffeus*.

Genus *Detracia* Gray 1840

DETRACIA BULLAOIDES Montagu 1808
Bulla Melampus **pl. 22**
Florida to the West Indies.
Description: Length about ½ inch. Colour greyish-brown, with white spiral bands. The shell tapers towards both ends, and the aperture extends less than half way up the shell. The spire is extended, and the shell relatively narrow.
Occurrence: Common under debris above the high tide mark at Runaway Bay, and on the harbour side of the Palisadoes. Dead shells are found on beaches in close proximity to swamps and salt marshes all around Jamaica.

Suborder Patelliformia

FAMILY SIPHONARIIDAE

Members of this family have limpet-shaped shells which closely resemble, and are often mistaken for *Acmaeidae*. The muscle scars are horseshoe shaped, and there is usually a distinct siphonal groove on the right side.

The animals possess both lungs and gills, and occur in the inter-tidal zone, often in very large numbers.

In the genus *Siphonaria*, the molluscs are true air-breathers, and are more closely related to land snails than any marine group.

Genus *Siphonaria* Sowerby 1824

SIPHONARIA PECTINATA Linné 1758

Striped False Limpet **pl. 22**

Florida to Texas and the West Indies. Also the Old World.

Description: Length $\frac{3}{4}$ to 1 inch. Colour whitish, with fine bifurcating brown lines, repeated in the interior. Sculpture of numerous fine radial threads; occasionally the surface may be smooth. There are three distinct swellings on the muscle scar.

Occurrence: This is an uncommon species in Jamaica. It occurs in the inter-tidal zone at Runaway Bay and Ocho Rios, and dead shells are found there in drift sand. There is some evidence that the species has only recently been introduced to the Island, but this has not yet been substantiated.

Other Jamaican Gastropods

Order ARCHAEOGASTROPODA

Superfamily PLEUROTOMARIACEA

Family Scissurellidae
Schismope cingulata O. G. Costa 1861

Family Pleurotomariidae
Perotrochus quoyana Fischer and Bernardi 1885

Family Fissurellidae
Emarginula tuberculosa Libassi 1859
Rimula aequisculpta Dall 1927
Fissurella barbouri Clench
Diodora jaumei Aguayo and Rehder 1936

Superfamily TROCHACEA

Family Trochidae
Euchelus guttarosea Dall 1889
Solariella amabilis Jeffreys 1865
Microgaza rotella Dall 1889
Gaza superba Dall 1889
Gaza fischeri Dall 1889
Synaptocochlea picta Orbigny 1842
Pseudostomatella coccinea A. Adams 1850
Pseudostomatella erythrocoma Dall 1889
Calliostoma torrei Clench and Aguayo 1940

Family Turbinidae
Arene tricarinata Stearns 1872
Homalopoma linnei Dall 1889

Family Phasianellidae
Tricolia adamsi Philippi 1853
Tricolia thalassicolia Robertson 1958
Tricolia bella M. Smith 1937
Tricolia tessellata Potiez and Michaud 1838

Superfamily NERITACEA

Family Phenacolepadidae
Phenacolepas hamillei Fischer 1856

Order MESOGASTROPODA

Superfamily RISSOACEA

Family Hydrobiidae
Truncatella caribaeensis Reeve 1842
Truncatella scalaris Michaud 1830
Truncatella scalaris clathrus Lowe 1832
Truncatella succinea C. B. Adams 1845
Truncatella bilabiata Pfeiffer 1840

Family Rissoidae
Rissoina bryerea Montagu 1803
Rissoina fischeri Desjardin 1949
Rissoina multicostata C. B. Adams 1850
Rissoina chesneli Michaud 1830
Rissoina striatacostata Orbigny 1842
Rissoina decussata Montagu 1803 (Plate 22)
Rissoina cancellata Philippi 1847
Rissoina striosa C. B. Adams 1850
Zebina browniana Orbigny 1842
Microdochus floridanus Rehder 1943
Alvania auberiana Orbigny 1842
Alvania gradata Orbigny 1842
Alvania aberrans C. B. Adams 1850

Family Vitrinellidae
Vitrinella helicoidea C. B. Adams 1850
Vitrinella multistriata Verrill 1884
Cyclostremiscus supranitidos Wood
Cyclostremiscus trilix Bush 1897
Teinostoma obtectum Pilsbry and McGinty 1945
Teinostoma clavium Pilsbry and McGinty 1945
Teinostoma parvicallum Pilsbry and McGinty 1945
Episcynia inornata Orbigny 1842
Cochliolepis parasitica Stimpson 1858
Parviturboides comptus Woodring 1928
Pleuromalaxis balesi Pilsbry and McGinty 1945

Superfamily CERITHIACEA

Family Architectonicidae
Spirolaxis exquisita Dall and Simpson 1901
Architectonica peracuta Dall

Family Caecidae
Caecum floridanum Stimpson 1851
Caecum pulchellum Stimpson 1851
Caecum imbricatum Carpenter 1858
Caecum insigne de Folin 1867

Caecum plicatum Carpenter 1858
Caecum coronellum Dall 1892
Caecum clava de Folin 1867
Caecum nebulosum Rehder 1943
Caecum nitidum Stimpson 1851

Family Planaxidae
Planaxis lineatus da Costa 1778

Family Cerithiidae
Bittium varium Pfeiffer 1840
Alabina adamsi Dall 1889
Alabina cerithidioides Dall 1859
Litiopa melanostoma Rang 1829
Alaba incerta Orbigny 1842
Cerithiopsis greeni C. B. Adams 1839
Cerithiopsis emersoni C. B. Adams 1838
Cerithiopsis latum C. B. Adams 1850
Seila adamsi H. C. Lea 1845

Family Mathildidae
Mathilda barbadensis Dall 1889

Family Triphoridae
Triphora turris-thomae Holten 1802
Triphora melanura C. B. Adams 1850
Triphora nigrocincta C. B. Adams 1839
Triphora ornata Deshayes 1832

Superfamily EPITONIACEA

Family Epitoniidae
Amaea retifera Dall 1889
Opalia hotessieriana Orbigny 1842
Epitonium echinaticostum Orbigny 1842
Epitonium krebsi Mörch 1874
Epitonium foliaceicostum Orbigny 1842
Epitonium venosum Sowerby 1844
Epitonium unifasciatum Sowerby 1844
Epitonium turritellulum Mörch 1874
Epitonium frieli Dall 1889
Epitonium denticulatum Sowerby 1844
Epitonium sericifilum Dall 1899
Epitonium tiburonense Clench and Turner 1952
Depressicala nautlae Mörch 1874
Depressicala nitidella Dall 1889

Superfamily AGLOSSA

Family Eulimidae
Balcis intermedia Cantraine 1835
Balcis conoidea Kurtz and Stimpson 1851

Balcis jamaicensis C. B. Adams 1845
Eulima bifasciata Orbigny 1842
Eulima patula Dall and Simpson 1901
Eulima auricincta Abbott 1958

Family Stiliferidae
Stilifer subulatus Broderick and Sowerby 1832 (Plate 22)

Superfamily HIPPONICACEA

Family Hipponix
Hipponix benthophilus Dall 1889

Family Fossaridae
Fossarus orbignyi Fischer 1864
Iselica anomala C. B. Adams 1850

Family Vanikoroidae
Vanikoro sulcata Orbigny 1842
Vanikoro oxychone Mörch 1877

Superfamily CALYPTRAEACEA

Family Calyptraeidae
Calyptraea centralis Conrad 1841
Crepidula convexa Say 1822
Crepidula glauca Say 1822
Crepidula protea Orbigny 1845

Superfamily CYPRAEACEA

Family Lamellariidae
Lamellaria rangi Bergh 1853

Family Eratoidae
Erato maugeriae Gray 1832

Family Ovulidae
Cyphoma intermedium Sowerby 1828

Superfamily NATICACEA

Family Naticidae
Polinices nubilus Dall 1889
Natica sagraiana Orbigny 1842
Tectonatica pusilla Say 1822

Order NEOGASTROPODA

Superfamily MURICACEA

Family Muricidae
Typhis longicornis Dall 1889

Superfamily BUCCINACEA

Family Columbellidae

Anachis subcostulata C. B. Adams 1845
Anachis catenata Sowerby 1844
Anachis sparsa Reeve 1859
Anachis mangeloides Reeve 1859
Anachis iontha Ravenel 1883
Anachis obesa C. B. Adams 1845
Anachis crassilabris Reeve 1859
Anachis hotessieriana Orbigny 1842
Nitidella dichroa Sowerby 1858
Mitrella lunata Say 1826
Mitrella fenestrata C. B. Adams 1850
Mitrella idalina Duclos 1840
Psarostola monilifera Sowerby 1844
Psarostola minor C. B. Adams 1845
Pleurotoma minor C. B. Adams 1850

Family Nassariidae

Nassarius consensa Ravenel
Nassarius hotessieri Orbigny
Nassarina metabrunnea Dall and Simpson 1901

Family Columbariidae

Columbarium bartletti Clench and Aguayo 1940

Superfamily VOLUTACEA

Family Volutidae

Scaphella neptunia Clench and Aguayo 1940

Family Olividae

Olivella dealbata Reeve 1850
Olivella perplexa Olsson 1956
Olivella acteocina Olsson 1956
Olivella floralia Duclos

Family Mitridae

Pusia hanleyi Dohrn 1861
Pusia histrio Reeve 1854

Family Marginellidae

Marginella haematita Kiener 1841
Marginella denticulata Conrad 1830
Marginella lactea Kiener 1841
Marginella striata Sowerby 1846
Marginella evadne Dall and Simpson 1901
Marginella interrupte-lineata Mühlfeld 1818

Marginella torticula Dall 1881
Marginella subtriplicata Orbigny 1845
Marginella fusca Sowerby 1846
Marginella gracilis C. B. Adams 1851
Bullata ovuliformis Orbigny 1842
Persicula pulcherrima Gaskoin 1849
Persicula lavalleeana Orbigny 1842
Persicula fluctuata C. B. Adams 1850
Hyalina albolineata Orbigny 1842
Hyalina avenacea Deshayes 1844
Prunum pruniosum Hinds 1844

Family Turridae
Crassispira nigrescens C. B. Adams 1845
Crassispira ebenina Dall 1883
Clathrodrillia melonesiana Dall and Simpson 1901
Leptadrillia splendida Bartsch 1934
Carinodrillia liella Corea 1934
Mangelia asarca Dall and Simpson 1901
Mangelia melanitica Dall 1885
Mangelia balteata Reeve 1845 (Plate 22)
Mangelia morra Dall 1881
Mangelia fusca C. B. Adams 1845
Mangelia elata Dall 1889
Mangelia biconica C. B. Adams 1850
Mangelia aguadillana Dall and Simpson 1901
Mangelia bartletti Dall 1889
Mangelia luctuosa Orbigny 1845
Mangelia trilineata C. B. Adams 1845
Mangelia quadrata Reeve 1845
Mangelia quadrilineata C. B. Adams 1850
Mangelia lavalleana Orbigny 1845
Borsonia rouaulti Dall 1889
Mitromorpha biplicata Dall 1889
Ithycythara parkeri Abbott 1958
Vitricythara metria Dall 1903
Glyphoturris quadrata rugirima Dall 1889
Pyrgocythara coxi Fargo 1953
Pyrgocythara densestriata C. B. Adams 1850
Drillia ponciana Dall and Simpson 1901
Drillia aepynota Dall 1889
Drillia lissotropis Dall 1881
Drillia albicoma Dall 1889
Drillia eucosmia Dall 1889
Drillia actinocycla Dall and Simpson 1901
Drillia gundlachi Dall and Simpson 1901
Drillia thea Dall 1883

Order TECTIBRANCHIA

Family Atyidae
Haminoea succinea Conrad 1846
Haminoea petiti Orbigny 1841
Atys guildingi Sowerby 1869
Atys riiseana Mörch 1875
Atys caribaea Orbigny 1841
Atys lineata Usticke 1959

Family Retusidae
Retusa candei Orbigny 1841
Retusa bullata Kiener 1834
Rhizorus oxytatus Bush 1885
Rhizorus acutus Orbigny 1841

Family Acteocinidae
Cylichna bidentata Orbigny 1841

Family Philinidae
Philine sagra Orbigny 1841

Superfamily APLYSIACEA

Family Pyramidellidae
Pyramidella fusca C. B. Adams 1839
Pyramidella candida Mörch 1875
Triptychus niveus Mörch, 1875
Odostomia laevigata Orbigny 1842
Odostomia seminuda C. B. Adams 1839
Odostomia solidula C. B. Adams 1850
Odostomia canaliculata C. B. Adams 1850
Odostomia gemmulosa C. B. Adams 1850
Turbonilla pupoides Orbigny 1842
Turbonilla unilirata Bush 1899
Turbonilla haycocki Dall and Bartsch 1911
Turbonilla abrupta Bush 1889
Turbonilla reticulata C. B. Adams
Turbonilla elegans Orbigny 1842
Turbonilla interrupta Totten 1835
Cingulina babylonia C. B. Adams 1845

Order SACOGLOSSA

Family Oxynoeidae
Oxynoe antillarum Mörch 1863

Order PTEROPODA

Family Spiratellidae
Spiratella lesueri Orbigny
Spiratella inflata Orbigny

Family Cavolinidae
Cavolina longirostris Lesueur 1821
Cavolina inflexa Lesueur 1813
Cavolina gibbosa Rang 1836
Cavolina trispinosa Lesueur 1821
Cavolina uncinata Rang 1836
Clio pyramidata Linné 1758
Creseis acicula Rang 1828
Styliola subula Quoy and Gaimard 1827
Cuvierina columnella Rang 1827

Order BASOMMATOPHORA

Family Ellobiidae
Pedipes mirabilis Mühlfeld 1818

Family Trimusculidae
Trimusculus goesi Hubendick 1946

Family Siphonariidae
Williamia krebsi Mörch 1877

Note
There are a number of common Jamaican *Opistobranchia* without external shells. The large ones (tectibranchs) are referred to locally as Sea Hares or Sea Slugs, though many West Indians are probably more familiar with the much less polite – if more descriptive – colloquial names.

Many of these molluscs are unusually beautiful, and there is a selection of North American nudibranchs illustrated in colour in *American Seashells* by R. Tucker Abbott.

Unfortunately, these animals are virtually impossible to preserve attractively and few amateur conchologists try to collect them. No attempt is made in this book to list the various genera.

CLASS PELECYPODA

Order Protobranchia

SUPERFAMILY NUCULACEA

FAMILY NUCULIDAE

This is a small, very primitive family which includes the little Nut Clams and the larger and less common Yoldias. The shells are characterised by a straight series of teeth on each valve which make an angle at a central pit.

Nuculidae are often found in the stomachs of fish, and their widespread distribution is partly attributed to the fact that some species are also a staple food of wild duck. Most Nut Clams occur in deep water, often in large colonies.

Genus *Yoldia* Möller 1842

Subgenus *Adrana* H. and A. Adams 1858

YOLDIA PERPROTRACTA Dall 1912 Long Yoldia **pl. 23**
Panama, Puerto Rico, Jamaica and Brazil.
Description: Length about 1 inch. Colour, glossy white. The shell is elongate, bluntly pointed at the posterior end and rounded at the anterior end. The hinge has about 38 anterior, and 48 posterior teeth, separated by a sub-triangular pit. The shell is thin and moderately fragile.
Occurrence: This is a rare species in Jamaica. Our few specimens were dredged alive, in about 100 feet, off Negril. We can trace no other local records. The shell lives just beneath the surface of soft, muddy substrate.

Order Filibranchia

Suborder Taxodonta

SUPERFAMILY ARCACEA

FAMILY ARCIDAE

This is the well-known family of Ark Shells, the 200 species of which occur mainly in shallow, tropical seas. The robust shells are characterised by thick periostracums and a long series of small teeth on the hinge, and the animals by greatly reduced siphons.

Most genera are anchored by a byssus of hair-like threads, but *Anadara* remain free. In the West Indies, several species are commonly used as fish bait.

SUBFAMILY ARCINAE

Genus *Arca* Linné 1758

ARCA ZEBRA Swainson 1883 Turkey Wing **pl. 23**
Bermuda, and North Carolina to the West Indies.
Description: Length 2 to 3 inches. Colour tan, with reddish-brown flecks and zebra-stripe markings. Sculpture of several low, irregular ribs. The shell has a very thick and persistent periostracum.
Occurrence: This is a very abundant species. It lives in 1 to 20 feet, attached to rocks, in Kingston harbour, Montego Bay, and in rocky areas all around Jamaica. The animal is widely used by fishermen as bait, and heaps of the discarded shells are common on all fishing beaches.

ARCA IMBRICATA Bruguière 1789 Mossy Ark **pl. 23**
North Carolina to the West Indies.
Description: Length 1 to $2\frac{1}{2}$ inches. Colour brown. This shell is similar to *A. zebra*, but has a very much larger byssal opening, finely beaded ribs, and no zebra stripes. The periostracum is thick and foliated.
Occurrence: Firmly attached to rocks in 2 to 15 feet, in Kingston harbour, Discovery Bay, and in rocky areas all around Jamaica. Dead shells, usually single valves, are often washed ashore. It is a locally common species.

Plate 19 OLIVES, MITRES

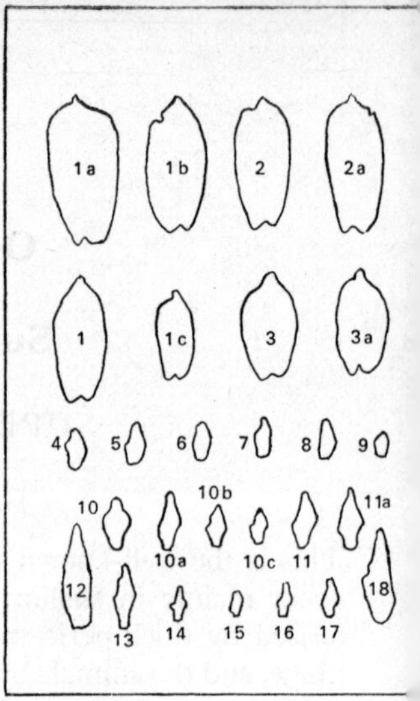

Scale: reduced by 15%

1. **Oliva reticularis** 162
Lamarck Netted Olive
Typical colour variety.
1a, *Oliva reticularis greenwayi* Caribbean Golden Olive
This is a variety of **1.** **1b,** dorsal view.
1c, *Oliva reticularis olorinella* Duclos White Olive
This is a variety of **1.**

2. **Oliva caribaeensis** 163
Dall and Simpson
Caribbean Olive
2a, ventral view.

3. **Oliva fusiformis** 163
Lamarck
3a, ventral view.

4. **Olivella minuta** Link 164
Minute Dwarf Olive

5. **Olivella petiolita** 164
Duclos Caribbean Dwarf Olive

6. **Olivella rotunda** 165
Dall Stout Dwarf Olive

7. **Olivella nivea** Gmelin 164
West Indian Dwarf Olive

8. **Jaspidella jaspidea** 165
Gmelin Jasper Dwarf Olive

9. **Olivella verreauxi** 164
Duclos Verreaux's Dwarf Olive

10. **Pusia pulchella** 167
Reeve Beautiful Mitre
10a, colour variety. **10b,** typical colour variety.
10c, deeper water variety (from about 200 feet).

11. **Pusia albocincta** 168
C. B. Adams
White-lined Mitre
11a, colour variety.

12. **Mitra barbadensis** 166
Gmelin Barbados Mitre

13. **Mitra fluviimaris** 166
Pilsbry and McGinty
Gulf Stream Mitre

14. **Pusia staminea** 167
A. Adams

15. **Pusia puella** Reeve 167
Girl Mitre
Early whorls usually missing in adult shells.

16. **Pusia gemmata** 167
Sowerby Little Gem Mitre

17. **Pusia cubana** Aguayo 166
and Rehder Cuban Mitre

18. **Mitra nodulosa** 166
Gmelin Beaded Mitre

Plate 20 **VOLUTES, NUTMEGS, MARGINELLAS, AUGERS, TURRET SHELLS**

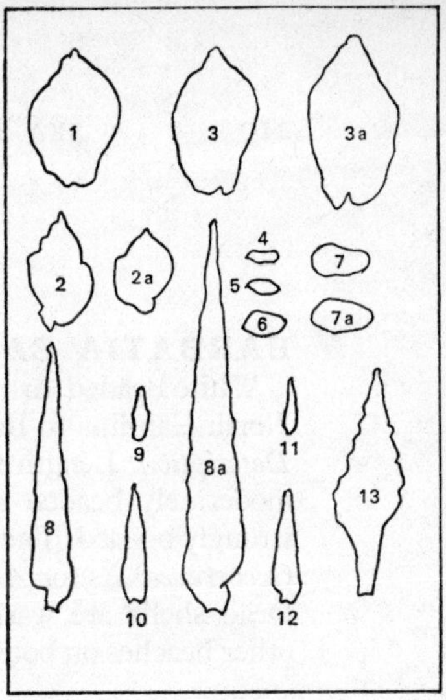

Scale: natural size

1. **Cancellaria reticulata** Linné Common Nutmeg 169

2. **Cancellaria conradiana** Dall 169
This 'species' is probably a form of **1. 2a,** dorsal view.

3. **Voluta musica** Linné 168
Music Volute
3a, ventral view.

4. **Hyalina avena** Kiener 170
Orange-banded Marginella

5. **Hyalina tenuilabra** 171
Tomlin Pale Marginella

6. **Prunum apicinum** 170
Menke Common Atlantic Marginella

7. **Prunum guttatum** 170
Dillwyn White-spotted Marginella
7a, ventral view.

8. **Terebra taurinus** 179
Solander Flame Auger
8a, dorsal view.

9. **Terebra hastata** 180
Gmelin Shiny Atlantic Auger

10. **Terebra cinerea** Born 180
Gray Auger

11. **Terebra protexta** 180
Conrad Fine-ribbed Auger

12. **Terebra dislocata** 180
Say Atlantic Auger

13. **Polystira albida** 181
Perry White Giant Turret

Genus *Barbatia* Gray 1847

Subgenus *Barbatia* Gray 1847

BARBATIA CANDIDA Helbling 1779
White Beaded Ark **pl. 23**
North Carolina to Brazil.
Description: Length 1 to 2 inches. Colour white. Sculpture of numerous moderately beaded ribs; those on the posterior dorsal area are very strongly beaded. The shell has a brown periostracum.
Occurrence: As for *Arca imbricata*, and it is an equally common species. Dead shells are washed ashore at Negril, Discovery Bay, and many other beaches on both sides of the Island.

BARBATIA CANCELLARIA Lamarck 1819
Red-brown Ark **pl. 23**
South Florida and the West Indies.
Description: Length 1 to 1½ inches. Colour dark red-brown, sometimes with a lighter area above the byssal opening. The surface of the valves is finely beaded.
Occurrence: Very commonly found attached to rocks in Kingston harbour, Port Antonio, and in rocky areas all around Jamaica. Dead shells are washed ashore on most beaches, often with both valves intact. The live shell occurs from 1 to 25 feet.

Subgenus *Acar* Gray 1857

BARBATIA DOMINGENSIS Lamarck 1819
White Reticulated Ark **pl. 23**
South-east U.S. and the Lesser Antilles.
Description: Length ½ to ¾ inch. Colour cream. Surface of the valves coarsely reticulated, with prominent concentric cords. The base of the shell is broad and slightly concave.
Occurrence: Very commonly found attached to the under side of rocks, in 1 to 10 feet, at Port Henderson, Runaway Bay and in rocky areas all around the Island. Dead shells are washed ashore on most beaches.
Note: Do not confuse with *Arcopsis adamsi*, which is usually less than ⅓ inch long and very common under rocks in shallow water.

Subgenus *Fugleria* Reinhart 1937

BARBATIA TENERA C. B. Adams 1845
Doc Bales Ark **pl. 23**
Southern Florida to Texas, and the Caribbean.
Description: Length 1 to 1½ inches. Colour white, with a thin brown periostracum. Sculpture of many fine ribs. The shell is thin and rather flat.
Occurrence: Commonly attached to rocks in 1 to 20 feet in Kingston harbour, Discovery Bay, and in most rocky areas all around the Island. Single valves are often washed ashore.

SUBFAMILY ANADARINAE

Genus *Anadara* Deshayes 1830

Subgenus *Larkinia* Reinhart 1935

ANADARA NOTABILIS Röding 1798 Eared Ark **pl. 23**
Florida to the Caribbean and Brazil.
Description: Length 1½ to 3½ inches. Colour white, with a thick brown periostracum. Sculpture of about 26 strong ribs, crossed by fine concentric threads. Specimens less than 2 inches long tend to have a prominent dorsal wing and a quadrate shape; the shell becomes more elongate as it grows larger.
Occurrence: This is a very common species in 1 to 6 feet, on mud or grass bottoms, in Kingston harbour, Runaway Bay, and in similar areas all around Jamaica. Dead shells are often cast ashore with the hinge ligament intact.

ANADARA LIENOSA FLORIDANA Conrad 1869
Cut-ribbed Ark **pl. 23**
South-east U.S., Texas, and the Greater Antilles.
Description: Length 2½ to 5 inches. Colour white, with a thick brown periostracum. This species is very similar to *A. notabilis*, but differs in having a single groove in the centre of each rib.
Occurrence: As for *A. notabilis*, but it is not as common. The two species often occur side by side.

ANDARA TRANSVERSA Say 1822 Transverse Ark
Eastern U.S., Texas and the West Indies.
Description: Length ½ to 1½ inches. Colour white, with a grey-brown

periostracum. Sculpture of about 33 ribs on each valve. There are usually beads on the ribs of the left valve, but rarely on the right one. The left valve overlaps the right.
Occurrence: This is an uncommon species found in about 3 feet of water in muddy areas along the harbour side of the Palisadoes, and in deeper water off the Bogue Islands. Beach specimens are rare in most areas, but single valves are commonly found along the causeway in Kingston harbour.

Subgenus *Lunarca* Gray 1847

ANADARA OVALIS Bruguière 1789 Blood Ark **pl. 23**
Cape Cod to the West Indies, and the Gulf States.
Description: Length 1½ to 2 inches. Colour white, with a dark-brown periostracum. Sculpture of 25 to 36 square ribs on each valve. The shell is roughly oval, and the beaks are close together.
Occurrence: As for *A. transversa*. It is not a very common species in Jamaica.

Subgenus *Cunearca* Dall 1898

ANADARA BRASILIANA Lamarck 1819
Incongruous Ark **pl. 23**
South-east U.S., and the West Indies to Brazil.
Description: Length 1 to 2½ inches. Colour white, occasionally with a very faint purplish cast on the umbones, which are well separated. Sculpture of 26 to 28 broad ribs per whorl, and many concentric grooves. The shell is inflated, and almost as high as it is long. There is a thin, light-brown periostracum.
Occurrence: This is a relatively common species, which prefers deeper water than most other members of the family. Many of our specimens came from about 30 feet off the Bogue Islands and in Ocho Rios harbour. Beach specimens are seasonally abundant near the Electric Power Station at Old Harbour Bay, and the species also occurs in Kingston harbour. The live unattached shells lie on the substrate.

ANADARA CHEMNITZI Philippi 1851
Chemnitz's Ark **pl. 23**
Florida, Texas and the West Indies.
Description: Length about ¾ inch. Colour white. This species is very similar to *A. brasiliana*, but smaller and with a thicker shell. The beaks are slightly forward of the ligament area.

Occurrence: As for *A. brasiliana,* but it is much less common. Beach specimens can usually be found at Old Harbour Bay, and the shell has been dredged off the Bogue Islands in 35 feet.

FAMILY GLYCYMERIDIDAE

These are smooth, solid, equivalve shells characterised by thin, velvety periostracums. They have wide hinges with equal-sized teeth, and powerful adductor muscles. The ligament is external.

All species live in sandy or muddy areas, and are widely distributed, mostly in tropical waters. *Glycymeris* bear light-receptive 'eyes'.

Genus *Glycymeris* da Costa 1778

GLYCYMERIS DECUSSATA Linné 1758
Decussate Bittersweet **pl. 23**
South-east Florida and the West Indies.
Description: Length about 2 inches. Colour cream or white, with irregular brown or orange markings. Sculpture of numerous radial lines. The hinge is curved, and there are many small, oblique teeth.
Occurrence: Common in about 30 feet off the Bogue Islands, and in Ocho Rios harbour. Single valves are frequently washed ashore at Old Harbour Bay and Negril. The shell lives just beneath the surface of soft sand or mud substrate.

GLYCYMERIS UNDATA Linné 1758
Atlantic Bittersweet **pl. 23**
South-east U.S. and the West Indies.
Description: Length about 2 inches. This species is very similar in most respects to *G. decussata,* but differs in having the beaks centrally located and pointing towards each other.
Occurrence: We have found this shell only on the beach at Negril and in Ocho Rios harbour. It probably occurs in sandy areas, in moderately shallow water, at other places around Jamaica.

GLYCYMERIS PECTINATA Gmelin 1791
Comb Bittersweet **pl. 23**
South-east U.S. and the West Indies.
Description: Length ½ to 1 inch. Colour cream or grey, with irregular patches of brown or purple. Sculpture of 20 to 40 radial ribs. The shell is oval and strong, and unusually attractive.

Occurrence: Not uncommon in about 15 feet at Negril, and in Ocho Rios harbour; it prefers sandy or muddy substrate. Dead shells are occasionally cast ashore, rarely with the hinge ligament intact.

Suborder Anisomyaria

SUPERFAMILY MYTILACEA

FAMILY MYTILIDAE

This is the familiar family of Mussels whose thin, streamlined shells are distributed in both fresh and salt water all over the world. They are probably the most abundant of all molluscs, and in the West Indies dense colonies are found on rocky shores attached by byssal threads. A few genera burrow into rock, coral or other molluscs.

In Europe mussels are commonly eaten; the West Indian species are edible but tasteless, and they are used mainly as fish bait.

Genus *Modiolus* Lamarck 1799

MODIOLUS AMERICANUS Leach 1815
Tulip Mussel **pl. 24**
North Carolina to the West Indies.
Description: Length 1 to 4 inches. Colour mahogany, orange, light purple or rose, arranged in rays. There is a thick, brown, hairy periostracum, under which the shell is relatively thin, smooth and polished. This is an unusually beautiful species.
Occurrence: Very commonly attached to thick algae or mossy rocks in 1 to 10 feet, in Kingston harbour, Discovery Bay, and in rocky areas all around the Island. Dead shells are washed ashore on many beaches. The living shell, with its dull algae-covered periostracum, blends perfectly with its background and is easily overlooked.

Genus *Brachidontes* Swainson 1840

Subgenus *Brachidontes* Swainson 1840

BRACHIDONTES CITRINUS Röding 1798
Yellow Mussel **pl. 24**
Southern Florida and the West Indies.
Description: Length 1 to $1\frac{1}{2}$ inches. Colour silvery, with purple umbones,

and a persistent yellow periostracum. Sculpture of many fine radial lines. There are about 30 very small teeth of equal size at the hinge. The shell is thin but strong.
Occurrence: As for *Modiolus americanus.* It is an abundant species, which occurs in large colonies in very shallow water. It is particularly common in, and outside, the pool at the Police Training School, Port Royal, and at Pimento Cove (Hanover). Dead shells are found on the beaches in Kingston harbour.

Subgenus *Hormomya* Mörch 1853

BRACHIDONTES EXUSTUS Linné 1758
Scorched Mussel **pl. 24**
South-east U.S. to the West Indies.
Description: Length ¾ inch. Colour dark brown, sometimes with a yellow border to the valves. The interior is mottled with a metallic purple and white. Sculpture of numerous thin radial ribs. The species is similar to *B. citrinus*, but it is smaller, broader and darker in colour.
Occurrence: As for *B. citrinus*. Dead shells are common on the beaches along the harbour side of the Palisadoes.

Subgenus *Ischadium* Jukes-Brown 1905

BRACHIDONTES RECURVUS Rafinesque 1820
Hooked Mussel **pl. 24**
Cape Cod to the West Indies.
Description: Length 1 to 2 inches. Colour greyish-brown to black; interior purplish to rosy-brown. Sculpture of numerous, wavy radial ribs. The shell is triangular, broad, and obliquely curved at the anterior end.
Occurrence: Commonly found attached to rocks and wharf pilings in about 2 feet of water in Kingston harbour, Port Antonio, and in many other areas all around the Island. Dead shells are abundant on many beaches.

Genus *Botula* Mörch 1853

BOTULA FUSCA Gmelin 1791
Cinamon Chestnut Mussel **pl. 24**
South-east U.S. and the West Indies.
Description: Length about ¾ inch. Colour greyish-black with a shiny

black periostracum. Sculpture of strong periodic growth lines. The beaks are anterior, the umbones prominent. Rare specimens are entirely smooth.

Occurrence: This species is an uncommon rock-borer. It can be found with the tips of the valves protruding from small holes in dead coral in Kingston harbour and Discovery Bay, in 6 to 20 feet of water. It is occasionally seen in the rocks that fringe the walls of the pool at the Police Training School, Port Royal.

Genus *Lioberus* Dall 1898

LIOBERUS CASTANEUS Say 1822

Say's Chestnut Mussel

Florida and the West Indies.

Description: Length ½ to 1 inch. Colour, glossy chestnut brown; interior iridescent bluish-white. The shell is thin and inflated, and the posterior end is usually covered by a periostracum. This species is similar to *Modiolus americanus*, but smaller.

Occurrence: This is not a common species in Jamaica. Our specimens come from Kingston harbour in about 6 feet of water. Beach specimens are occasionally found at Negril and Old Harbour Bay.

Genus *Lithophaga* Röding 1798

LITHOPHAGA NIGRA Orbigny 1842

Black Date Mussel **pl. 24**

South-east Florida and the West Indies.

Description: Length 1 to 2 inches. Colour black-brown, with an iridescent interior. Sculpture of vertical markings on the anterior part of each valve; there are also several irregular growth lines.

Occurrence: This is a relatively common rock borer, which prefers to live in soft, dead coral. It is found in Kingston harbour, at Runaway Bay, at Negril, and in coral-producing areas all around the Island. Coral fragments picked up on the beach sometimes contain dead shells of this species still firmly wedged in their holes.

LITHOPHAGA ANTILLARUM Orbigny 1842

Antillean Date Mussel **pl. 24**

Southern Florida and the West Indies.

Description: Length 2 to 5 inches. Colour light brown. This species is very similar to *L. nigra*, except for its larger size and lighter colour. The growth lines are usually prominent.

Occurrence: As for *L. nigra*. Dead shells are occasionally freed from their surrounding coral or stone, and then washed ashore.

Subgenus *Diberius* Dall 1898

LITHOPHAGA BISULCATA Orbigny 1842
Mahogany Date Mussel **pl. 24**
North Carolina, the Gulf of Mexico and the West Indies.
Description: Length 1 to 1½ inches. Colour mahogany-brown, with an iridescent interior. The surface of the shell is commonly encrusted with calcareous deposits. There is a sharp, oblique line which divides each valve into two sections. The shell is cylindrical in shape, and comes to a blunt point at the posterior end.
Occurrence: As for *L. nigra*. We have found both species occurring in the same block of coral.

Genus *Gregariella* Monterosato 1884

GREGARIELLA CORALLIOPHAGA Gmelin 1791
Ridged Mussel **pl. 24**
West Indies.
Description: Length ½ to ¾ inch. Colour: mahogany-brown at the anterior end; yellowish at the posterior end. There is a thick, hairy periostracum on the posterior dorsal area. Sculpture of growth lines, and thin radiating striae on the posterior two-thirds of the shell.
Occurrence: This is a relatively uncommon rock-borer. Our specimens are from dead coral, in about 6 feet, at Negril and in Kingston harbour.

SUPERFAMILY PTERIACEA

FAMILY ISOGNOMONIDAE

These are very thin, flat shells with pearly interiors. The anterior margin has a narrow byssal gape near the dorsal margin of the valves. The shells live colonially in shallow, protected water, where they occur in large numbers, and are anchored by a silky byssus to some firm surface.

Genus *Isognomon* Solander 1786

ISOGNOMON ALATUS Gmelin 1791

Flat Tree Oyster **pl. 24**

South Florida and the West Indies.

Description: Length 2 to 3 inches. Colour dull grey or purplish; interior pearly. Sculpture of many coarse growth lines. The shell is fan-shaped, with very flat valves. There are 8 to 12 parallel grooves on the hinge.

Occurrence: Abundant on mangrove roots in about 6 inches of water in Kingston harbour, at Falmouth, and wherever there are mangrove swamps. In the absence of mangroves, the shells will attach themselves to any solid object in the inter-tidal area of protected bays.

ISOGNOMON BICOLOR C. B. Adams 1845

Two-toned Tree Oyster **pl. 24**

Florida Keys, Bermuda and the West Indies.

Description: Length 1 to 1½ inches. This species is similar in all respects to *I. alatus*, but it is smaller, heavier and has stronger scale-like lamellations on the exterior of the valves.

Occurrence: As for *I. alatus*. It is equally common, and small specimens are often concealed in rock crevices

ISOGNOMON RADIATUS Anton 1839

Lister's Tree Oyster **pl. 24**

South-east Florida and the West Indies.

Description: Length 1 to 2 inches. Colour very pale reddish-brown, with irregular white radial streaks. The hinge is short, with 4 to 8 grooves. The shell is thin, flat, and more or less elongate; it is brittle and easily fractured.

Occurrence: Commonly found attached to the underside of rocks in 1 to 10 feet in Kingston harbour, at Negril, and in rocky areas all around the Island. The species thrives in many environments, but it is not really colonial in habit, as are the two other members of the genus.

Genus *Malleus* Lamarck

Subgenus *Parimalleus* Iredale 1884

MALLEUS CANDEANUS Orbigny 1842

Jamaica and the Greater Antilles (?).

Description: Length 1 to 2½ inches. Colour brown to black; interior

pearly. This species is superficially very similar to *Isognomon bicolor*, but the genus has an entirely different hinge structure which lacks *Isognomon's* characteristic parallel grooves.
Occurrences: Relatively common in many areas under rocks, in 3 to 1600 feet. Our specimens come from Discovery Bay and Negril. Young shells attached to living *Spondylus americanus* often adapt, as they grow, to the contours of their host's spines, though they are in fact secured by a byssus like all members of the family.

FAMILY PTERIIDAE

These are flat, thin, relatively delicate shells, with beautiful nacreous interiors. They are widely distributed in warmer waters, and attach themselves by a silky byssus to some firm surface.

Commercial pearls come from the large Pacific species; pearls are occasionally found in the Caribbean members of the family, but they are rarely of gem quality. The Jamaican *Pteriidae* are edible, but few people find them palatable.

Genus *Pteria* Scopoli 1777

PTERIA COLYMBUS Röding 1798
Atlantic Wing Oyster **pl. 24**
South-east U.S. and the West Indies.
Description: Length 1½ to 3 inches. Colour dark brown with lighter radial lines which are usually broken; interior pearly. Characterised by the distinctive elongate posterior wing. It is a beautiful shell.
Occurrence: Relatively common in 1 to 30 feet, attached to rocks, wharf pilings and sea fans, all around the Island. Excellent examples are found on the sea walls at Port Royal.

Genus *Pinctada* Röding 1798

PINCTADA RADIATA Leach 1814
Atlantic Pearl Oyster **pl. 24**
Southern Florida and the West Indies.
Description: Length 1½ to 3 inches. Colour variable, but generally either tan, reddish-brown, or greenish-yellow; the interior is pearly. Sculpture of delicate, overlapping scales, which are spiny in young specimens. The shell is usually slightly inflated; the valves are thin and

brittle. Pearls are occasionally found in this species, but they are very rarely of gem quality.
Occurrence: As for *Pteria colymbus*. The two species are often found living side by side, with *P. radiata* being considerably more common.

FAMILY PINNIDAE

This is a distinctive family of large, fragile, triangular shells, which live buried upright in soft sand or mud with only the tips exposed. The shell is anchored by a silky byssus, which protrudes from the base, and the animals are unique in possessing a pallial organ and waste canal which remove debris from the mantle cavity.

The translucent, amber coloured varieties are sometimes used as lamp shades, and shed an unusual pale and restful light. The family is widespread in warmer waters.

Genus *Pinna* Linné 1758

PINNA CARNEA Gmelin 1791 Amber Pen Shell **pl. 25**
South-east Florida and the West Indies.
Description: Length 4 to 11 inches. Colour light to dark orange. Sculpture of about 10 longitudinal ribs, which may be either smooth or spiny. There is a central radial ridge, which is most conspicuous at the pointed end. The shell is brittle and easily fractured.
Occurrence: Relatively common on eel grass beds and in sandy areas in 1 to 15 feet, at Negril, Discovery Bay, and off the Port Royal Cays. Dead and damaged shells are cast ashore on many other beaches, particularly after stormy weather.

Genus *Atrina* Gray 1840

Subgenus *Servatrina* Iredale 1939

ATRINA SEMINUDA Lamarck 1819 Spiny Pen Shell **pl. 25**
Eastern U.S. to Brazil.
Description: Length 5 to 9 inches. Colour light to dark brown. Sculpture of between 5 and 16 spiny, longitudinal ridges, which are occasionally smoothish. The shell is proportionately much broader than *Pinna carnea*.
Occurrence: Very common in shallow, protected muddy water, par-

ticularly in the vicinity of mangrove swamps. It occurs in large numbers in 1 to 6 feet along the harbour side of the Palisadoes, and in a similar environment all around the Island.

ATRINA SERRATA Sowerby 1825 Saw-toothed Pen Shell
North Carolina to the West Indies.
Description: Length 5 to 9 inches. Colour light brown. This species is very similar to *A. seminuda*, but the spines are smaller, more numerous, and imbricated. The shell is usually thinner and more fragile.
Occurrence: As for *A. seminuda*, but it is less common. Our specimens come from the harbour side of the Palisadoes.

SUPERFAMILY PECTINACEA

FAMILY PLICATULIDAE

Shells of this family, aptly known as Kittens Paws, are very similar to those of the following group (*Pectinidae*), and some authorities do not consider that the *Plicatulidae* merit true family status.

They differ from Scallops in possessing heavy, strongly hinged valves, and lack the ability to swim. The single Jamaican species is an abundant shell in many protected areas.

Genus *Plicatula* Lamarck 1801

PLICATULA GIBBOSA Lamarck 1801 Kittens Paw **pl. 27**
South-east U.S., the Gulf States and the West Indies.
Description: Length about 1 inch. Colour generally cream or very light brown, with red or brown markings. Sculpture of 5 to 9 strong, smooth radial ribs, which produce fluted margins on the valves. The shell is thick and strong.
Occurrence: Abundant in about 1 foot of water all along the harbour side of the Palisadoes, where it is usually found attached to some solid object. Less common in deeper water at Negril, where many specimens are all white in colour. Beach specimens are abundant at both places.

FAMILY PECTINIDAE

This is the family of beautiful Scallop shells, whose graceful form has appealed to man's aesthetic instincts for countless centuries. The edge of the mantle bears efficient and colourful 'eyes'.

The family is characterised by the familiar fan-like shape, and the ability of the molluscs to 'swim' in an effective but ungainly fashion.

Scallops are widely distributed in warm and temperate seas, and several large species are fished commercially, for the succulent adductor muscle. The Lion's Paw (*Lyropecten nodosus*) is a Caribbean species considered by many collectors to be the most beautiful scallop of them all. At rest, all *Pectinidae* lie on their right sides.

Genus *Pecten* Müller 1776

Subgenus *Amusium* Röding 1798

PECTEN LAURENTI Gmelin 1791 Laurent's Scallop **pl. 27**
West Indies.
Description: Length 2½ to 3½ inches. Colour: convex valve cream to tan, sometimes with light brown radial lines; flat valve reddish-brown flecked with light and dark patches. Sculpture: convex valve exterior smooth or with very lightly incised radial lines; interior with 30 to 40 fine ribs, commonly arranged in pairs; flat valve with numerous rounded radial ribs. The shell is relatively thin but strong.
Occurrence: This is a rare species, and should not be confused with the similar *Pecten ziczac* which is found all around Jamaica. Our specimens of *laurenti* came from about 30 feet off the Bogue Islands and in Ocho Rios harbour. Beach specimens are occasionally found at both places, and the species probably occurs in moderately deep water in other areas along both north and south coasts.

PECTEN PAPYRACEUS Gabb Paper Scallop **pl. 27**
Gulf of Mexico to the West Indies.
Description: Diameter about 2½ inches. Colour: upper valve light mauve to reddish-brown; lower valve white in the centre, darker at the margins. The shell is smooth, polished and rather delicate. Internally, there are about 22 fine ribs, usually arranged in pairs.
Occurrence: Relatively common in certain areas of the West Indies in about 250 feet. It has been dredged off Montego Bay at this depth, and is quite often brought up by shrimp trawlers off Barbados and Trinidad.

Subgenus *Euvola* Dall 1898

PECTEN ZICZAC Linné 1758 Zigzag Scallop **pl. 27**
South-east U.S. and the West Indies.
Description: Length 2 to 4 inches. Colour: convex valve tan to brownish-

red; flat valve maroon, mottled with brown and white. Sculpture: convex valve with 18 to 22 broad, low ribs on the exterior; flat valve with a number of straight radial grooves.
Occurrence: This is a relatively common species which prefers calm, protected, shallow water. Live specimens can be collected in 3 feet on the eel grass beds along the harbour side of the Palisadoes, and dead shells are common on the beaches at Negril.

PECTEN CHAZALIEI Dautzenberg 1900
Chazalie Scallop **pl. 27**
Southern Florida, the Gulf of Mexico, Puerto Rico and Jamaica.
Description: Length about 1 inch. Colour light tan with reddish-brown flecks. Sculpture of about 20 small narrow ribs on the flat upper (left) valve; the convex (right) valve has a number of low ribs. The shell is light and relatively fragile.
Occurrence: This is a rare species in Jamaica. It occurs in about 40 feet at Orange Bay and in Ocho Rios harbour. We have found no beach specimens.

Genus *Chlamys* Röding 1798

CHLAMYS SENTIS Reeve 1853 Sentis Scallop **pl. 26**
South-east U.S. and the West Indies.
Description: Length 1 to 1½ inches. Colour white, orange, vermilion, or purple-red, sometimes mottled near the beaks. The valves are nearly flat, and one hinge ear is twice as large as the other. Sculpture of about 50 ribs of various sizes.
Occurrence: Relatively common down to 50 feet at Prospect, Port Antonio, Orange Bay, and many other areas around the Island. Fine live specimens can be collected in a few feet of water in the sandy depressions in the eel grass beds at Prospect; usually they are found attached by the byssus to the underside of rocks.

CHLAMYS ORNATA Lamarck 1819 Ornate Scallop **pl. 26**
South-east Florida to the West Indies.
Description: Length 1 to 1½ inches. Colour variable, but generally cream with strong, light or dark maroon maculations. This species is very similar to *C. sentis*, but the sculpture consists of 18 groups of three closely spaced, scaly riblets, of which the centre rib is usually the largest.
Occurrence: Common in 1 to 20 feet at Negril, Prospect, the Port Royal Cays, and in many areas around Jamaica. It is often found attached by the byssus to the underside of rocks and coral slabs in sandy localities, and single valves are common on the beaches.

Plate 21 CONE SHELLS

Scale: reduced by 35%

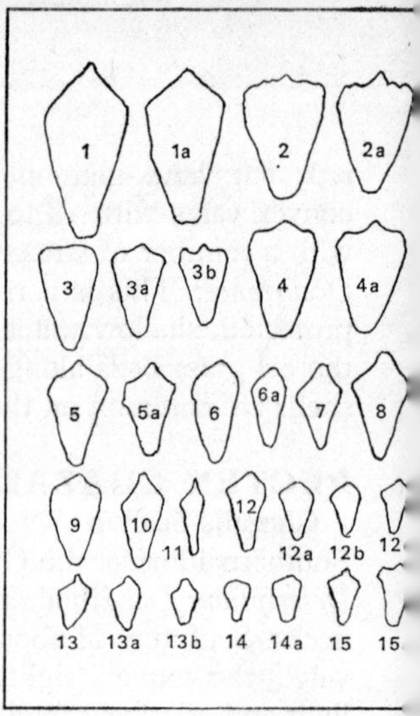

1. Conus ranunculus Hwass Atlantic Agate Cone 174
1a, typical colour variety.

2. Conus regius Gmelin 171
Crown Cone
Typical colour variety.
2a, *Conus regius citrinus* Gmelin. This is an uncommon colour form.

3. Conus daucus Hwass 173
Carrot Cone
Unusual banded colour variety. **3a,** common colour variety. **3b,** common colour variety known as *luteus* Krebs.

4. Conus spurius Gmelin 172
Alphabet Cone
4a, common colour variety.

5. Conus centurio Born 173
Centurion Cone
5a, colour variety.

6. Conus granulatus Linné 175
Glory of the Atlantic Cone
6a, colour variety.

7. Conus clarki Rehder and Abbott Clark's Cone 178
See text note on relationship between this and **8.**

8. Conus austini Rehder and Abbott Austin's Cone 175
Neither this species nor **7** has yet been found in Jamaican waters.

9. Conus mus Hwass 178
Mouse Cone

10. Conus dominicanus Hwass Antillean Cone 179

11. Conus mazei Deshayes 178
Maze's Cone

12. Conus juliae Clench 174
Julia's Cone
Typical colour variety. **12a, 12b and 12c,** colour varieties.

13. Conus jaspideus Gmelin 172
Jasper Cone
Typical colour variety. **13a,** example with thin periostracum still in place. **13b,** example with pustuled surface. Sometimes classified as *C. verrucosus*, a separate species. See text for note on this subject.

14. Conus jaspideus havanensis Aguayo and Farfante Havana Cone 172
14a, typical colour variety. Note the nodules at the shoulder of the body whorl which distinguish it from true *jaspideus*.

15. Conus patae Abbott 175
Note the characteristic axial folds of this very recently described species. **15a,** rare pink colour variety.

Plate 22 TRIVIAS, TURRETS, DRILLIAS, MELAMPUS SHELLS, PYRAMS, BUBBLE SHELLS, SEA BUTTERFLIES AND OTHERS

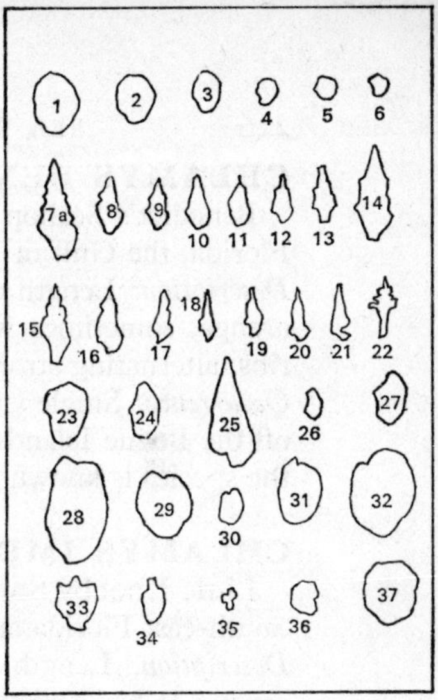

Scale: natural size

1. **Trivia pediculus** Linné 103
Coffee Bean Trivia
2. **Trivia nix** Schilder White 104
Globe Trivia
3. **Trivia suffusa** Gray Suffuse 103
Trivia
4. **Trivia quadripunctata** Gray 104
Four-spotted Trivia
5. **Trivia leucosphaera** Gray 104
Little White Trivia
6. **Trivia antillarum** Schilder 104
Antillean Trivia
7. **Fenimorea janetae** Bartsch 184
Janet's Turret
8. **Drillia cydia** Bartsch Cydia 183
Drillia
9. **Crassispira leucocyma** Dall 183
White-knobbed Turret
10. **Drillia interpleura** Dall and 184
Simpson Interpleura Drillia
11. **Drillia albinodata** Reeve 184
White-noded Turret
12. **Crassispira** species 183
13. **Crassispira** species 182
14. **Crassispira fuscescens** 182
Reeve Ebony Turret
15. **Daphnella stegeri** McGinty 186
Steger's Turret
16. **Daphnella lymneiformis** 186
Kiener Volute Turret
17. **Kurtziella limonitella** Dall 185
Punctate Mangelia
18. **Ithycythara lanceolata** 184
C. B. Adams Spear Turret
19. **Mangelia balteata** Reeve 203
20. **Rissoina decussata** Montagu 199
Decussate Risso
21. **Stilifer subulatus** Broderip 201
and Sowerby Thomas' Stilifer
22. **Ancistrosyrinx radiata** Dall 181
Common Star Turret
23. **Melampus monile** Bruguière 196
Yellowish Melampus
24. **Tralia ovula** Bruguière Egg 195
Tralia
25. **Pyramidella dolabrata** 190
Linné Giant Atlantic Pyram
26. **Detracia bullaoides** 196
Montagu Bulla Melampus
27. **Melampus coffeus** Linné 196
Coffee Melampus
28. **Bulla striata** Bruguière 189
Striated-bubble
29. **Haminoea antillarum** 190
Orbigny Antillean Paper-bubble
30. **Micromelo undata** Bruguière 188
Miniature Melo
31. **Haminoea elegans** Gray 189
Elegant Paper-bubble
32. **Hydatina vesicaria** Solander 188
Brown-lined Paper-bubble
33. **Cavolina tridentata** Forskäl 191
Three-toothed Cavolina
34. **Acteon punctostriatus** C. B. 187
Adams Adams Baby-bubble
35. **Murex tristichus** Dall 136
Tristichus Murex
36. **Atlanta peroni** Lesueur 109
Peron's Atlanta
37. **Siphonaria pectinata** Linné 197
Striped False Limpet

CHLAMYS BENEDICTI Verrill and Bush 1897
Benedict's Scallop **pl. 26**
Florida, the Gulf of Mexico, Jamaica and Puerto Rico.
Description: Length about ½ inch. Colour yellow, purple, pink, red, or orange, sometimes with white zigzag stripes. Sculpture of about 45 ribs, alternating strong and weak. The shell is thin and light.
Occurrence: Single valves are uncommon on the beaches at Negril, and off the Bogue Islands; we have seen no live specimens in Jamaica, but the species is known to occur from 30 to 80 feet in Puerto Rico.

CHLAMYS IMBRICATA Gmelin 1791
Little Knobby Scallop **pl. 26**
South-east Florida and the West Indies.
Description: Length 1 to 1¾ inches. Colour and shape similar to *C. ornata*. It has a distinctive sculpture of prominent, cup-shaped scales all along the 8 to 10 radial ribs. There are smaller cords between the main ribs.
Occurrence: Relatively common in 3 to 25 feet at Montego Bay, Priory, Yallahs (St Thomas), and in many similar areas around the Island. Beach specimens usually have lost their delicate scales, and this may complicate identification. The live shell is often found attached to the underside of large rocks or dead coral.

CHLAMYS EFFLUENS Dall 1886 **pl. 26**
West Indies.
Description: Length 1 to 2 inches. Colour red, yellow or maroon. Sculpture of numerous faint radial threads which bear minute scales on the lower two thirds of the valves. The shell has an oily sheen. Both valves are thin, translucent, fragile and nearly flat. It is a delicately beautiful species.
Occurrence: This is a very rare shell in Jamaica. It occurs in 50 to 160 feet at Orange Bay, Discovery Bay and in Ocho Rios harbour. Our specimens were found on sandy substrate. We have no other relevant information about the species, and, apparently, it has not previously been reported from Jamaica.

CHLAMYS species **pl. 26**
Jamaica.
Description: Length ¾ to 1¼ inches. Reddish-brown or yellow, alone or in various combinations, often with extensive white speckling on the ribs. Occasionally the darker colour extends to the limits of a particular growth stage and is then replaced by yellow to the margins, producing a ringed effect. Sculpture of 10 to 15 broad rounded ribs which have

squared edges on the interior; there are fine radial threads between the ribs and these continue over them in some specimens. The interior is polished and slightly pearly, particularly towards the umbones. Both valves are equally, and only slightly, inflated. The shell is thin but surprisingly strong.
Occurrence: Common on mud or sand substrate in about 30 feet at the Bogue Islands, and at Negril. Dead shells, occasionally with both valves intact, are sometimes found on the Long Bay beach at Negril after heavy weather. We have no other records for this species.
Note: At the time of writing, this shell is being examined by Dr R. T. Abbott to ascertain whether it has previously been described. We have been unable to find any trace of it in the records available.

Genus *Leptopecten* Verrill 1897

LEPTOPECTEN BAVAYI Dautzenberg 1900
Bavay's Scallop **pl. 26**
West Indies.
Description: Length ½ inch or less. Colour white, usually streaked with pale yellow or brown. The shell is thin, fragile, and sculptured with about 20 sharp ribs and numerous microscopic, wavy growth lines.
Occurrence: This is a rare species in Jamaica. Our specimens came from the lagoon in the centre of the Palisadoes, just outside Port Royal, and in Ocho Rios harbour in 20 feet. We can trace no other Jamaican records.

Genus *Lyropecten* Conrad 1862

LYROPECTEN ANTILLARUM Récluz 1853
Antillean Scallop **pl. 26**
South Florida and the West Indies.
Description: Length ½ to ¾ inch. Colour white, light yellow, orange, or light brown, usually mottled with white. The valves are fragile and nearly flat. Sculpture of 10 to 15 broad, slightly rounded ribs, and exceedingly fine growth lines. This is an exquisite little shell.
Occurrence: Relatively uncommon in about 25 feet off the Bogue Islands, and in Ocho Rios harbour. Dead shells are occasionally found on the beaches at Negril, especially after rough weather.
Note: A bright red colour form (usually flecked with white) occurs at Negril.

Subgenus *Nodipecten* Dall 1898

LYROPECTEN NODOSUS Linné 1758

Lion's Paw **pls. 1, 27**

South-east U.S. and the West Indies.

Description: Length 3 to 6 inches. Colour variable; usually purplish-red, maroon or (less frequently) orange. The shell is heavy and strong. Sculpture of 7 to 9 large, nodulous ribs, and many fine riblets. This is an unusually beautiful shell, both in form and colour.

Occurrence: It is a comparatively rare species in Jamaica. It occurs in approximately 50 feet off the Bogue Islands, at Negril, and at the mouth of Kingston harbour. Beach specimens are rare and usually damaged. In the southern Caribbean, fine, live examples are occasionally dredged by shrimp trawlers in about 150 feet. The species appears to seek deeper water as it matures.

Genus *Aequipecten* P. Fischer 1886

Subgenus *Aequipecten* P. Fischer 1886

AEQUIPECTEN MUSCOSUS Wood 1828

Rough Scallop **pl. 26**

South-east U.S. and the West Indies.

Description: Length $\frac{3}{4}$ to $1\frac{1}{4}$ inches. Colour bright yellow, orange, red, or orange-brown, sometimes mottled with white. Both valves are moderately inflated. Sculpture of 18 to 20 ribs, each bearing 3 rows of very small, sharp scales. The surface is rough to the touch in fresh specimens, but beach-worn valves may be quite smooth.

Occurrence: Relatively common in about 15 feet at Negril, and in many areas along the north coast. Dead shells are common on beaches west of Montego Bay, and along the southern Palisadoes.

AEQUIPECTEN LINEOLARIS Lamarck 1819

Wavy-lined Scallop **pl. 26**

Florida Keys to the West Indies.

Description: Length about 1 inch. Colour rosy tan, with numerous wavy brown concentric lines. There are generally about 4 small patches of solid colour; the bottom valve is white, occasionally flecked with brown on the ribs. Sculpture of about 18 ribs which are angular at the summit (in fresh specimens). The interior edge of both valves is strongly crenulated. The shell is solid and strong, with moderately inflated valves.

Occurrence: This is a rare and choice species, from moderately deep water. We have found it only at Runaway Bay and in Ocho Rios harbour, in about 40 feet. There are indications that it occurs in limited numbers in other areas along the north coast, and we found single valves in the extensive commercial dredgings at the Bogue Islands.

Subgenus *Argopecten* Monterosato 1889

AEQUIPECTEN GIBBUS Linné 1758

Calico Scallop **pl. 26**

Eastern U.S., the Gulf of Mexico and the West Indies.

Description: Length 1 to 2 inches. Colour: upper valve with pink, purple, or brown mottlings on a white background; lower valve whitish. The shell is strong, with moderately inflated valves. Sculpture of about 20 ribs and very fine concentric lamellations; the surface is smooth to the touch.

Occurrence: This is the most common member of the family in Jamaica. Live specimens can be collected in 1 to 15 feet on the eel grass beds along the harbour side of the Palisadoes, and single valves are found on many beaches all around the Island.

FAMILY SPONDYLIDAE

These are the beautiful and spectacular 'Chrysanthemum Shells', well-known and admired by collectors everywhere. The heavy valves are covered with long, graceful spines, and many species are bright red or orange in colour. The shells are attached by the right (lower) valve.

The family is characterised by the 'ball and socket' joint at the hinge, which incidentally serves as a quick means of identifying juvenile specimens (which may resemble *Pectinidae*). The mantle-edge is studded with 'eyes'.

The shape of *Spondylidae* influenced the art of Central and South American Indian civilisations as far back as 900 B.C., and examples of ceramic vases from Peru with stylised *Spondylus* motifs can be seen in many large museums.

Genus *Spondylus* Linné 1758

SPONDYLUS AMERICANUS Hermann 1781

Atlantic Thorny Oyster **pl. 25**

Florida and the West Indies.

Description: Length 3 to 6 inches. Colour variable: it may be all white

all red, or all orange; or any combination of these colours. The most common Jamaican variety is white, with red or orange on the umbones. The adult shell is very solid and heavy. Most examples have many long spines, but these vary greatly in length between individuals. Generally, specimens which occur in open muddy or sandy areas are smaller and much more spinose than those from coral reefs. The presence of the characteristic 'ball and socket' hinge quickly identifies the species.
Occurrence: This is a common and widespread species, found all around the Island in depths of 10 to 60 feet. At Cow Bay, small but exquisite specimens occur in quantity at about 35 feet, attached to small stones lying in the mud. In Discovery Bay and off Port Royal, massive and colourful examples are accidentally brought up, tangled in seine nets or fishing lines. Worn, single valves (usually the upper, unattached ones) are common on beaches everywhere in Jamaica. The living shell is always attached by the lower valve to some hard and firm surface.

FAMILY LIMIDAE

This is a distinctive family of compressed, equivalve shells, with straight anterior, and rounded posterior, ends. The group is characterised by the ability to 'swim' in a clumsy but effective fashion (unlike the *Pectinidae*, they do so in an upright position).

Limidae occasionally spin a silky byssus to attach themselves to some firm surface, and adults can construct rudimentary 'nests' from fragments of shell and coral. There are about 100 species worldwide.

The living animal of most species is a spectacular scarlet-red colour, and they make fascinating subjects for study in a marine aquarium.

Genus *Lima* Bruguière 1797

Subgenus *Lima* Bruguière 1796

LIMA LIMA Linné 1758 Spiny Lima **pl. 27**
South-east Florida and the West Indies.
Description: Length 1 to 1½ inches. Colour white. Sculpture of numerous, spiny, radial ribs. The hinge line is oblique, and the anterior ear is much smaller than the posterior one.
Occurrence: Relatively uncommon under rocks in 1 to 12 feet at Prospect, Port Royal and Discovery Bay. Single valves are often washed ashore with the spines worn smooth. The live shell is usually secured by its byssus to some firm surface.

Subgenus *Limaria* Link 1807

LIMA PELLUCIDA C. B. Adams 1846 Antillean Lima **pl. 27**
South-east U.S. and the West Indies.
Description: Length $\frac{3}{4}$ to 1 inch. Colour white; shell semi-transparent. Sculpture of numerous, fine, irregular radial ribs. The shell is thin, oblique, and elongate; immature specimens are exceedingly delicate.
Occurrence: Relatively uncommon under rocks in 1 to 20 feet at Orange Bay, Ocho Rios and Montego Bay. Beach specimens are found near Fort Augusta, in Kingston harbour. Live shells will often 'swim' into rock crevices to avoid capture.

Subgenus *Ctenoides* Mörch 1853

LIMA SCABRA Born 1778 Rough Lima **pl. 27**
South-east Florida and the West Indies.
Description: Length 1 to 3 inches. Colour white. Sculpture of irregular radial rows of short ribs. The periostracum, when present, is light brown.
Occurrence: Common in 1 to 20 feet at Runaway Bay, Prospect, Port Royal, and in rocky areas all around the Island. The live shell may temporarily attach itself by the byssus to the underside of rocks. When unattached, it swims in a clumsy but effective fashion, apparently using its long, flexible tentacles to assist it through the water.
Forms: Lima scabra form *tenera* Sowerby 1843 is similar, but the shell is thinner and more delicate, with a satin-smooth surface (pl. 27). It occurs in the same environment as typical *scabra*, but is less common.

SUPERFAMILY ANOMIACEA

FAMILY ANOMIIDAE

These are thin, translucent, slightly pearly shells, of variable shape and size. The family is characterised by the distinctive notch in the lower (concave) valve, through which passes the byssus attaching the shell to some solid surface.

Anomiidae are a highly specialised group of bivalves, and the position of the byssus has resulted in the displacement of other organs.

Genus *Anomia* Linné 1758

ANOMIA SIMPLEX Orbigny 1842
Common Jingle Shell **pl. 28**
Eastern U.S., the Gulf of Mexico and the West Indies.
Description: Length 1 to 2 inches. Colour yellow, orange, or silvery, sometimes translucent. The shell is thin, light, and irregular in shape. The lower valve is flat, with a large notch, through which the byssus passes to anchor it to some firm surface. The upper valve has 1 large and 2 small muscle scars.
Occurrence: Common in about 3 feet of water attached to rocks, shells, or any other firm surface, in Discovery Bay, Kingston harbour, and rocky areas all around Jamaica. We have seen the deck of a large barge, submerged for eighteen months and then raised, entirely covered with these shells like a coat of shimmering silver paint.

Genus *Pododesmus* Philippi 1837

PODODESMUS RUDIS Broderip 1834
False Jingle Shell **pl. 28**
Florida and the West Indies.
Description: Length 1 to 4 inches. Colour whitish to grey-brown. Sculpture of irregular, broken ridges, and scaly growth lines. The upper valve (without the hole) has only 2 muscle scars.
Occurrence: As for *Anomia simplex*, but less common. It is often found attached to the bottoms of boats which have been moored for some time in the same place.

SUPERFAMILY OSTREACEA

FAMILY OSTREIDAE

Members of this widely distributed family have irregular, inequivalve shells, of variable shape. The larger (lower) valve adheres to some solid object, and adapts to its contours. Sex reversal is common.

The family includes the oviparous Caribbean Oyster (*Crassostrea rhizophorae*), which is gathered commercially throughout the West Indies, and brings joy to discriminating palates.

Genus *Ostrea* Linné 1758

OSTREA EQUESTRIS Say 1834 Crested Oyster **pl. 28**
South-east U.S., the Gulf of Mexico and the West Indies.
Description: Length 1 to 2 inches. Colour dirty white; interior grey with an irregular patch of greenish-brown. The shell is more or less oval, with heavily crenulated margins. It lacks the clasping projections of *O. frons* (the following species).
Occurrence: Common in 1 to 6 feet on rocks, piles, the bottoms of boats, and other firm surfaces, in Kingston harbour, at Negril and in many areas all around the Island. Single valves are found on most beaches.

OSTREA FRONS Linné 1758 'Coon Oyster **pl. 28**
South-east U.S. and the West Indies.
Description: Length 1 to 2½ inches. Colour purplish-brown, occasionally with patches of gold. Sculpture of radial plications and sharply folded valve margins. There is a series of characteristic clasping projections, which fasten the shell to the surface on which it is found. Large specimens are very beautiful shells.
Occurrence: Very common in 2 to 15 feet on mangrove roots, sea fan stems, and the wire of fish pots, in Kingston harbour, Falmouth, Discovery Bay, and in many other areas all around the Island. Perfect specimens can be recovered from old fish pots along the beach at Por Henderson.

Genus *Crassostrea* Sacco 1897

CRASSOSTREA RHIZOPHORAE Guilding 1828
Caribbean Oyster **pl. 28**
Caribbean.
Description: Length 2 to 6 inches. Colour greyish-white. The shell is very variable in shape, but the upper valve is generally flat and fits deeply into the lower one; the margins are straight, sharp, and smooth. This is the common edible Oyster.
Occurrence: Very common on mangrove roots (and sometimes on the backs of other shells), in 1 to 10 feet in Kingston harbour, Falmouth, Runaway Bay, and in quiet, protected waters all around the Island. Dead shells are often washed ashore.

Order Eulamellibranchia

Suborder Heterodonta

SUPERFAMILY CARDITACEA

FAMILY CARDITIDAE

Genus *Cardita* Bruguière 1792

CARDITA GRACILIS Shuttleworth 1856
West Indian Cardita **pl. 28**
Mexico to the West Indies.
Description: Length 1 to $1\frac{1}{2}$ inches. Colour whitish, with brown overtones. Sculpture of about 17 flat, wrinkled ribs; on the posterior slope these are more elevated and almost spinose.
Occurrence: This is a rare species in Jamaica if, in fact, it occurs at all. Our specimens were reportedly dredged in about 15 feet at Negril, but the record needs confirmation. Elsewhere in the West Indies it is a relatively common shell.

SUPERFAMILY DREISSENACEA

FAMILY DREISSENIDAE

Genus *Mytilopsis* Conrad 1857

MYTILOPSIS DOMINGENSIS Rècluz 1852
False Mussel **pl. 28**
West Indies.
Description: Length $\frac{3}{4}$ to $1\frac{1}{4}$ inches. Colour greyish-white, covered with a brown periostracum. This shell closely resembles a Mussel (*Mytilidae*) in shape, but there are no radial markings and *Mytilopsis* possesses a distinctive shelf beneath the ligament.
Occurrence: Common in 1 to 8 feet, attached by a short byssus to rocks or mangrove roots, in Kingston harbour and brackish, protected bays all around the Island. Beach specimens (rarely with the hinge ligament intact), are very common in these areas.

FAMILY TRAPEZIIDAE

Genus *Coralliophaga* Blainville 1824

CORALLIOPHAGA CORALLIOPHAGA Gmelin 1791
Coral Clam
Florida, Texas and the West Indies.
Description: Length ½ to 1¾ inches. Colour yellowish white. Sculpture of radial threads and concentric ridges. The umbones are at the anterior end. It can be separated from the superficially similar *Lithophaga antillarum* by the presence of teeth at the hinge.
Occurrence: This species lives in the burrows of other rock-boring bivalves. It is uncommon in about 6 feet in Kingston harbour, at Negril and Montego Bay. Dead shells are occasionally washed ashore.

SUPERFAMILY LUCINACEA

FAMILY DIPLODONTIDAE

These are relatively small, white, inflated, subcircular shells, which are closely related to the *Lucinidae* (the following family). The group is characterised by the simple shape of the valves, and the lateral teeth of the hinge, which are either obscure or entirely absent. All species are infaunal, i.e. live within the substrate, but lack siphons. The sexes are separate, and the ligament is external.

Genus *Diplodonta* Bronn 1831

DIPLODONTA PUNCTATA Say 1822
Common Atlantic Diplodon **pl. 29**
South-east U.S. and the West Indies.
Description: Length ¾ inch. Colour white. Shell inflated. Sculpture of many fine concentric threads and coarse growth lines. This is the largest of the Jamaican *Diplodonta*.
Occurrence: Common from 3 to 12 feet in Kingston harbour, at Robins Bay and at Negril. It is usually found on soft sand or mud bottoms. Beach specimens are common along the harbour side of the Palisadoes.

FAMILY LUCINIDAE

The 200 members of this family are widely distributed in tropical seas, and many of the Jamaican species occur in large numbers in shallow water all around the Island. The sexes are separate.

Lucinidae are distinguished by their orbicular shape, long anterior muscle scars, and (in the West Indies) their white exteriors. The ligament is external. The live shells are lightly buried in soft sand or mud substrate, and construct an anterior exhalant shaft from sand and mucus.

Genus *Lucina* Bruguière 1797

LUCINA PENSYLVANICA Linné, 1758
Pennsylvania Lucina **pl. 29**
South-east U.S. and the West Indies.
Description: Length 1 to 2 inches. Colour white, with a thin yellow periostracum. There is a distinctive furrow on each valve, from the beak to the posterior ventral edge. Sculpture of thin, sharp, concentric ridges in fresh specimens; in beach-worn examples, these are usually worn away and the shell surface is smooth.
Occurrence: Common in 4 to 10 feet at Negril, Discovery Bay, off the Port Royal Cays, and in sandy areas all around the Island. Single valves are found on most beaches.

LUCINA PHENAX Dall and Simpson 1901
Mock Lucina **pl. 29**
Puerto Rico and Jamaica.
Description: Length about ½ inch. Colour white, with a thin, grey periostracum. Sculpture very similar to that of *L. pensylvanica* (the preceding species), but more pronounced. The shell is inflated and lacks hinge teeth.
Occurrence: Common in sand at about 35 feet in Ocho Rios harbour and at Negril, but we have found it nowhere else. Beach specimens are rare.

Genus *Phacoides* Gray 1847

PHACOIDES PECTINATUS Gmelin 1791
Thick Lucina **pl. 29**
South-east U.S., Texas and the West Indies.
Description: Length 1 to 2½ inches. Colour white, usually tinged with

orange. Sculpture of numerous thin, sharp, concentric ridges. The shell is solid and ovate.
Occurrence: This is a relatively uncommon species in Jamaica. It occurs in 3 to 20 feet in Kingston and Ocho Rios harbours, and off the Bogue Islands. Beach specimens are rarely found with the hinge ligament intact, but single valves are common at Farquhars Beach (Clarendon). The species prefers mud to sand, and is most often found in the vicinity of mangrove swamps.

Subgenus *Lucinisca* Dall 1901

PHACOIDES MURICATUS Spengler 1798
Spinose Lucina **pl. 29**
Florida Keys and the West Indies.
Description: Length about ½ inch. Colour white. Sculpture of many radial ribs which bear small, sharp spines; these are often worn smooth in beach specimens.
Occurrence: As for *Lucina pensylvanica*. It is an abundant species in most sandy areas, and single valves are very common on the beaches near Montego Bay.

Subgenus *Callucina* Dall 1901

PHACOIDES RADIANS Conrad 1841 Dosinia-like Lucina
South-east U.S., Puerto Rico and Jamaica.
Description: Length ¾ to 1 inch. Colour white. Sculpture of fine concentric lines and very faint radial threads. The lunule is deep and the ventral margin is crenulated. Superficially this species resembles *Dosinia*.
Occurrence: This is not a common species in Jamaica. It is found in sand, at about 15 feet, in Ocho Rios harbour and at Negril. Beach specimens are rare.

Genus *Anodontia* Link 1807

ANODONTIA ALBA Link 1807 Buttercup Lucina **pl. 29**
South-east U.S. and the West Indies.
Description: Length 1½ to 2 inches. Colour white, more or less tinged with yellow-orange. Sculpture of weak, irregular, concentric growth lines. The shell is circular, inflated and relatively thin.

Occurrence: Common in 3 to 15 feet in sandy areas, everywhere around Jamaica. Dead shells are frequently washed ashore, and between December and March they are abundant on the beaches at Negril.

ANODONTIA PHILIPPIANA Reeve, 1850

Chalky Buttercup

South-east U.S. and the West Indies.

Description: Length 2 to 4 inches. This species is very similar to *A. alba*, but it is all-white in colour, and generally has a pustulose interior.

Occurrence: As for *A. alba*, but it is far less common.

Genus *Codakia* Scopoli 1777

Subgenus *Codakia* Scopoli 1777

CODAKIA ORBICULARIS Linné 1758

Tiger Lucina **pl. 29**

Florida to Texas, and the West Indies.

Description: Length 2 to 3½ inches. Colour white, often tinged with rose at the ends of the hinge and the margins of the valves. Sculpture of numerous, thin, radial cords, and finer concentric threads. The shell is compressed and nearly circular.

Occurrence: This is an abundant species in 1 to 10 feet, in muddy or sandy areas, all around Jamaica. Live shells are often exposed on the mud flats at low tide, and dead shells are common on the beaches along both sides of the Palisadoes. It is possibly the most common of the larger Jamaican bivalves.

Subgenus *Ctena* Mörch 1860

CODAKIA ORBICULATA Montagu 1808

Dwarf Tiger Lucina **pl. 29**

South-east U.S. and the West Indies.

Description: Length about 1 inch. Colour white. This species is very similar to *C. orbicularis*, but it is smaller, pure white, and has stronger radial sculpture.

Occurrence: As for *C. orbicularis*, but it is not quite as common. Dead shells are found on many beaches around the Island. Live shells are often found on eel grass beds.

CODAKIA COSTATA Orbigny 1842 Costate Lucina **pl. 29**
South-east U.S. and the West Indies.
Description: Length about ½ inch. Colour white, occasionally yellowish. Sculpture of fine radial threads crossed by even finer concentric threads. There may be an occasional coarse growth line.
Occurrence: Common in sand at depths of 3 to 15 feet, at Negril, Discovery Bay and Prospect. Beach specimens often have the fine sculpture worn smooth, and this is true of most examples from the Palisadoes.

Genus *Divaricella* von Martens 1880

DIVARICELLA QUADRISULCATA Orbigny 1842
Cross-hatched Lucina **pl. 29**
Eastern U.S. and the West Indies.
Description: Length ¾ to 1 inch. Colour white. Sculpture of thin lines which rise from the margins to a blunt apex, roughly in the centre of the shell, and together strongly resemble a fingerprint. There are also several coarse concentric growth lines. The shell is almost circular, inflated, thin and glossy.
Occurrence: Single valves are relatively common on the beaches at Negril, Montego Bay and the Port Royal Cays. The live shell occurs in about 6 feet, in many sandy areas all around Jamaica.

FAMILY CHAMIDAE

This is a small family (about 20 species) of brilliantly coloured, tropical shells, which is well represented in the West Indies.

The shells are permanently attached, usually by the left valve, to a solid surface, and the family is characterised by the prominent scaly foliations on both valves. The ligament is external.

Genus *Chama* Linné 1758

CHAMA MACEROPHYLLA Gmelin, 1791
Leafy Jewel Box **pl. 29**
South-east U.S. and the West Indies.
Description: Length 1 to 3 inches. Colour white, yellow, or purple; sometimes in combinations of all three colours. Sculpture of numerous, thin, overlapping scales. The inner margins of both valves are crenulated. The shell is solid and irregular in shape.
Occurrence: This colourful and attractive shell is common in many

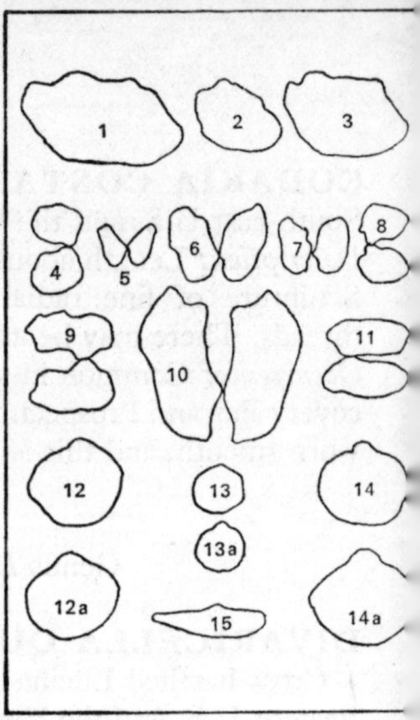

Plate 23 ARKS, BITTERSWEETS, YOLDIAS

Scale: reduced by 40%

1. **Anadara lienosa floridana** Conrad Cut-ribbed Ark 211

2. **Anadara brasiliana** Lamarck Incongruous Ark 212

3. **Anadara notabilis** Röding Eared Ark 211

4. **Anadara ovalis** Bruguière Blood Ark 212

5. **Barbatia domingensis** Lamarck White Reticulated Ark 210

6. **Barbatia candida** Helbling White Beaded Ark 210

7. **Barbatia tenera** C. B. Adams Doc Bales' Ark 211

8. **Anadara chemnitzi** Philippi Chemnitz's Ark 212

9. **Arca imbricata** Bruguière Mossy Ark 207

10. **Arca zebra** Swainson Turkey Wing 207

11. **Barbatia cancellaria** Lamarck Red-brown Ark 210

12 and **12a. Glycymeris decussata** Linné Decussate Bittersweet 213
Compare the position of the beaks in this species with **14 and 14a.**

13 and **13a. Glycymeris pectinata** Gmelin Comb Bittersweet 213

14 and **14a. Glycymeris undata** Linné Atlantic Bittersweet 213

15. **Yoldia perprotracta** Dall Long Yoldia 206

Plate 24 MUSSELS, OYSTERS

Scale: reduced by 35%

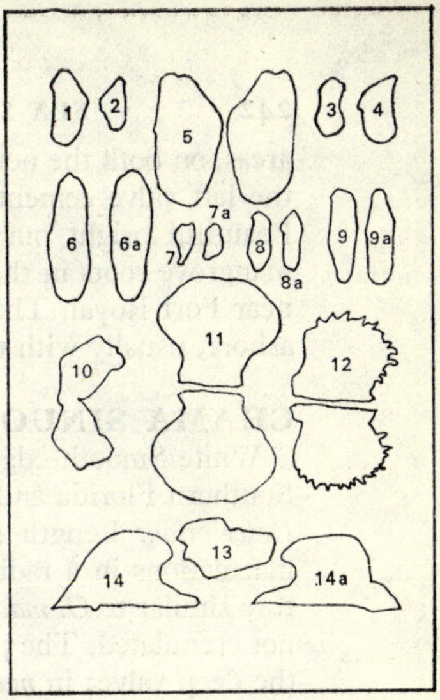

1. Brachidontes citrinus Röding Yellow Mussel 214

2. Brachidontes exustus Linné Scorched Mussel 215

3. Botula fusca Gmelin Cinnamon Chestnut Mussel 215

4. Brachidontes recurvus Rafinesque Hooked Mussel 215

5. Modiolus americanus Leach Tulip Mussel 214

6 and **6a. Lithophaga antillarum** Orbigny Antillean Date Mussel 216

7 and **7a. Gregariella coralliophaga** Gmelin Ridged Mussel 217

8 and **8a. Lithophaga nigra** Orbigny Black Date Mussel 216

9 and **9a. Lithophaga bisulcata** Orbigny Mahogany Date Mussel 217

10. Isognomon bicolor C. B. Adams Two-toned Tree Oyster 218
See text for difference between this and the superficially similar *Malleus candeanus.*

11. Isognomon alatus Gmelin Flat Tree Oyster 218

12. Pinctada radiata Leach Atlantic Pearl Oyster 219

13. Isognomon radiatus Anton Lister's Tree Oyster 218

14 and **14a. Pteria colymbus** Röding Atlantic Wing Oyster 219 areas, on both the north and south coasts. It occurs in 1 to 30 feet, with the left valve cemented firmly to a rock or some other hard surface. Beautiful bright purple or yellow specimens can be taken from the mangrove roots in the 'rivers' which connect the lagoons with the sea, near Port Royal. The right valve of dead shells is commonly washed ashore, usually with the scales worn smooth.

CHAMA SINUOSA Broderip
White Smooth-edged Jewel Box **pl. 28**
Southern Florida and the West Indies.
Description: Length 1 to 3 inches. Colour white, usually with brown maculations in a radial pattern; interior occasionally greenish. Sculpture similar to *C. macerophylla*, but the inner margins of the valves are not crenulated. The pallial line joins the bottom of the muscle scars of the deep valve; in *macerophylla* it joins at the upper outside edges.
Occurrence: Uncommon in 1 to 3 feet, on mangrove roots, in the 'rivers' which connect the lagoons to the sea just east of Port Royal. It is also reported from Negril. Abbott (*American Seashells*) says that it is a reef species, and its occurrence on mangrove roots may be very unusual.

CHAMA CONGREGATA Conrad 1833
Little Corrugated Jewel Box **pl. 28**
South-east U.S. and the West Indies.
Description: Length ½ to 1½ inches. Colour white, usually with radial rows of reddish-brown speckling. Sculpture of wavy, concentric cords, and axial folds. It is attached by the left valve to some firm surface; the right valve may have a few poorly developed scales. The shell is solid and irregular in shape.
Occurrence: Relatively common in 1 to 50 feet on both sides of the Palisadoes, at Runaway Bay, and in rocky areas all around Jamaica. The right valve is often washed ashore, usually worn smooth by the waves. In its live state, the shell may be covered with green algae which efficiently camouflages it.

CHAMA SARDA Reeve 1847 Red Jewel Box **pl. 28**
Florida Keys and the West Indies.
Description: Length ¾ to 1 inch. Colour bright red, sometimes with lighter streaks. Sculpture of irregular foliations; the internal margins are crenulated. The left valve, which is attached to a solid surface, is deeply cupped; the right valve may be almost flat.
Occurrence: As for *C. congregata*. Right valves are common on many beaches, and are conspicuous by their vivid red colour.

CHAMA FLORIDA Lamarck 1819 Florida Jewel Box **pl. 28**

South Florida to the West Indies.

Description: Length about 1 inch. Colour white, with radial rows of pink speckling. Sculpture of small, irregular scales. The lower (left) valve, which is attached to some firm surface, is much deeper than the upper valve, which may be almost flat. The shell is solid and subcircular in shape. There is a thin, yellow periostracum.

Occurrence: Very common in 2 to 50 feet of water, attached to rocks or coral, on both sides of the Palisadoes, at Discovery Bay, and in many other areas all around Jamaica. Single valves (usually the unattached ones) are found on most beaches.

Genus *Pseudochama* Odhner 1917

PSEUDOCHAMA RADIANS Lamarck 1819

Atlantic Left-handed Jewel Box **pl. 28**

Southern Florida and the West Indies.

Description: Length 1 to 3 inches. Colour white, with brown shadings. This species is a mirror image of *Chama macerophylla*, but the scale-like lamellations are not as pronounced.

Occurrence: Relatively common in 1 to 50 feet, attached to a solid surface, along the harbour side of the Palisadoes, and in protected areas near Falmouth and Negril. Our live-collected specimens came from mangrove roots in the 'rivers' connecting the lagoons with the sea near Port Royal.

Genus *Echinochama* P. Fischer 1887

ECHINOCHAMA ARCINELLA Linné 1767

True Spiny Jewel Box **pl. 28**

West Indies to Brazil.

Description: Length 1 to $1\frac{1}{2}$ inches. Colour white, with the interior of some specimens stained purplish-pink. Sculpture of 16 to 35 spiny radial ribs; in beach specimens these are usually worn smooth. The surface is pitted. The shell is solid and strong, with a heart-shaped lunule. Undamaged adult specimens are beautiful shells.

Occurrence: Relatively common in about 8 feet in Kingston harbour, near Old Harbour Bay, at Montego Bay, and in many other areas all around the Island. The shell may be attached by the right valve to some firm surface, or in muddy or sandy places it may be free.

SUPERFAMILY CARDIACEA

FAMILY CARDIIDAE

This is one of the most important, and most beautiful, of the bivalve families, with a world-wide distribution in both very deep and very shallow water. There are more than 200 species.

The family is distinguished by the graceful heart-shape of the double valves, and the mollusc's powerful foot with which it can leap several inches through the water when disturbed from its bed in the substrate. In parts of Europe 'Cockles' are an important article of food, but the West Indian species are not generally eaten. The ligament in all species is external.

SUBFAMILY TRACHYCARDIINAE

Genus *Trachycardium* Mörch 1853

TRACHYCARDIUM MAGNUM Linné 1758
Magnum Cockle **pl. 29**
Lower Florida Keys and the West Indies.
Description: Length 2 to $3\frac{1}{2}$ inches. Colour: exterior cream, with irregular patches of reddish-brown or shades of yellow; interior white, with yellow in the umbonal area. The posterior margin is yellow, merging into purple at the extreme edge. Sculpture of 32 to 35 smooth radiating ribs; those at the posterior end have small scales.
Occurrence: This is a very uncommon species in Jamaica. We have found it only at Runaway Bay, Ocho Rios and Negril, where it occurs in sand at about 20 feet. Beach specimens are rare. We have seen one live specimen trapped in a fish pot, but this must be a very unusual accident.

TRACHYCARDIUM MURICATUM Linné 1758
Yellow Cockle **pl. 29**
South-east U.S. and the West Indies.
Description: Length $1\frac{1}{2}$ to 2 inches. Colour: exterior cream, with irregular patches of brownish-red, and shades of yellow; interior usually white, sometimes extensively tinted with yellow and purple. Sculpture of 30 to 40 radiating ribs with small, tooth-like scales.
Occurrence: This is the most common member of the family in Jamaica. It occurs in 1 to 10 feet in sandy or muddy areas all around the Island, and dead shells are found on most beaches. Fine, live specimens can be collected in the 'rivers' which connect the lagoons with the sea near

Port Royal – a vigorous raking of the bottom quickly brings them to the surface of the mud.

TRACHYCARDIUM ISOCARDIA Linné 1758
Prickly Cockle **pl. 29**
West Indies to the northern coast of South America.
Description: Length 2 to 3 inches. Colour: exterior cream, with irregular reddish-brown mottling; interior white with a broad purple band leading to the umbones. Sculpture of 31 to 37 radiating ribs, covered with sharp, imbricated scales. In general shape the shell is very similar to *T. muricatum*. It is an unusually beautiful species.
Occurrence: As for *T. muricatum*, but it is relatively uncommon, and its distribution is not as widespread. We have found it in fair numbers in shallow water at the northern end of the beach at Discovery Bay.

Genus *Papyridea* Swainson 1840

PAPYRIDEA SOLENIFORMIS Bruguière 1789
Spiny Paper Cockle **pl. 29**
South-east U.S. and the West Indies.
Description: Length 1 to $1\frac{3}{4}$ inches. Colour cream, with irregular violet or pinkish mottling. Sculpture of 40 to 55 very fine, oblique, radiating ribs, which are covered with minute spines. The shell is fragile and gapes at both ends. There are about 10 serrations on the posterior half of the valves. This is a delicate and attractive species.
Occurrence: Relatively common in 1 to 6 feet in Kingston harbour, at Negril, and in sandy areas all around the Island. Beautiful specimens can be found in the 'rivers' which connect the lagoons with the sea near Port Royal.
Forms: There is a rare yellowish-orange colour form, which occurs at Port Royal, and in the Great Salt Pond (St Catherine).

PAPYRIDEA SEMISULCATA Gray 1825
Frilled Paper Cockle **p. 29**
Southern Florida to the West Indies.
Description: Length about $\frac{1}{2}$ inch. Colour variable, but usually white, pink, or orange-yellow. Sculpture of about 30 fine, oblique, radiating ribs; there are a number of sharp serrations on the posterior of both valves. The shell is thin and very fragile.
Occurrence: Relatively uncommon in 6 to 25 feet at Negril, Montego Bay, Ocho Rios harbour, and in sandy areas all around the Island. Single valves occasionally wash ashore, and can be found in drift sand.

SUBFAMILY FRAGINAE

Genus *Americardia* Stewart 1830

AMERICARDIA GUPPYI Thiele 1910

Guppy's Strawberry Cockle **pl. 29**

Bahamas, Florida and the Caribbean.

Description: Length about ½ inch. Colour cream, usually with brown or purple maculations. Sculpture of 26 to 29 radial ribs. The shell is small but robust.

Occurrence: Uncommon in Kingston and Ocho Rios harbours, in mud and sand, at about 20 feet. Dead shells are washed ashore on the eastern side of the causeway across Kingston harbour.

AMERICARDIA MEDIA Linné 1758

Atlantic Strawberry Cockle **pl. 29**

South-east U.S. and the West Indies.

Description: Length 1 to 2 inches. Colour white, with irregular orange-brown markings. Sculpture of 33 to 37 slightly scaly radial ribs. The posterior slope of the valves is flattened and concave. The shell is strong, inflated, and has a quadrate outline.

Occurrence: This is a relatively uncommon species in Jamaican waters. It occurs in 1 to 15 feet at Port Royal, Negril, and in Ocho Rios harbour. Live specimens can sometimes be found in the 'rivers' which connect the lagoons with the sea near Port Royal.

SUBFAMILY LAEVICARDIINAE

Genus *Laevicardium* Swainson 1840

LAEVICARDIUM LAEVIGATUM Linné 1758

Common Egg Cockle **pl. 29**

South-east U.S. and the West Indies.

Description: Length 1 to 2 inches. Colour variable, but usually yellow with orange; many specimens have purple or brown mottling. The shell is smooth, polished, and light in structure. It is an unusually attractive species.

Occurrence: Abundant on the beaches at Negril. It is less common along the north coast, and at Old Harbour Bay. Live shells occur in sand or mud at about 10 feet, often in beds of eel grass.

Forms: There is a distinct, larger, dull brown form with a sculpture of fine radial lines. It is not uncommon at Farquhars Beach (Clarendon), Old Harbour Bay and in Kingston and Ocho Rios harbours, in depths up to 20 feet. This form may merit sub-specific status.

LAEVICARDIUM PICTUM Ravenel 1861
Ravenel's Egg Cockle
South-east U.S. and the West Indies.
Description: Length ½ to 1 inch. Colour white or cream with delicate rose shadings; occasionally with brown zigzag lines. Shell smooth and polished with very low umbones set near the anterior end.
Occurrence: This is a rare species in Jamaica. It occurs in sand in about 25 feet in Ocho Rios harbour and possibly at Negril. It is known to occur also in deep water.

LAEVICARDIUM SYBARITICUM Dall 1886
Dall's Egg Cockle
South-east U.S. and the West Indies.
Description: Length ½ to 1 inch. Colour brownish-white, with deep pink umbones. Shell smooth and polished; it is squarish in shape and inflated.
Occurrence: This is an uncommon species in Jamaica. It occurs in about 20 feet in Kingston harbour and at Negril. Beach specimens are occasionally found at Fort Augusta.

SUPERFAMILY VENERACEA

FAMILY VENERIDAE

This is a major group of strong, brightly coloured, equivalve shells, some of which are edible. There are over 400 species with a world-wide distribution. The family is characterised by the distinctive hinge, which bears interlocking lateral and cardinal teeth.

Many Jamaican species occur abundantly in sand or mud all around the Island, and the unusual and beautiful Royal Comb Venus (*Pitar dione*) is commonly found on the south coast.

SUBFAMILY VENERINAE

Genus *Antigona* Schumacher 1817

Subgenus *Periglypta* Jukes-Brown 1914

ANTIGONA LISTERI Gray 1838 Princess Venus **pl. 30**
South-east Florida and the West Indies.
Description: Length 2 to 4 inches. Colour cream, with irregular brown maculations; the cream interior usually has a purplish-brown stain on the posterior muscle scar. Sculpture of many radial riblets, crossed by sharp concentric cords, giving a rough, cancellate surface. The shell is heavy and strong.
Occurrence: Relatively common on mud or sand bottoms, in Kingston harbour, Discovery Bay, at Negril, and many other areas, in depths of 1 to 15 feet. Fine specimens can be collected in the 'rivers' connecting the lagoons with the sea, just east of Port Royal. The big shells which occur near Port Rhoades in Discovery Bay are (with the exception of the sessile *Spondylus americanus*) the heaviest Jamaican bivalves.

Subgenus *Ventricolaria* Keen 1954

ANTIGONA RIGIDA Dillwyn 1817 Rigid Venus **pl. 30**
Florida Keys to Brazil.
Description: Length 2 to 2½ inches. Colour cream with brown mottling. Sculpture of numerous, raised, concentric ribs, between which are 1 to 3 finer concentric threads. The lunule is deeply impressed, and the shell is strong, inflated, and almost circular in outline.
Occurrence: As for *A. listeri*. The two species often occur together, and both can be collected in the 'rivers' which connect the lagoons to the sea, just east of Port Royal.

Genus *Chione* Mühlfeld 1811

Subgenus *Chione* Mühlfeld 1811

CHIONE CANCELLATA Linné 1767
Cross-barred Venus **pl. 30**
South-east U.S., Texas and the West Indies.
Description: Length 1 to 1¾ inches. Colour cream, extensively overlaid with irregular brown rays; some specimens have a purplish cast. The

interior is white, usually with deep purple patches. Sculpture of sharp, raised, concentric ridges, and very weak radial riblets.

Occurrence: Very common in mud or sand, in 1 to 6 feet at Old Harbour Bay, Discovery Bay, in Kingston harbour, and in many other areas around the Island. Dead shells are found on the beaches along the harbour side of the Palisadoes; live specimens often come to the surface when the mud flats are exposed by low tides at Old Harbour Bay.

CHIONE GRANULATA Gmelin 1791 Beaded Venus **pl. 30**

West Indies.

Description: Length ¾ to 1¼ inches. Colour variable, usually yellowish-cream, with extensive brown or purple maculations. Some specimens may be all cream, and others all brown; most have interior purple markings. Sculpture of numerous radiating ribs, crossed by concentric ridges, giving a granulated effect.

Occurrence: As for *C. cancellata*, with which it is often found. Both species are equally common in Jamaica, and occur in large numbers in Kingston harbour.

CHIONE INTAPURPUREA Conrad 1849

Lady-in-waiting Venus **pl. 30**

North Carolina, the Gulf States, and the West Indies.

Description: Length 1 to 1½ inches. Colour white, usually with irregular brown or orange maculations; interior sometimes with reddish-purple splotches. Sculpture of numerous sharp, concentric ridges, which are serrated on the lower edge.

Occurrence: This is an uncommon species in Jamaica. Our specimens come from Old Harbour Bay and Kingston harbour; it occurs in moderate numbers, in about 15 feet off Fort Augusta, and dead valves are occasionally washed ashore. We did not find it in the extensive commercial dredgings off the Bogue Islands and in Ocho Rios harbour.

Subgenus *Timoclea* Brown 1827

CHIONE PYGMAEA Lamarck 1818

White Pygmy Venus **pl. 30**

Florida and the West Indies.

Description: Length about ½ inch. Colour white, speckled with brown; the beaks are usually pink. Sculpture coarsely cancellate. The shell is not unlike a miniature *Antigona listeri*.

Occurrence: Relatively common in sand, in 3 to 6 feet, at Negril, the Port Royal Cays, Prospect, and in sandy areas all around the Island. Dead shells are found on many beaches, particularly at Robins Bay.

Subgenus *Lirophora* Conrad 1862

CHIONE PAPHIA Linné 1767 King Venus **pl. 30**
Lower Florida Keys and the West Indies.
Description: Length 1 to 1½ inches. Colour white or cream, mottled with purple-brown. Interior white, usually with purple splotches. Sculpture of 10 to 12 broad, concentric ribs, which are thin and sharp at both ends.
Occurrence: This is a common species in Jamaica. Our specimens came from about 3 feet of water in Kingston harbour (in the eel grass beds off the old runway), and single valves are occasionally found on the beaches at Negril and Old Harbour Bay. Unusually large and colourful examples are often washed ashore on the Bogue Islands.

CHIONE LATILIRATA Conrad 1841 Imperial Venus **pl. 30**
North Carolina to Brazil.
Description: Length 1 to 1½ inches. This species is very similar to *C. paphia*, but has 5 to 7 broad, bulbous, concentric ribs, which are not flattened at the ends.
Occurrence: This species is reported from about 80 feet off the Morant Cays. The record needs confirmation, but as the species is known from as close as Puerto Rico, it may well be accurate. Our specimens come from the southern Caribbean in about 120 feet, and were recovered by shrimp trawlers.

Genus *Anomalocardia* Schumacher 1817

ANOMALOCARDIA BRASILIANA Gmelin 1791
West Indian Pointed Venus **pl. 30**
West Indies to Brazil.
Description: Length ¾ to 1¼ inches. Colour yellowish-cream, generally strongly speckled with purple or brown; the interior may bear extensive patches of deep purple. Sculpture of numerous concentric ribs. The inner margins of the valves are crenulated. The shell is distinguished by the unusual extension of the posterior edge of the valves.
Occurrence: This species is locally common, but not widespread in its distribution. It occurs lightly buried in the mud in 3 to 15 feet in Kingston harbour, and dead shells are very common on the beach between Port Henderson and Fort Augusta, and also in the Great Salt Pond (St Catherine).

SUBFAMILY MERETRICINAE

Genus *Transennella* Dall 1883

TRANSENNELLA CONRADINA Dall 1883
Conrad's Transennella
Southern Florida and the West Indies.
Description: Length about ½ inch. Colour white, with zigzag markings of pale brown or orange. Sculpture of fine growth lines; the surface of the shell is polished. There are three cardinal teeth in each valve, and the inner ventral margin has oblique grooves. The shell is relatively thin but strong.
Occurrence: Moderately common in Ocho Rios harbour and at Negril, in about 20 feet in sand. Single valves are occasionally washed ashore on the Port Royal Cays, but it is not a common species on the south coast.
Note: All Transennellas are distinguished from other similar bivalves by the oblique grooves on the inner ventral margins.

Genus *Tivela* Link 1807

TIVELA MACTROIDES Born 1778 Trigonal Tivela **pl. 30**
West Indies to Brazil.
Description: Length 1 to 1½ inches. Colour white, with broad rays of orange-brown; near the umbones, the colour is purple-blue. The interior is occasionally flushed with purple, particularly towards the beaks. Sculpture of fine, irregularly spaced growth lines; the surface is smooth and polished. The beaks are central and prominent.
Occurrence: This species is locally common, but restricted in its distribution. Dead shells are often found on the beaches at Negril, where the live shells occur in sand in about 6 feet of water, and it also occurs in Kingston harbour and west to Alligator Pond.
Note: (i) There is an all-brown (occasionally greenish) variety which occurs at Ocho Rios and in Kingston harbour. Apart from the difference in colour, these shells are narrower than typical *mactroides* and may be a distinct form.

(ii) The many shells we have found at Negril are all less than ¾ inch long, and very brightly coloured in shades of purple, brown and white. They do not appear to be juveniles, and may represent a distinct sub-species. They may be identical with *T. floridana* Rehder.

TIVELA ABACONIS Dall 1902 Abaco Tivela
Bahamas, The Caribbean and Vera Cruz.
Description: Length ½ to ¾ inch. Very similar to *T. mactroides*, but it is white with pink towards the umbones, semi-translucent, and has coarser growth lines.
Occurrence: This is a relatively rare species in Jamaica. Our few specimens come from dredgings in about 35 feet, off the Bogue Islands and in Ocho Rios harbour.

SUBFAMILY PITARINAE

Genus *Pitar* Römer 1857

Subgenus *Pitar* Römer 1857

PITAR ARESTA Dall and Simpson 1901
West Indian Venus **pl. 30**
West Indies.
Description: Length 1½ to 2 inches. Colour white, sometimes with a greyish tinge. Sculpture of numerous, irregular growth lines, which become more prominent towards the margins of the valves. The shell is solid, inflated and inequilateral.
Occurrence: This is a very uncommon species in Jamaica. It occurs in mud at about 20 feet off the Bogue Islands, and we found it in the recent extensive commercial dredgings in Kingston harbour. Dead shells are reported from Farquhars Beach (Clarendon).

PITAR ALBIDA Gmelin 1791 White Venus **pl. 30**
West Indies.
Description: Length 1 to 2 inches. Colour white, very occasionally with orange rays. Sculpture of numerous, fine, irregular growth lines. The anterior end is slightly narrowed, and smoothly rounded; the posterior end is much broader but also rounded. The surface of fresh specimens is relatively smooth and polished.
Occurrence: Common at low tide exposed on the grass flats on the harbour side of the Palisadoes. Elsewhere, it occurs in mud or sand, at about 8 feet in Old Harbour Bay and at Negril, and beach specimens are often found there.

PITAR FULMINATA Menke 1828 Lightning Venus **pl. 30**
South-east U.S. and the West Indies.
Description: Length ½ to 1½ inches. Colour white, with brown or orange zigzag markings, or spots of colour usually in radial bars. Sculpture of

numerous, fine, raised concentric cords. The large lunule is bordered by an impressed line. Some specimens have a purple line on one or both sides of each umbone.
Occurrence: Common in mud or sand at 10 to 40 feet in Kingston harbour, Old Harbour Bay, Discovery Bay and at Ocho Rios, often within the *thalassia* beds. Single valves are found on the beach at Negril.
Note: We have found no Jamaican specimens as large as ¾ inch, and most have been under ½ inch. Possibly they represent a dwarf race of the typical variety.

Subgenus *Hysteroconcha* P. Fischer 1887

PITAR DIONE Linné 1758 Royal Comb Venus **pl. 30**
Texas to Panama, and the West Indies.
Description: Length 1 to 1¾ inches. Colour purplish-white. Sculpture of numerous sharp, concentric ridges. The shell is characterised by two radial rows of long spines at the posterior end of each valve. This species is perhaps the most beautiful of an unusually attractive family.
Occurrence: This is an uncommon species, with a relatively widespread distribution in Jamaica. Single valves are found on the beaches of the southern Palisadoes, at Alligator Pond, Black River, Port Maria, and several other places on both the north and south coasts. It is very much more common at Farquhars Beach (Clarendon) where the live shell lives buried in the mud at about 8 feet, and dead but complete shells are often washed ashore.
Forms: There is a white and pale yellow colour form which we have found only in Kingston harbour. Our specimens were dredged in about 25 feet just east of Fort Augusta.

PITAR CIRCINATA Born 1778 Purple Venus **pl. 30**
West Indies.
Description: Length 1 to 1¾ inches. Colour all-white, all-purple, or a combination of both. Sculpture and shape are very similar to *P. dione*, but it lacks the spines on the posterior ends of the valves. It is a strikingly attractive species.
Occurrence: This is an uncommon shell in Jamaica. It occurs in sand, in about 20 feet, in Ocho Rios and Kingston harbours, and dead shells have been reported from the beaches at Old Harbour Bay, Black River and Negril. Our specimens were dredged just east of Fort Augusta (in Kingston harbour), and collected on the shore at Farquhars Beach (Clarendon), where it is relatively common.

SUBFAMILY CIRCINAE

Genus *Macrocallista* Meek 1876

MACROCALLISTA MACULATA Linné 1758

Calico Clam **pl. 30**

South-east U.S. and the West Indies.

Description: Length $1\frac{1}{2}$ to $3\frac{1}{2}$ inches. Colour cream, with reddish-brown markings, usually in a checkerboard pattern. The shell is solid and strong with a smooth, highly polished surface. A thin, pale yellow periostracum is present in fresh specimens. It is a colourful and attractive species.

Occurrence: Relatively common in mud or sand in about 6 feet, at Negril, Discovery Bay, in Kingston harbour, and many other areas all around the Island. The stout ligament often keeps both valves together when the shell is washed ashore.

Forms: There is a rare, all-white albino form which we have found only at Negril. Occasional shells are all-brown.

SUBFAMILY DOSINIINAE

Genus *Dosinia* Scopoli 1777

Subgenus *Dosinidia* Dall 1902

DOSINIA CONCENTRICA Born 1778

Southern Dosinia **pl. 30**

Cuba south to Brazil.

Description: Length 2 to 3 inches. Colour yellowish-white. Sculpture of numerous, regular concentric ridges. The surface is relatively smooth, and the valves are compressed and glossy.

Occurrence: This is not a common species in Jamaica. It occurs in 4 to 30 feet in Kingston harbour, off the Bogue Islands, and in Ocho Rios harbour. Single valves are rarely washed ashore. In the southern Caribbean, shrimp trawlers bring them up from depths down to 150 feet.

Genus *Cyclinella* Dall 1902

CYCLINELLA TENUIS Récluz 1852

Atlantic Cyclinella **pl. 30**

Eastern U.S. and the West Indies.

Description: Length 1 to 2 inches. Colour dull white. Sculpture of

numerous irregular growth lines. The shell is thin and almost circular in outline.
Occurrence: As for *Dosinia concentrica* (the preceding species). It is relatively uncommon in about 6 feet of water, along the harbour side of the Palisadoes. Beach specimens are very uncommon.

FAMILY PETRICOLIDAE

Juvenile members of this family bore their way into hard substances such as coral, rock, or clay, and remain in these burrows for the rest of their lives. The shells are thin, white, and more or less oval.

Most species are relatively common, and the family is widely distributed in shallow, tropical seas.

SUBFAMILY PETRICOLINAE

Genus *Petricola* Lamarck 1801

PETRICOLA LAPICIDA Gmelin 1791 Boring Petricola
Southern Florida and the West Indies.
Description: Length ½ to 1 inch. Colour white. Sculpture of numerous, branching, radial lines, and concentric growth lines. Abbott (*Caribbean Seashells*) points out that this is not sculpture on the true surface of the shell, but is part of a calcareous coating. The shell is ovate and moderately inflated.
Occurrence: This species is a coral borer, and occurs from 1 to 20 feet, wherever coral is found around Jamaica. Our specimens come from Kingston harbour, just west of Port Royal.

Genus *Rupellaria* Fleuriau 1802

RUPELLARIA TYPICA Jonas 1844
Atlantic Rupellaria **pl. 30**
South-east U.S. and the West Indies.
Description: Length about 1 inch. Colour: exterior greyish-white; interior stained with buff or brown. Sculpture of numerous coarse, irregularly-spaced radial ribs. The shell is moderately compressed and gapes posteriorly.
Occurrence: As for *Petricola lapicida* (the preceding species). It is less common, but has an equally wide distribution.

Plate 25 THORNY OYSTERS, PEN SHELLS

Scale: reduced by 65%

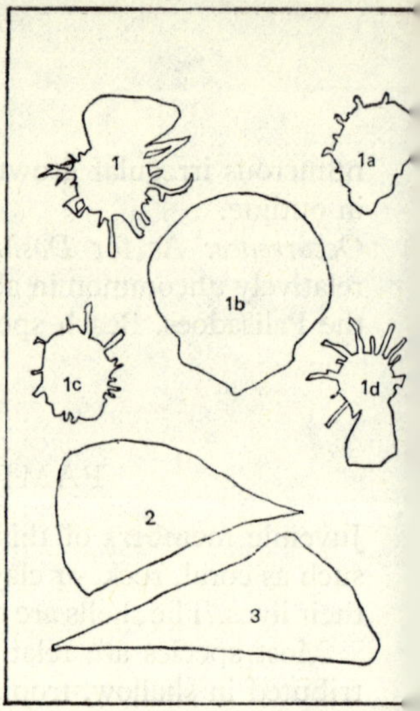

1. **Spondylus americanus** Hermann 229
Atlantic Thorny Oyster
Note attachment of lower valve to pebble. **1a,** dark red colour variety. **1b,** large, short-spined variety taken from a coral reef. **1c,** yellow and white colour variety. **1d,** young specimen taken from a pebble-strewn mud slope exposed to the open sea in 40 feet.

2. **Atrina seminuda** Lamarck 220
Spiny Pen Shell
Note that the profile is proportionately broader than **3.**

3. **Pinna carnea** Gmelin 220
Amber Pen Shell
This species may, or may not, be spinose.

Plate 26 SCALLOPS

Scale: reduced by 15%

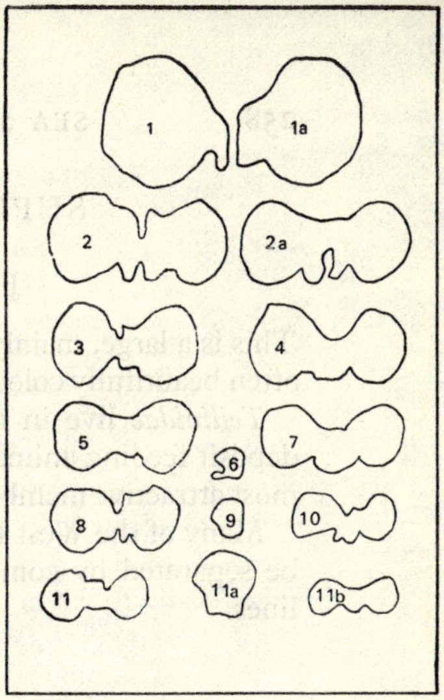

1. **Chlamys effluens** 226
Dall
Note young worm shells on valve at right.

2. **Aequipecten muscosus** 228
Wood Rough Scallop
2a, orange colour variety

3. **Aequipecten lineolaris** 228
Lamarck Wavy-lined Scallop

4. **Chlamys sentis** Reeve 223
Sentis Scallop

5. **Chlamys** species 226
We have been unable to identify this shell positively; it could be a form of **11.**

6. **Leptopecten bavayi** 227
Dautzenberg Bavay's Scallop

7. **Chlamys ornata** 223
Lamarck Ornate Scallop

8. **Aequipecten gibbus** 229
Linné Calico Scallop
Typical colour variety.

9. **Chlamys benedicti** 226
Verrill and Bush
Benedict's Scallop

10. **Chlamys imbricata** 226
Gmelin Little Knobby Scallop

11. **Lyropecten antillarum** Récluz 227
Antillean Scallop
Red and white colour variety. **11a,** 'Peacock's tail' colour variety. **11b,** orange, black and white colour variety.

SUPERFAMILY TELLINACEA

FAMILY TELLINIDAE

This is a large, mainly tropical family, with elongate, rather fragile, and often beautifully coloured shells. The ligaments are external.

Tellinidae live in the substrate of sandy or muddy areas, and the deposit feeding animals have two long, separate siphons. Several of the most attractive members of the family occur in Jamaican waters.

Many of the West Indian species closely resemble each other, but can be separated by comparison of their internal muscle scars and pallial lines.

Genus *Tellina* Linné 1758

Subgenus *Tellina* Linné 1758

TELLINA RADIATA Linné 1758 Sunrise Tellin **pl. 31**
South-east U.S. and the West Indies.
Description: Length 2 to 4 inches. Colour white, with rays of pink, orange-red or yellow. The umbones are usually red, and the interior is stained yellow. The shell is smooth and highly polished. This is an exceptionally beautiful species, which is well known to collectors all over the world; it is widely used in the manufacture of shell jewellery.
Occurrence: Relatively common in 5 to 15 feet, in sandy areas all along the north and west coasts. We have found it in Ocho Rios and Discovery Bay harbours, at Braco and Florida (Trelawny) beaches, and at Montego Bay and Negril. (It is possible to collect them at these places by diving to the bottom and vigorously raking the sand with one's fingers.) Beach specimens are not uncommon, but exposure to the sun fades their exquisite colours.
Forms: There is a yellowish-white colour form with red umbones called *T. radiata unimaculata* Lamarck. In some areas (such as Negril) this form outnumbers the typical variety.

TELLINA LAEVIGATA Linné 1758 Smooth Tellin **pl. 31**
Southern Florida and the West Indies.
Description: Length 2 to 3 inches. Colour white, with faint rays or bands of yellow or orange. Sculpture of microscopic growth lines; the surface is smooth to the touch.
Occurrence: This is a rare species in Jamaica. We have found it (dead) off Rackhams Cay (Port Royal Cays) in 10 feet, and on the beach at

Negril. Extensive dredging at the Bogue Islands and Ocho Rios suggests that it does not occur in those areas.

Subgenus *Tellinella* Mörch 1853

TELLINA LISTERI Röding 1798 Speckled Tellin **pl. 31**
South-east U.S. and the West Indies.
Description: Length 2 to 3½ inches. Colour whitish, with irregular specklings of brown or purple-brown; interior flushed with yellow. Sculpture of numerous, evenly spaced, concentric threads. This is an unusually attractive species.
Occurrence: Relatively common in sand or mud, in 1 to 150 feet, at Priory, Runaway Bay, Negril, the Port Royal Cays, and many other areas all around the Island. Beach specimens are not uncommon in sandy places.

Subgenus *Angulus* Mühlfeld 1811

TELLINA MERA Say 1834 Mera Tellin
Florida, the Bahamas, Puerto Rico and Jamaica.
Description: Length about ¾ inch. Colour white. Sculpture of fine irregular concentric lines most noticeable near the margins. The shell has a distinctive blunt posterior end.
Occurrence: This is an uncommon species in Jamaica. It occurs in sand at depths of about ten feet at Robins Bay and in Ocho Rios harbour. We have found one specimen living in only six inches of water in The Great Salt Pond (St Catherine), but this is most unusual.

Subgenus *Scissula* Dall 1900

TELLINA SIMILIS Sowerby 1806 Candy Stick Tellin
Florida and the West Indies.
Description: Length about 1 inch. Colour white, often tinted with yellow. Sculpture of fine, concentric threads, which cross the growth lines at an oblique angle.
Occurrence: Uncommon in about 6 feet at Old Harbour Bay, off the Port Royal Cays, and in the vicinity of the Bogue Islands. We have found beach specimens at Negril.

TELLINA CARIBAEA Orbigny 1842 Caribbean Tellin **pl. 31**
West Indies.
Description: Length about 1 inch. Colour white, pink, or yellowish. Sculpture of minute concentric lines, and fine oblique grooves. The shell is thin, polished, and moderately inflated.
Occurrence: Beach specimens are relatively uncommon at Ocho Rios, Discovery Bay and Negril. We have found no living shells, but there are indications that it occurs in about 10 feet, in sandy areas, in several places on the north and south coasts.

TELLINA EXILIS Lamarck 1818 Pink Thin Tellin **pl. 31**
Bermuda and the West Indies.
Description: Length $\frac{1}{2}$ to $\frac{3}{4}$ inch. Colour pink. There is an external oblique sculpture, and at the posterior end, there are two weak, white internal radial ribs. The shell is thin and translucent.
Occurrence: As for *T. caribaea.* It is more common, however, and its distribution in Jamaica appears to be widespread on all coasts. Perfect beach specimens are abundant at Port Morant.

Subgenus *Scrobiculina* Dall 1900

TELLINA MAGNA Spengler 1798 Great Tellin **pl. 31**
South-east U.S. and the West Indies.
Description: Length 3 to $4\frac{1}{2}$ inches. Colour: left valve white, rarely yellowish; right valve orange or pink. The shell is compressed, smooth, and glossy. This beautiful species is easily the largest of our local Tellins.
Occurrence: This is a very rare species in Jamaica, and far from common in any other area. Our specimens were dredged in about 30 feet in Ocho Rios harbour: we found no trace of it in the large commercial dredgings from the same area 1 year later, and have no south coast records. It occurs in mud at Montego Bay, in 40 feet.

Subgenus *Eurytellina* P. Fischer 1887

TELLINA GUILDINGII Hanley 1844
The Jamaican Tellin **pl. 31**
Bermuda, the Bahamas, and south to Barbados.
Description: Length 1 to $1\frac{3}{4}$ inches. Colour white, usually with red rays, which become more distinct near the margin. The upper half of the shell is faintly suffused with yellow. Sculpture of numerous, microscopic, concentric threads; the shell is smooth and quite well polished.

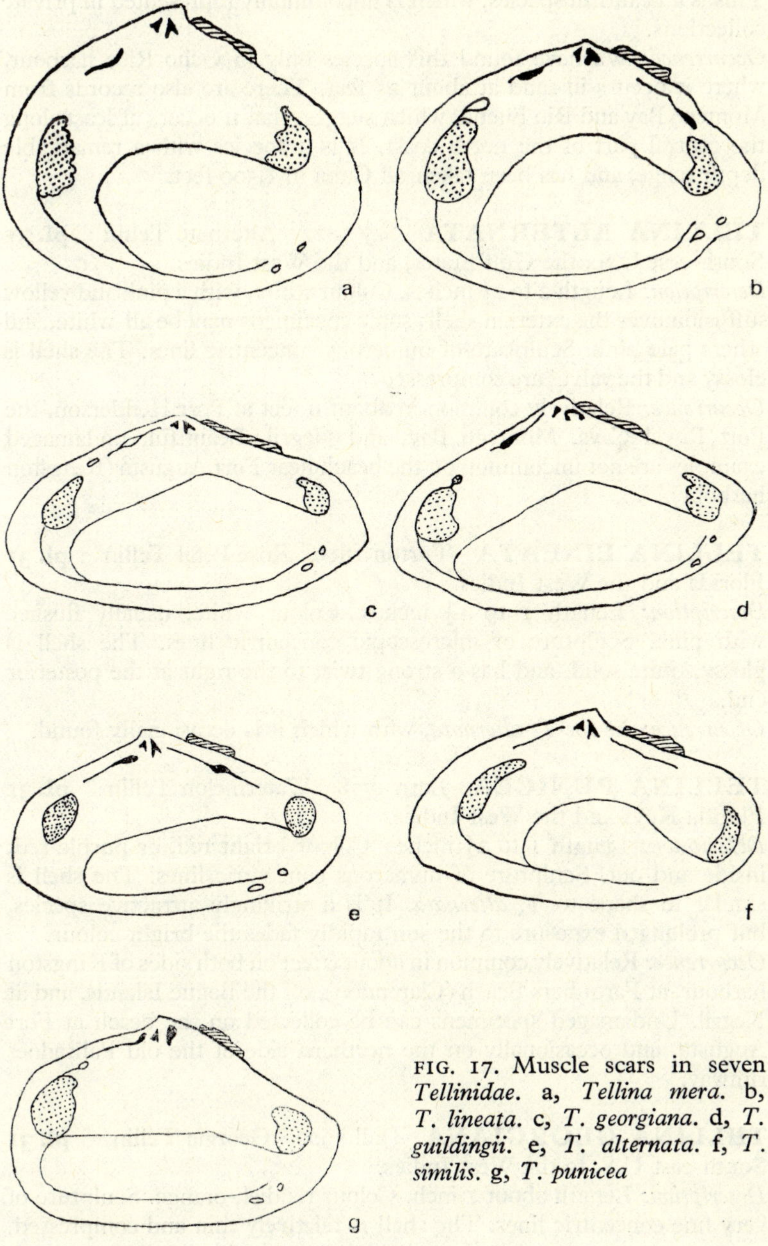

FIG. 17. Muscle scars in seven *Tellinidae*. a, *Tellina mera*. b, *T. lineata*. c, *T. georgiana*. d, *T. guildingii*. e, *T. alternata*. f, *T. similis*. g, *T. punicea*

This is a beautiful species, which is uncommonly represented in private collections.

Occurrence: We have found this species only in Ocho Rios harbour, where it occurs in sand at about 25 feet. There are also records from Montego Bay and Rio Bueno, which suggest that it occurs at least along the central part of the north coast. It is a species with a remarkable depth range, and has been taken off Cuba in 1,500 feet.

TELLINA ALTERNATA Say 1822 Alternate Tellin **pl. 31**

South-east U.S., the Gulf States, and the West Indies.

Description: Length 2 to $2\frac{1}{2}$ inches. Colour white, with a pink and yellow suffusion over the external shell; some specimens may be all-white, and others pale pink. Sculpture of numerous concentric lines. The shell is glossy and the valves are compressed.

Occurrence: Relatively common in about 6 feet at Port Henderson, the Port Royal Cays, Montego Bay, and Negril. Beautiful, undamaged examples are not uncommon on the beach near Fort Augusta (Kingston harbour).

TELLINA LINEATA Turton 1819 Rose Petal Tellin **pl. 31**

Florida and the West Indies.

Description: Length 1 to $1\frac{1}{2}$ inches. Colour white, usually flushed with pink. Sculpture of microscopic concentric lines. The shell is glossy, quite solid, and has a strong twist to the right at the posterior end.

Occurrence: As for *T. alternata*, with which it is occasionally found.

TELLINA PUNICEA Born 1778 Watermelon Tellin **pl. 31**

Florida Keys and the West Indies.

Description: Length 1 to $2\frac{1}{4}$ inches. Colour bright red, or purple-red, inside and out. Sculpture of numerous concentric lines. The shell is similar in shape to *T. alternata.* It is a strikingly attractive species, but prolonged exposure to the sun rapidly fades the bright colour.

Occurrence: Relatively common in about 6 feet on both sides of Kingston harbour, at Farquhars Beach (Clarendon), off the Bogue Islands, and at Negril. Undamaged specimens can be collected on the beach at Fort Augusta, and occasionally on the northern side of the old Palisadoes runway.

TELLINA GEORGIANA Dall 1900 Georgia Tellin **pl. 31**

South-east U.S. to the West Indies.

Description: Length about 1 inch. Colour reddish-orange. Sculpture of very fine concentric lines. The shell is relatively thin and compressed.

It can be separated from *T. punicea* and *T. alternata* by its colour and smaller size.
Occurrence: As for *T. punicea.* It is an uncommon species in Jamaica, except in Ocho Rios harbour, and we have found only dead shells.

Genus *Arcopagia* Brown 1827

ARCOPAGIA FAUSTA Pulteney 1799 Faust Tellin **pl. 31**
South-east U.S. and the West Indies.
Description: Length 2 to 4 inches. Colour yellowish-white; the yellow coloration is more pronounced on the inside. Sculpture of irregular growth lines. The shell is strong and heavy with a yellow-brown periostracum, which quickly flakes off when the animal dies.
Occurrence: Very common in 1 to 6 feet, in sandy or muddy areas, all around Jamaica. Dead but undamaged shells are found on the beaches on both sides of Kingston harbour, at Negril, and in many areas along the north coast. In Barbados, single valves are mounted on sharks' vertebrae, and brightly coloured pictures of local scenes are painted on the interior surface of the shells. They are then sold in the form of ash trays, as souvenirs of the Island.
Note: This is a favourite food of *Octopus vulgaris,* and empty shells surround the entrances to octopus holes.

Genus *Strigilla* Turton 1822

STRIGILLA CARNARIA Linné 1758 Large Strigilla **pl. 31**
South-east U.S. and the West Indies.
Description: Length $\frac{3}{4}$ to 1 inch. Colour usually deep pink inside and out; the exterior may be very pale in some specimens, and is usually white if the shell has been exposed to the sun for some weeks. Sculpture of oblique, radial lines in the central and posterior regions of the valves; at the anterior end, the lines are wavy, and run in the opposite direction. The shell is thin and relatively fragile.
Occurrence: Very common in about 6 feet, in sandy areas, all around Jamaica. Dead shells are found along the southern side of the Palisadoes, at Port Antonio, Discovery Bay and Negril. This species is widely used in the manufacture of shell jewellery.
Note: S. rombergi Mörch 1853 is very similar, but is less pink and the pallial sinus does not reach the anterior adductor muscle as is the case in *carnaria.*

Genus *Macoma* Leach 1819

Note: Macomas are modified Tellins, but can be distinguished by their chalky shells and lack of lateral teeth.

MACOMA CONSTRICTA Bruguière 1792
Constricted Macoma **pl. 31**
Florida to Texas and the West Indies.
Description: Length 1 to 2½ inches. Colour white; there is a thin, grey periostracum. Sculpture of fine concentric growth lines. The posterior end is twisted to the right.
Occurrence: Relatively common in 1 to 6 feet in Kingston harbour, at Runaway Bay, and at Negril. Dead shells are uncommonly washed ashore on other beaches on the north and south coasts and the species is particularly common in the Great Salt Pond (St Catherine).

MACOMA BREVIFRONS Say 1834 Short Macoma
New Jersey to Brazil.
Description: Length 1 to 1½ inches. Colour white with an orange flush on the umbones, which are also iridescent. Sculpture of numerous fine growth lines. This species is very similar to *M. constricta*.
Occurrence: As for *M. constricta* (the preceding species). This is not a common shell in Jamaica. Beach specimens are found at Farquhars Beach (Clarendon), and the live shell occurs in mud at about 25 feet in Kingston harbour.

MACOMA TAGELIFORMIS Dall 1900
Tagelus-like Macoma **pl. 31**
West Indies.
Description: Length 1 to 2 inches. Colour dull white. Sculpture of numerous fine growth lines, which start below the umbones. The shell is relatively thick and robust.
Occurrence: As for *M. constricta*, with which it is occasionally found, but it is a less common species.

MACOMA AURORA Hanley 1844 Hanley's Macoma **pl. 31**
Florida to Brazil.
Description: Length about ¾ inch. Colour white, flushed with orange. The shell is thin but quite strong, smooth and polished. The posterior end is short and slightly twisted to the right.
Occurrence: Uncommon in 5 to 15 feet in Kingston and Ocho Rios harbours. Dead shells are occasionally found on the beach along the

eastern side of the causeway in Kingston harbour, especially after heavy weather.

MACOMA TENTA Say 1834 Tenta Macoma **pl. 31**
Cape Cod to Florida, and the West Indies.
Description: Length ½ to 1 inch. Colour white, with a delicate iridescence. The shell is smooth and fragile.
Occurrence: Common in muddy areas, in 5 to 20 feet, in Kingston harbour, Ocho Rios, and off the Bogue Islands. We have found beach specimens only at Farquhars Beach (Clarendon).

MACOMA PSEUDOMERA Dall and Simpson 1901
Puerto Rico Macoma **pl. 31**
Puerto Rico and Jamaica.
Description: Length about ¾ inch. Colour white. The shell is smooth and characterised by its short, truncate posterior end. In this respect, it resembles *Tellina mera*.
Occurrence: Uncommon in sand or mud, in about 6 feet, at Montego Bay and Ocho Rios. Beach specimens are occasionally found along the southern side of the Palisadoes and at Robins Bay.

Genus *Apolymetis* Salisbury 1929

APOLYMETIS INTASTRIATA Say 1826
Atlantic Grooved Macoma **pl. 31**
Southern Florida and the Caribbean.
Description: Length 2 to 3 inches. Colour white. There is a strong radial rib at the posterior end of the right valve, and a radial groove at the same end of the opposite valve. The shell is thin and brittle.
Occurrence: Relatively common lightly buried in the substrate, in about 6 feet, in Kingston harbour, Ocho Rios, Negril, and many other muddy or sandy areas all around the Island. Dead shells are not uncommon in shallow water off the old Palisadoes runway.

FAMILY SEMELIDAE

These are rather small, rounded, thin-shelled bivalves, some of which are delicately tinted with a variety of attractive colours. Most West Indian species prefer shallow, protected water, where they burrow into the mud or sand of the sea floor. All have external ligaments.

Genus *Semele* Schumacher 1817

SEMELE PROFICUA Pulteney 1794
White Atlantic Semele **pl. 32**
South-east U.S. and the West Indies.
Description: Length $\frac{3}{4}$ to $1\frac{1}{2}$ inches. Colour yellowish-white on the exterior; the light yellow interior often has sparse, purplish markings, in a continuous V-shaped pattern. Sculpture of fine concentric lines, and microscopic radial striations.
Occurrence: Relatively common along the harbour side of the Palisadoes, in about 3 feet of water. Dead shells are washed ashore at Negril, Runaway Bay, and Folly (Port Antonio). Excellent specimens can be collected in the 'rivers' which connect the lagoons with the sea just east of Port Royal.
Forms: There is an uncommon colour form, with a few pink rays on the exterior surface, known as *S. proficua radiata* (pl. 32). We have found it only at Negril and off the Bogue Islands.

SEMELE PURPURASCENS Gmelin 1791
Purplish Semele **pl. 32**
South-east U.S. and the West Indies.
Description: Length 1 to $1\frac{1}{2}$ inches. Colour yellowish-cream, heavily speckled with purple (and occasionally orange) markings. Sculpture of fine concentric lines, and microscopic oblique threads. This is an unusually attractive species.
Occurrence: As for *S. proficua*, with which it is found in the 'rivers' connecting the lagoons to the sea, just east of Port Royal.

FAMILY DONACIDAE

These are small, active, wedge-shaped bivalves, which live in the sand of many tropical and temperate beaches. Most of the 50 species are brilliantly coloured in an endless variety of attractive patterns.

An indication of their abundance is the result of a quantitive study of *Donax fossor* in New Jersey, U.S.A., which produced 1,510 live shells in one square foot of sand. A similar study of *D. denticulatus* on certain Jamaican beaches would no doubt produce comparable results.

Genus *Donax* Linné 1758

DONAX DENTICULATUS Linné 1758
Common Caribbean Donax **pl. 32**
Caribbean.

Description: Length ¾ to 1 inch. Colour, almost unbelievably varied: usually a combination of white, blue, purple, brown, orange and tan, generally in a radial pattern. Many specimens are solid orange or purple. Sculpture of minute radial grooves; the zigzag lines on the posterior slope of each valve separate it from the other similar species of the same genus. The bright colours, infinite variety of pattern, and the graceful 'butterfly' shape of the open valves, make this one of the most attractive of West Indian bivalves. The ligament is sturdy, and dead shells are very often found intact on the beaches.

Occurrence: This is one of the most abundant Caribbean species. They live in migrant colonies of hundreds of thousands on sandy beaches, just where the waves break. As each wave recedes, they can be seen frantically burrowing their way beneath the sand again, and this is the moment to capture them. They occur all around the Island, but are particularly numerous along the beach from Port Henderson to Fort Augusta (Kingston harbour), just east of Black River, and at Negril. They are edible, and make a delicious broth or chowder.

DONAX TUMIDUS Philippi 1849 Fat Little Donax **pl. 32**
Gulf of Mexico to the West Indies.

Description: Length ⅓ to ½ inch. Colour yellowish white. It is very similar to *D. denticulatus*, but smaller, fatter, and less colourful. There are heavily beaded radial threads on the posterior slope of each valve.

Occurrence: This species is neither as common nor as widely distributed as *denticulatus*. It is abundant, however, on the Long Bay beach at Negril, particularly at the northern end. The remarks on the occurrence of *denticulatus* apply otherwise to this species.

DONAX STRIATUS Linné 1767 Striate Donax **pl. 32**
West Indies.

Description: Length ¾ to 1 inch. Colour usually a combination of cream and bluish-purple; occasional specimens are pink. The posterior slope is flat or slightly concave, and bears numerous fine radial threads. The shell is otherwise very similar to *D. denticulatus*.

Occurrence: This is a relatively common species in Jamaica, but with a limited distribution. Perfect specimens can often be found on the beach at Port Henderson, and occasionally at Old Harbour Bay. It also occurs in mud, at about 6 feet, off Newport West and along the causeway in Kingston harbour. It apparently prefers a muddy rather than sandy environment, unlike the two preceding species, and it is abundant on the beach just east of Black River.

Genus *Iphigenia* Schumacher 1817

IPHIGENIA BRASILIANA Lamarck 1818

Giant False Donax **pl. 32**

Southern Florida and the West Indies.

Description: Length 2 to $2\frac{1}{2}$ inches. Colour cream, with purple or orange on the umbones. Sculpture of irregular growth lines. There is a thin, brown periostracum. The shell is solid and strong.

Occurrence: Dead shells are relatively uncommon in shallow water along the harbour side of the Palisadoes, and near Fort Augusta. It has also been dredged in about 20 feet in Ocho Rios harbour and off the Bogue Islands. Beach specimens can be found just east of Black River, and at Farquhars Beach (Clarendon).

FAMILY SANGUINOLARIIDAE

Members of this family are closely related to the *Tellinidae* and are generally characterised by thin, translucent shells and a preference for warm, shallow, muddy water. They live just beneath the surface of the substrate.

There are about 100 species with a world-wide distribution in tropical seas. All have large, external ligaments, and many are considered edible.

Genus *Sanguinolaria* Lamarck 1799

SANGUINOLARIA CRUENTA Solander 1786

Atlantic Sanguin **pl. 32**

South Florida, the Gulf States, and the West Indies.

Description: Length $1\frac{1}{2}$ to 2 inches. Colour white, with a pink flush on the upper half of the valves towards the umbones. The shell is smooth, but not polished; the valves are thin and slightly inflated.

Occurrence: This species is relatively uncommon, and has a limited distribution in Jamaica. Dead, but perfect, specimens are often washed ashore at Prospect and on the beach between Port Henderson and Fort Augusta (in Kingston harbour), and occasionally at Negril. We have rarely found it on the north coast, but it does occur in Ocho Rios harbour in 3 to 30 feet.

SANGUINOLARIA OPERCULATA Gmelin
Operculate Sanguin **pl. 32**
Caribbean to Brazil.
Description: Length about 2 inches. Colour rose. Sculpture of very fine concentric lines, and even finer radial threads. The left valve is nearly flat, the right inflated. The shell is more robust than the preceding species (*S. cruenta*).
Occurrence: This attractive shell is reported to occur in mud at about 10 feet near the mouth of the Milk River (Clarendon). We have not found it there or elsewhere in Jamaica, and the record needs confirmation.

Genus *Asaphis* Modeer 1793

ASAPHIS DEFLORATA Linné 1758 Gaudy Asaphis **pl. 32**
South-east Florida and the West Indies.
Description: Length 1½ to 2½ inches. Colour yellowish-white, often stained with purple or rose; the interior is usually the same colour as the exterior but brighter. Sculpture of numerous radial threads, fine concentric lines, and occasional coarse growth lines.
Occurrence: Common just below the surface of the substrate in 5 feet, all along the south coast, and in muddy areas on the north coast. Beautiful specimens are washed ashore between Gunboat Beach and Harbour View (in Kingston harbour), in the shallow areas on both sides of Falmouth, and in many other relatively protected places all around Jamaica. We recently found a live colony at the water's edge in Kingston harbour, with the posterior ends of the shells just breaking the surface of the substrate.

Genus *Tagelus* Gray 1847

TAGELUS DIVISUS Spengler 1794 Purplish Tagelus **pl. 32**
Eastern U.S., the Gulf States and the West Indies.
Description: Length 1 to 1½ inches. Colour whitish, with a faint purple tinge, and a distinctive radial purple streak starting at the umbones. There is a thin, yellowish periostracum. The shell is thin, fragile and smooth-surfaced.

Occurrence: Locally quite common in muddy areas, in about 4 feet of water, but it is limited in distribution. Dead but perfect specimens are seasonally common on the beach, and in very shallow water, at the old Naval Dockyard, Port Royal, and (to a lesser extent) across the harbour at Fort Augusta. Broken valves are frequently washed ashore at Negril.

TAGELUS PLEBEIUS Solander 1786 Stout Tagelus **pl. 32**
Cape Cod to Florida and the West Indies.
Description: Length 2 to 3½ inches. Colour white. Sculpture of concentric growth lines. This species is similar to *T. divisus* in shape, but is larger, heavier, and lacks coloration. The periostracum is thin, dark yellow and persistent.
Occurrence: This is a relatively uncommon species in Jamaica. It occurs in mud in 1 to 20 feet in Kingston harbour, and dead shells are found along the causeway beach. It also lives off the Bogue Islands, near the mouth of the Black River, and in Ocho Rios harbour at approximately the same depth. It is most common at Farquhars Beach, where double valves are often washed ashore.

Genus *Heterodonax* Mörch 1853

HETERODONAX BIMACULATUS Linné 1758
Small False Donax **pl. 32**
Southern Florida and the West Indies.
Description: Length ½ to 1 inch. Colour very variable: it may be white, purple, orange, or red; or in any combination of these colours. Sculpture of microscopic growth lines. The shell is thin and relatively fragile.
Occurrence: Very common lightly buried in mud, in about 4 feet, all along the harbour side of the Palisadoes, and at Folly (Port Antonio), Runaway Bay, Negril, and many other areas all around Jamaica. Dead but perfect specimens are commonly washed ashore on the beach between the Gypsum Pier and Gunboat Beach (in Kingston harbour). The hinge ligament is strong, and usually keeps the valves together when they are washed up the beach.

Genus *Solecurtus* Blainville 1825

SOLECURTUS CUMINGIANUS Dunker 1861

Corrugated Razor Clam **pl. 32**

South-east U.S. and the West Indies.

Description: Length 1 to 2 inches. Colour white, with a yellowish periostracum. Sculpture of coarse, concentric growth lines, and fine, oblique, wavy threads.

Occurrence: This is a rare shell in Jamaica. It occurs in mud or sand, in about 20 feet, in Kingston and Ocho Rios harbours, and dead shells are found on the beach just east of Black River.

SOLECURTUS SANCTAEMARTHAE Orbigny 1842

St Martha's Razor Clam **pl. 32**

South-east U.S. and the West Indies.

Description: Length ½ to 1½ inches. Colour white. Sculpture similar to *S. cumingianus* but more pronounced. It is proportionately shorter and more inflated than *cumingianus*, and considerably more fragile.

Occurrence: This is a relatively rare species in Jamaica. It occurs in sand at about 10 feet in Bloody Bay, Negril. Perfect specimens are occasionally washed ashore there, but we have found it nowhere else.

Suborder Adapedonta

SUPERFAMILY SOLENACEA

FAMILY SOLENIDAE

This is the family of long, slender Jacknife Clams, or Razor Shells, and includes a total of about 40 species. It is a remarkable and distinctive group, whose members burrow deep into the substrate of relatively shallow water, and when disturbed, use their muscular foot to execute astonishing leaps. Some species can bury themselves in the mud faster than one can dig with a spade.

There are no common Jamaican species, and those which do occur here are usually collected by dredging in depths of about 20 feet.

Plate 27 **SCALLOPS, KITTEN'S PAWS, FILE SHELLS**

Scale: reduced by 35%

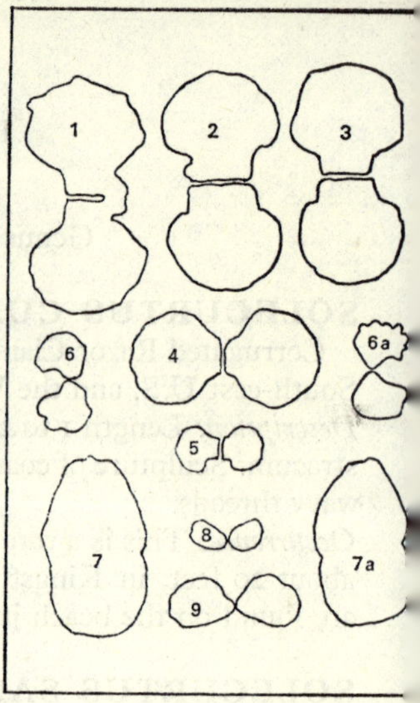

1. **Lyropecten nodosus** Linné Lion's Paw 228
The prominent nodules are filled with an aqueous solution.

2. **Pecten ziczac** Linné Zigzag Scallop 222
The upper valve is often pinkish.

3. **Pecten laurenti** Gmelin Laurent's Scallop 222

4. **Pecten papyraceus** Gabb Paper Scallop 222

5. **Pecten chazaliei** Dautzenberg Chazalie Scallop 223
This shell is easily confused with juvenile specimens of **2.**

6. **Plicatula gibbosa** Lamarck Kitten's Paw 221
Colour variety usually associated with sandy areas. **6a,** colour variety usually associated with muddy areas.

7. **Lima scabra** Born Rough Lima 231
This is the typical rough-surfaced form. **7a,** *Lima scabra tenera* Sowerby. This is a smooth-surfaced form.

8. **Lima pellucida** C. B. Adams Antillean Lima 231
The red stains on the shell illustrated are coralline deposits. The shell itself is translucent.

9. **Lima lima** Linné Spiny Lima 230

Plate 28 OYSTERS, JINGLE SHELLS, FALSE MUSSELS, JEWEL BOXES, CARDITAS

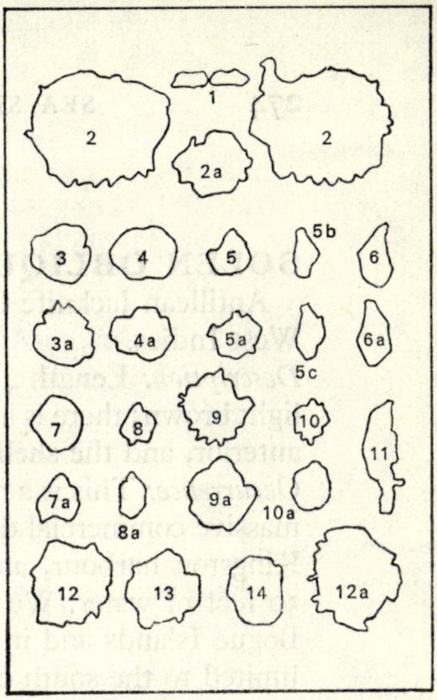

Scale: reduced by 40%

1. Cardita gracilis Shuttleworth West Indian Cardita 234
See text for note on the need for confirmation that this species occurs in Jamaica.

2. Ostrea frons Linné 'Coon Oyster 233
Interior of both valves shown; note *Petaloconchus mcgintyi* on exterior surface of valve at right. **2a,** exterior view. This very variable species takes its shape from the surface to which it adheres, and some specimens are long and narrow, unlike those illustrated here.

3 and 3a. Pododesmus rudis Broderip False Jingle Shell 232
Exterior and interior views.

4 and 4a. Anomia simplex Orbigny Common Jingle Shell 232
Exterior and interior views.

5 and 5a. Ostrea equestris Say Crested Oyster 233
Exterior and interior views. **5b and 5c,** unusual narrow specimen; exterior and interior views.

6 and 6a. Mytilopsis domingensis Récluz False Mussel 234
Exterior and interior views. Note the characteristic 'shelf' at the umbone of **6a.**

7 and 7a. Chama sada Reeve Red Jewel Box 242
Exterior and interior views.

8 and 8a. Chama congregata Conrad Little Corrugated Jewel Box 242
Exterior and interior views.

9 and 9a. Echinochama arcinella Linné True Spiny Jewel Box 243
Exterior and interior views.

10 and 10a. Chama florida Lamarck Florida Jewel Box 243
Exterior and interior views.

11. Crassostrea rhizophorae Guilding Caribbean Oyster 233
This species cements itself to some firm surface. The shell illustrated is attached to a thin mangrove root, which can be seen projecting below it. The picture is taken from the side, with the valves partially open.

12. Chama macerophylla Gmelin Leafy Jewel Box 239
The illustration shows two shells—the smaller, with valves open, attached to the upper valve of the larger. **12a,** two specimens—the smaller (purple) shell is attached to the upper (right) valve of the yellow specimen. Attachment in this species is by the left valve.

13. Chama sinuosa Broderip White Smooth-edged Jewel Box 242
See text for differences between this and the very similar *C. macerophylla.*

14. Pseudochama radians Lamarck Atlantic Left-handed Jewel Box 243
This species is a mirror image of **12.**

Genus *Solen* Linné 1758

SOLEN OBLIQUUS Spengler 1794
Antillean Jacknife Clam **pl. 32**
West Indies.
Description: Length 4 to 6 inches. Colour dull white, with faint rays of light brown; there is a persistent brown periostracum. The umbones are anterior, and the shell is relatively thick and strong.
Occurrence: This is a rare species in Jamaica. Our specimens come from massive commercial dredgings off Newport West and Fort Augusta, in Kingston harbour, and from Farquhars Beach (Clarendon), in 20 to 30 feet of water. We found no trace of it in similar dredgings at the Bogue Islands and in Ocho Rios harbour, and its distribution may be limited to the south coast.

SUPERFAMILY MACTRACEA

FAMILY MACTRIDAE

These are the well-known Surf Clams, which are found within the substrate of sandy beaches in tropical and temperate climates. The shells are characterised by a spoon-shaped depression on the hinge, and the few Jamaican species lack colour (unlike some attractive Pacific varieties).

The Atlantic Surf Clam (*Spisula solidissima*) of the eastern United States is the source of commercial clam chowder.

Genus *Mactra* Linné 1767

MACTRA ALATA Spengler 1802 Winged Mactra **pl. 32**
Caribbean to Brazil.
Description: Length 2 to 4 inches. Colour white, with a thin, yellowish periostracum. The posterior slope is flattened, and has an elevated ridge. The shell is relatively thin and inflated.
Occurrence: Uncommon in about 10 feet in Kingston harbour, at Farquhars Beach (Clarendon), Port Antonio, and at Negril. Single valves are occasionally washed ashore on the harbour side of the Palisadoes, particularly in the vicinity of the old runway.

MACTRA FRAGILIS Gmelin 1791
Fragile Atlantic Mactra **pl. 32**
South-east U.S., Texas and the West Indies.
Description: Length 2 to 3 inches. Colour white, with a thin, brown, rather silky periostracum. The posterior slope is rounded, with two radial lines. The shell is thin and relatively fragile.
Occurrence: Reasonably common in about 5 feet in Kingston harbour, Discovery Bay and at Negril. Dead but perfect specimens are often washed ashore on the harbour side of the Palisadoes, and on the beach near Fort Augusta (in Kingston harbour).

Genus *Mulinia* Gray 1837

MULINIA PORTORICENSIS Shuttleworth 1856
Puerto Rican Surf Clam **pl. 32**
Puerto Rico and Jamaica.
Description: Length 1 to 1½ inches. Colour white or cream, occasionally with orange stains. The shell is smooth, except for irregular growth lines; fresh specimens have a brown periostracum. The beaks are prominent and central.
Occurrence: Uncommon in mud in about 15 feet in Kingston harbour, off the Bogue Islands, and possibly in Ocho Rios harbour. Dead shells are occasionally washed ashore on the beach at Montego Freeport.

Genus *Labiosa* Möller 1832

LABIOSA LINEATA Say 1822 Smooth Duck Clam
Eastern U.S., Jamaica and Puerto Rico.
Description: Length 2 to 3 inches. Colour usually white, occasionally tan. Sculpture of fine growth lines, which become more irregular as they approach the ventral margin; there is a distinct radial rib at the posterior end. The valves gape, and have flaring edges. The shell is thin and translucent, but not as fragile as it looks.
Occurrence: This is a rare species in Jamaica. We found fragments of two valves in Kingston harbour at a depth of 20 feet on mud. The species may not have survived the recent pollution of the harbour water. It is not uncommon on the beach at Port Morant, but it has not been recorded previously from Jamaica.

Plate 29 LUCINAS, DIPLODONS, COCKLES

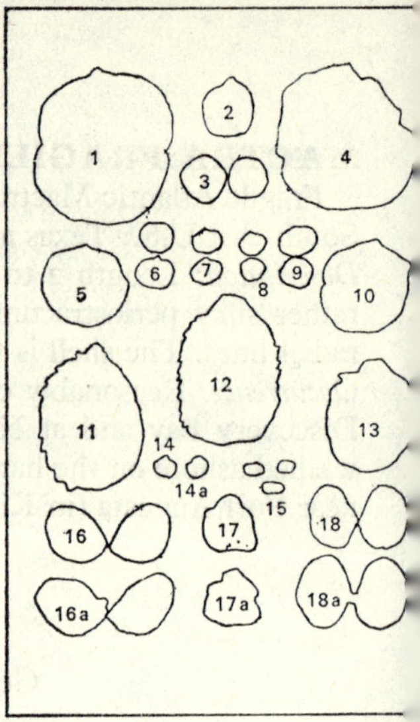

Scale: reduced by 40%

1. Codakia orbicularis Linné Tiger Lucina 238

2. Codakia orbiculata Montagu Dwarf Tiger Lucina 238

3. Divaricella quadrisulcata Orbigny Cross-hatched Lucina 239

4. Phacoides pectinatus Gmelin Thick Lucina 236
Many shells lack a yellow interior.

5. Anodontia alba Link Buttercup Lucina 237

6. Phacoides muricatus Spengler Spinose Lucina 237

7. Codakia costata Orbigny Costate Lucina 239

8. Lucina phenax Dall and Simpson Mock Lucina 236

9. Diplodonta punctata Say Common Atlantic Diplodon 235

10. Lucina pensylvanica Linné Pennsylvania Lucina 236

11. Trachycardium magnum Linné Magnum Cockle 244

12. Trachycardium isocardia Linné Prickly Cockle 245

13. Trachycardium muricatum Linné Yellow Cockle 244

14 and **14a. Americardia guppyi** Thiele Guppy's Strawberry Cockle Exterior and interior views. 246

15. Papyridea semisulcata Gray Frilled Paper Cockle 245
Three different colour varieties illustrated.

16. Laevicardium laevigatum Linné Common Egg Cockle 246
16a, common colour variety, from a sandy substrate.

17 and **17a. Americardia media** Linné Atlantic Strawberry Cockle Exterior and interior views. 246

18. Papyridea soleniformis Bruguière Spiny Paper Cockle 245
Unusual yellow/orange colour variety. **18a,** typical colour variety.

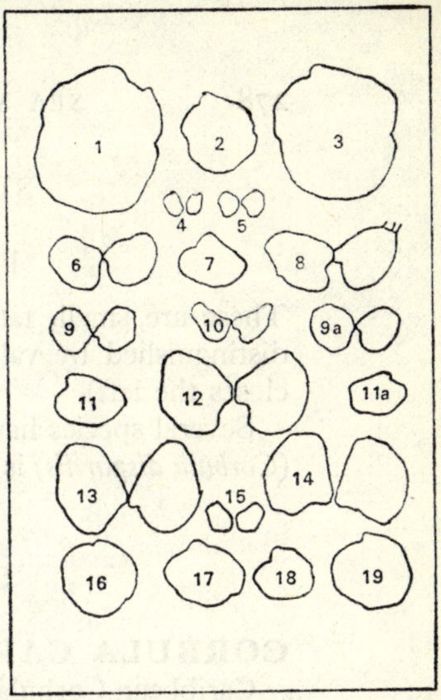

Plate 30 **VENUS SHELLS, RUPELLARIAS, CYCLINELLAS, DOSINIAS**

Scale: reduced by 50%

1. **Antigona rigida** 248
Dillwyn Rigid Venus

2. **Chione intapurpurea** 249
Conrad Lady-in-waiting Venus

3. **Antigona listeri** Gray 248
Princess Venus

4. **Chione pygmaea** 249
Lamarck White Pygmy Venus

5. **Pitar fulminata** 252
Menke Lightning Venus

6. **Chione cancellata** 248
Linné Cross-barred Venus

7. **Chione latilirata** 250
Conrad Imperial Venus

8. **Pitar dione** Linné 253
Royal Comb Venus
Young specimens usually possess a double row of spines on each valve.

9. **Chione granulata** 249
Gmelin Beaded Venus
9a, colour variation; occasional specimens are solid brown.

10. **Anomalocardia brasiliana** Gmelin 250
West Indian Pointed Venus
Typical colour variety; occasional specimens are solid brown.

11. **Pitar circinata** Born 253
Purple Venus
Typical colour variety.
11a, the rare (in Jamaica) purple colour variety.

12. **Chione paphia** Linné 250
King Venus

13. **Macrocallista maculata** Linné 254
Calico Clam Rare specimens are all-white.

14. **Tivela mactroides** 251
Born Trigonal Tivela

15. **Rupellaria typica** 255
Jonas Atlantic Rupellaria

16. **Cyclinella tenuis** 254
Récluz Atlantic Cyclinella

17. **Pitar albida** Gmelin 252
White Venus

18. **Pitar aresta** Dall and Simpson 252
West Indian Venus

19. **Dosinia concentrica** 254
Born Southern Dosinia

SUPERFAMILY MYACEA

FAMILY CORBULIDAE

These are small, fat, generally white shells; most species are easily distinguished by valves of unequal size (the right valve usually encloses the left).

Several species have an extraordinary depth range, and one of them (*Corbula disparilis*) is known to occur from 30 to over 4,000 feet.

Genus *Corbula* Bruguière 1792

CORBULA CARIBAEA Orbigny 1842
Caribbean Corbula **pl. 32**
Eastern U.S. and the West Indies.
Description: Length $\frac{1}{2}$ to $\frac{5}{8}$ inch. Colour whitish. Sculpture of irregular, concentric ridges. The right valve encloses the left, and the shell is greatly inflated.
Occurrence: Common in about 3 feet, lightly buried in the substrate, all along the harbour side of the Palisadoes, and in protected, muddy areas east of Falmouth. Dead shells are common on the beach between the Gypsum Pier and Gunboat Beach, in Kingston harbour.

CORBULA AEQUIVALVIS Philippi 1836
Equal-valved Corbula **pl. 32**
West Indies.
Description: Length $\frac{1}{2}$ to $\frac{3}{4}$ inch. Colour white. Sculpture of strong, regular concentric ridges. Unlike most Corbulas, both valves are very nearly of equal size. There is a strong, sharp posterior ridge.
Occurrence: Relatively common lightly buried in the mud, in about 6 feet in Kingston harbour, at Ocho Rios and off the Bogue Islands. Perfect specimens are washed ashore all along the harbour side of the Palisadoes, and single valves can be found on the beaches at Negril.

CORBULA DIETZIANA C. B. Adams 1852
Dietz's Corbula **pl. 32**
South-east U.S. to the West Indies.
Description: Length about $\frac{1}{2}$ inch. Colour cream with patches of red, especially near the umbones, and on the sloping ventral margins where the colour is usually in vertical bars. Interior often bright pink. The right valve encloses the left, and is characterised by the very sharply angled margin. This is an unusually attractive little shell.

Occurrence: Uncommon in about 4 feet along the northern side of the Palisadoes, particularly in the area between Gunboat Beach and the Gypsum pier, where dead shells can be found on the beach. Surprisingly, we have not found it on the opposite side of the harbour, but two dead valves were recovered on the southern side of the Palisadoes near the Light House.

SUPERFAMILY ADESMACEA

FAMILY PHOLADIDAE

This is the family of the notorious borer shells, and includes the delicate and beautiful Angel Wings (*Cyrtopleura*). The group is widely distributed in tropical and temperate seas, and the *Martesia*, which commonly bore into driftwood, may travel from one continent to another. Some species are hermaphroditic.

The fragile shells require careful removal from their burrows, and in the case of *Cyrtopleura*, which may dig two or three feet into hard clay, a certain skill and stamina are also necessary!

Genus *Cyrtopleura* Tryon 1862

CYRTOPLEURA COSTATA Linné 1758 Angel Wing
Eastern U.S.; Texas and the West Indies.
Description: Length 4 to 8 inches. Colour chalky white. Sculpture of about 30 beaded or scaly radial ribs. The shell is light and moderately fragile. This is a very beautiful species; it belongs to a family which is much sought after by collectors.
Occurrence: We have not found this species in Jamaica. It is included here because it is known to occur in Florida, Cuba, and Hispaniola, and this suggests that it may eventually be discovered locally.

Genus *Barnea* Risso 1826

Subgenus *Anchomasa* Leach 1852

BARNEA TRUNCATA Say 1822 Fallen Angel Wing
Maine to Brazil.
Description: Length $1\frac{1}{2}$ to $2\frac{1}{2}$ inches. Colour translucent white. Sculpture of weak, beaded, radial ribs most pronounced at the anterior end. There

is a characteristic, long thin condyle, and the fragile shell gapes at both ends. This is a delicately beautiful species.

Occurrence: We found several perfect valves of this species on the beach at Old Harbour Bay while this book was in proof. It is known to bore to a depth of 20 inches in wood, clay or soft rock, in shallow water. Our specimens probably came from a submerged bank of clay near the water's edge. We can trace no previous Jamaican records.

Genus *Martesia* Blainville 1824

MARTESIA STRIATA Linné 1767 Striate Martesia **pl. 32**

West Florida to Texas, and the West Indies.

Description: Length $\frac{3}{4}$ to $1\frac{1}{4}$ inches. Colour whitish. There is a faint radial groove on each valve extending from the beak to the ventral margin. Behind this groove, the sculpture is of concentric growth lines. The calcareous shield over the umbones is heart-shaped. The shell is thin, rather fragile, and gapes at the posterior end.

Occurrence: This is a very common wood-borer found in old pilings and water-logged tree trunks in Kingston harbour, and in many other areas all around the Island. A small piece of soft wood often plays host to an almost unbelievable concentration of these shells, as a carefully prepared cross-section will show.

MARTESIA CUNEIFORMIS Say 1822

Wedge-shaped Martesia

South-east U.S., Texas and the West Indies.

Description: Length $\frac{1}{2}$ to $\frac{3}{4}$ inches. Colour whitish. Very similar to the preceding species (*M. striata*), but smaller and chubbier. The calcareous shield over the umbones is shaped like the head of a spear, and usually has a crease running down the centre.

Occurrence: As for *M. striata*, but it is a far less common species in Jamaican waters.

Note: The two local members of this genus are generally considered to be entirely destructive. In fact, by constantly breaking down wooden debris in the sea, and particularly in congested harbour areas, the Martesias perform a most useful marine function.

Other Jamaican Pelecypods

Order PROTOBRANCHIA

Family Solemyacidae
Solemya occidentalis Deshayes 1857

Superfamily NUCULACEA

Family Nuculidae
Nucula aegeensis Jeffreys 1879
Nuculana acuta Conrad 1831

Family Malletiidae
Tindaria corpulenta Dall 1886

Order FILIBRANCHIA

Superfamily ARCACEA

Family Arcidae
Arcopsis adamsi Dall 1886

Family Glycymerididae
Glycymeris sericata Reeve 1843

Superfamily MYTILACEA

Family Mytilidae
Crenella divaricata Orbigny 1842
Crenella glandula Totten
Musculus lateralis Say 1822
Amygdalum dendriticum Mühlfeld 1811

Superfamily PECTINACEA

Family Pectinidae
Pecten raveneli Dall 1898
Chlamys nana Verrill and Bush 1897

Family Propeamussiidae
Propeamussium pourtalesianum Dall
Propeamussium dalli E. A. Smith

Family Limidae
Limatula subauriculata Montagu 1808

Superfamily OSTREACEA

Family Ostreidae
Ostrea permollis Sowerby

Order EULAMELLIBRANCHIA

Superfamily ASTARTACEA

Family Crassatellidae
Crassinella lunulata Conrad 1834
Crassinella martinicensis Orbigny 1842
Crassinella guadalupensis Orbigny 1842

Superfamily LUCINACEA

Family Diplodontidae
Diplodonta nucleiformis Wagner 1852
Diplodonta semiaspera Philippi 1836
Diplodonta gabbi Dall 1900
Diplodonta notata Dall and Simpson 1901
Thyasira trisinuata Orbigny 1842

Family Lucinidae
Lucina blanda Dall and Simpson 1901
Codakia pectinella C. B. Adams 1852
Myrtea pristiphora Dall and Simpson 1901

Superfamily ERYCINACEA

Family Erycinidae
Erycina periscopiana Dall 1899
Erycina emmonsi Dall 1899
Basterotia quadrata Hinds 1843
Basterotia elliptica Récluz 1850

Superfamily CARDIACEA

Family Cardiidae
Trigoniocardia ceramidum Dall 1886
Trigoniocardia antillarum Orbigny 1842
Microcardium tinctum Dall 1881
Microcardium peramabile Dall 1881

Superfamily VENERACEA

Family Veneridae
Antigona rugatina Heilprin 1887
Transennella stimpsoni Dall 1902
Transennella culebrana Dall and Simpson 1901
Transennella cubaniana Orbigny 1842
Gouldia cerina C. B. Adams 1845
Parastarte triquetra Conrad 1846

Family Petricolidae
Cooperella atlantica Rehder 1943

Superfamily TELLINACEA

Family Tellinidae

Tellina versicolor De Kay
Tellina candeana Orbigny 1842
Tellina sybaritica Dall 1881
Tellina pauperata Orbigny 1842
Tellina vitrea Orbigny 1842
Tellina vespusiana Orbigny 1842
Tellina cristallina Spengler 1798
Tellina aequistriata Say 1824
Tellina martinicensis Orbigny 1842
Strigilla pisiformis Linné 1758
Strigilla rombergi Mörch 1853
Strigilla mirabilis Philippi 1841
Quadrans lintea Conrad 1837
Macoma orientalis hendersoni Rehder 1939

Family Semelidae

Semele bellastriata Conrad 1837
Semele nuculoides Conrad 1841
Cumingia antillarum Orbigny 1842
Abra aequalis Say 1822
Abra lioica Dall 1881

Superfamily MACTRACEA

Family Mesodesmatidae

Ervilia nitens Montagu 1806
Ervilia rostratula Rehder 1943

Family Hiatellidae

Saxicavella sagrinata Dall and Simpson 1901

Superfamily MYACEA

Family Myacidae

Sphenia antillensis Dall and Simpson 1901

Family Corbulidae

Notocorbula operculata Philippi 1848
Corbula contracta Say 1822

Superfamily GASTROCHAENACEA

Family Gastrochaenidae

Gastrochaena hians Gmelin 1791
Spengleria rostrata Spengler 1793

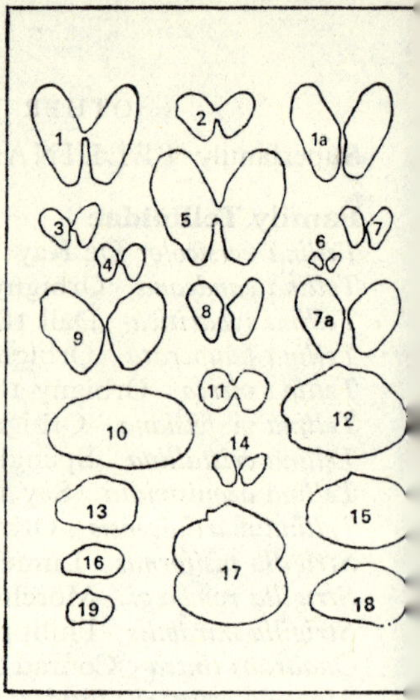

Plate 31 TELLINS, MACOMAS, STRIGILLAS

Scale: reduced by 50%

1. **Tellina radiata** Linné 258
Sunrise Tellin
Typical colour variety.
1a, *Tellina radiata unimaculata* Lamarck. This is the yellowish-white colour form.

2. **Tellina lineata** Turton 262
Rose Petal Tellin
Some shells may be all-white.

3. **Tellina georgiana** 262
Dall Georgia Tellin

4. **Tellina caribaea** 260
Orbigny Caribbean Tellin

5. **Tellina listeri** Röding 259
Speckled Tellin

6. **Tellina exilis** Lamarck 260
Pink Thin Tellin

7. **Tellina punicea** Born 262
Watermelon Tellin
7a, faded specimen. Exposure to the sun eventually produces a white shell.

8. **Tellina guildingii** 260
Hanley
The Jamaican Tellin

9. **Tellina alternata** Say 262
Alternate Tellin

10. **Tellina magna** 260
Spengler Great Tellin

11. **Strigilla carnaria** 263
Linné Large Strigilla
See the text note under this species concerning the very similar *S. rombergi*, with which it is easily confused.

12. **Apolymetis intastriata** Say 265
Atlantic Grooved Macoma

13. **Macoma constricta** 264
Bruguière Constricted Macoma

14. **Macoma aurora** 264
Hanley Hanley's Macoma

15. **Tellina laevigata** 258
Linné Smooth Tellin

16. **Macoma tenta** Say 265
Tenta Macoma

17. **Arcopagia fausta** 263
Pulteney Faust Tellin

18. **Macoma tageliformis** 264
Dall Tagelus-like Macoma

19. **Macoma pseudomera** 265
Dall and Simpson
Puerto Rico Macoma

Plate 32 **ASAPHIS SHELLS, SANGUINS, SEMELES, WEDGE SHELLS, CORBULAS, SURF CLAMS, MACTRAS, RAZOR SHELLS, MARTESIAS, TAGELUS CLAMS**

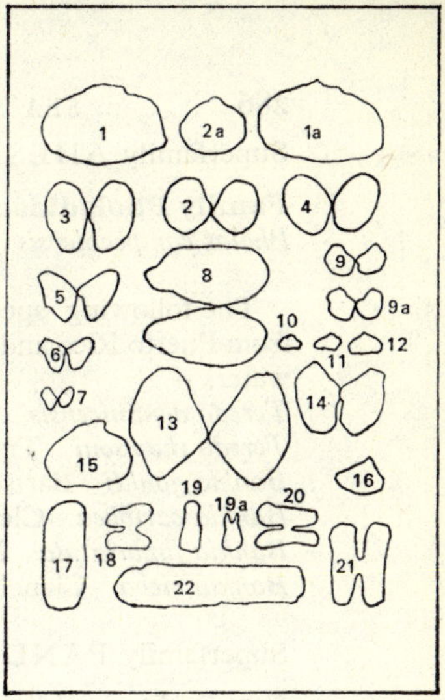

Scale: reduced by 45%

1 and 1a. Asaphis deflorata 269
Linné Gaudy Asaphis
Exterior and interior views.

2. Semele proficua Pulteney 266
White Atlantic Semele
2a, *Semele proficua radiata* Say.
This is the very uncommon rayed form.

3. Sanguinolaria cruenta 268
Solander Atlantic Sanguin

4. Semele purpurascens 266
Gmelin Purplish Semele

5. Donax striatus Linné 267
Striate Donax

6. Donax denticulatus Linné 267
Common Caribbean Donax

7. Donax tumidus Philippi 267
Fat Little Donax

8. Sanguinolaria operculata 269
Gmelin Operculate Sanguin
Note that the left valve is much flatter than the right.

9. Heterodonax bimaculatus 270
Linné Small False Donax
This is the typical colour variety. **9a,** pink colour variety.

10. Corbula dietziana C. B. 278
Adams Dietz's Corbula
Note the angled ventral margin. The right valve (illustrated) encloses the left.

11. Corbula caribaea Orbigny 278
Caribbean Corbula
The right valve (illustrated) encloses the left.

12. Corbula aequivalvis Philippi 278
Equal-valved Corbula

13. Iphigenia brasiliana 268
Lamarck Giant False Donax

14. Mactra fragilis Gmelin 275
Fragile Atlantic Mactra
Shell shown with periostracum in place.

15. Mactra alata Spengler 274
Winged Mactra

16. Mulinia portoricensis 275
Shuttleworth Puerto Rican Surf Clam

17. Solecurtus cumingianus 271
Dunker Corrugated Razor Clam

18. Solecurtus sanctaemarthae 271
Orbigny
St Martha's Razor Clam

19. Martesia striata Linné 280
Striate Martesia
19a, view showing the heart-shaped calcareous shield which distinguishes it from the very similar *Martesia cuneiformis* (see text).

20. Tagelus divisus Spengler 269
Purplish Tagelus
Shell shown with periostracum in place.

21. Tagelus plebeius Solander 270
Stout Tagelus
Shell shown with periostracum in place.

22. Solen obliquus Spengler 274
Antillean Jacknife Clam

Superfamily ADESMACEA

Family Pholadidae
Pholas campechiensis Gmelin 1790

The following species of the family *TEREDINIDAE* are reported from Puerto Rico and the Virgin Islands, and probably occur in Jamaican waters –
Teredo dominicensis Bartsch 1921
Teredo thomsoni Tryon 1863
Bankia gouldi Bartsch 1908
Bankia caribbea Clench and Turner 1946
Bankia fimbriatula Möll and Roch 1931
Bankia cieba Clench and Turner 1946

Superfamily PANDORACEA

Family Lyonsiidae
Lyonsia beana Orbigny 1842
Lyonsia hyalina floridana Conrad 1849

Family Pandoridae
Pandora bushiana Dall 1886

Family Thraciidae
Cyathodonta semirugosa Reeve 1859

Order SEPTIBRANCHIA

Family Verticordiidae
Verticordia ornata Orbigny 1842

Family Cuspidariidae
Cardiomya perrostrata Dall 1881
Cardiomya ornatissima Orbigny 1842

CLASS CEPHALOPODA

SUBCLASS COLEOIDEA

Order Decapoda

Members of this order, which include the squids, are characterised by the possession of ten arms, only two of which are long and tentacular. The small suckers on the arms are generally set on peduncles, and most genera have an internal shell.

FAMILY SPIRULIDAE

Genus *Spirula* Lamarck 1799

SPIRULA SPIRULA Linné 1758 Common Spirula **fig. 3**
World-wide.
Description: Diameter about 1 inch. Colour white. The interior of the shell is divided by concave, nacreous, septa into separate chambers. There is a narrow, siphonal tube, which runs back into the shell and pierces the septa. The shell is fragile, buoyant, and delicately beautiful.
Occurrence: This is the internal shell of a small, squid-like animal, which occurs at great depths in the Atlantic and Pacific Oceans. The living *Spirula* is rarely dredged alive, but the buoyant shell rises to the surface when the mollusc dies and decomposes. The shell is very common on most Jamaican beaches, and throughout the West Indies. Perfect specimens can usually be found along the southern side of the Palisadoes.

Order Octopoda

Members of this order, which include the familiar octopus, differ from the *Decapoda* in having only eight arms, and their suckers are not set on peduncles. In addition to this, none of the species possesses an internal (or external) shell, and the 'shells' of the family *Argonautidae* are, in fact, the egg cases of the female Argonaut.

The Argonauts are pelagic animals, and the dorsal arms of the female are modified to secrete and hold the delicate, calcareous 'shells'. The males of the genus are very much smaller than the females, and lack a shell of any description.

FAMILY ARGONAUTIDAE

Genus *Argonauta* Linné 1758

ARGONAUTA ARGO Linné 1758 Common Paper Nautilus
World-wide in tropical seas.
Description: Length 4 to 8 inches. Colour opaque, milky white; the nodules on the early part of the shell are dark purplish-brown. The shell is fragile and laterally compressed.
Occurrence: This is a very rare shell on Jamaican beaches. We have found one damaged specimen at Robins Bay, and can trace no other local records. In southern Florida they are washed ashore seasonally, and this may well be the case in the West Indies.
Note: The smaller but similar *A. hians* is illustrated in plate 1. We have not found it in Jamaican waters.

Other Jamaican Cephalopods

Order DECAPODA

Family Loliginidae
Lolliguncula brevis Blainville
Septioteuthis sepioidea Blainville
Doryteuthis plei Blainville

Family Ommastrephidae
Sthenoteuthis bartrami Lesueur

Order OCTOPODA

Family Argonautidae
Argonauta hians Solander **(pl. 1)**

Family Octopodidae
Octopus vulgaris Lamarck
Octopus briareus Robson
Octopus joubini Robson

CLASS AMPHINEURA

SUBCLASS POLYPLACOPHORA

Order Chitonida

There are approximately 20 known species of this order in Jamaican waters. They occur from the high-water mark down to several hundred fathoms.

The eight species described here are the larger and more common varieties; they can all be collected in water less than three feet deep.

FAMILY MOPALIIDAE

Genus *Ceratozona* Dall 1882

CERATOZONA SQUALIDA C. B. Adams 1845
Rough Girdle Chiton **fig. 18**
South-east Florida to the West Indies.
Description: Length 1 to 2 inches. Colour grey, with blue-green mottling; girdle yellowish brown. The surface of the valves is roughly sculptured, and frequently eroded. The foot of the living animal is bright orange. This species is readily identified by the girdle, which is covered with long, coarse 'hairs'.
Occurrence: Relatively common in the inter-tidal area of rocky coasts all around the Island. Our specimens were taken from rocks at Robins Bay, when exposed at low tide.

FAMILY CRYPTOPLACIDAE

Genus *Acanthochitona* Gray 1821

ACANTHOCHITONA HEMPHILLI Pilsbry 1893
Red Glass-haired Chiton **fig. 18**
Key West, Florida and the West Indies.
Description: Length about 1 inch. Colour rust-red, sometimes spotted

with white; the girdle, which covers most of the shell, is reddish-brown, with a spiculose fringe. There are 4 tufts of long glassy bristles near the anterior valve, and one on each side of the other valves (making a total of 18).

Occurrence: This is not a common species in Jamaica. Our specimens were taken under rocks in about 3 feet, near Port Rhoades at Discovery Bay. It is usually found on dead *Porites* coral and is also reported from the south coast.

FAMILY ISCHNOCHITONIDAE

Genus *Calloplax* Thiele 1909

CALLOPLAX JANEIRENSIS Gray 1828

Rio de Janeiro Chiton **fig. 18**

Lower Florida Keys, the West Indies and Brazil.

Description: Length $\frac{1}{2}$ to $\frac{3}{4}$ inch. Colour grey or greenish-brown, sometimes speckled with red; interior of valves white. Strongly sculptured with 3 or 4 coarsely beaded ribs on the lateral areas; the anterior valve has 12 to 18 ribs. The central ridge has longitudinal rows of fine beads; the apex is high, smooth, and rounded. There are usually a few fine, short hairs on the girdle.

Occurrence: This is a common species, which lives under rocks in 1 to 6 feet in many areas around the Island. Our specimens came from Robins Bay and Kingston harbour, in about 18 inches of water.

Genus *Ischnochiton* Gray 1847

Subgenus *Stenoplax* Carpenter 1878

ISCHNOCHITON LIMACIFORMIS Sowerby 1832

Slender Chiton **fig. 18**

Florida Keys, West Indies, Central America and Peru.

Description: Length 1 to 2 inches. Colour buff, grey, or green, often marbled with darker colour. Sculpture: central areas with fine, smooth longitudinal riblets, which become flat and broader on the lateral areas. The end valves have wavy, concentric ridges. The shell is narrow and elongate.

Occurrence: Relatively common on rocky shores facing the open sea, all around Jamaica. Our specimens were collected under rocks in 2 feet, at Runaway Bay and Port Henderson.

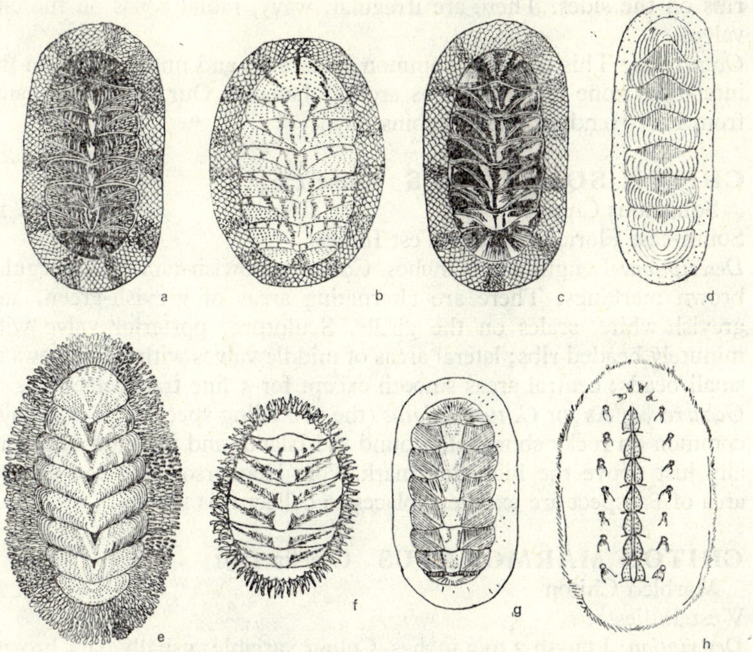

FIG. 18. Eight West Indian chitons. a, *Chiton tuberculatus*. b, *Chiton squamosus*. c, *Chiton marmoratus*. d, *Ischnochiton limaciformis* e, *Acanthopleura granulata*. f, *Ceratozona squalida*. g, *Calloplax janeirensis*. h, *Acanthochitona hemphilli*

FAMILY CHITONIDAE

SUBFAMILY CHITONINAE

Genus *Chiton* Linné 1758

CHITON TUBERCULATUS Linné 1758

Common West Indian Chiton **pl. 1, fig. 18**

West Indies.

Description: Length 2 to 3 inches. Colour greyish-green to olive green. There are alternating areas of dark green and white on the girdle. Sculpture: lateral areas with about 5 irregular radiating cords; central area smooth at the top, and with many long, coarse, wavy longitudinal

ribs on the sides. There are irregular, wavy, radial cords on the end valves.
Occurrence: This is a very common species on and under rocks, in the inter-tidal zone of many areas around Jamaica. Our specimens came from Port Henderson and Robins Bay.

CHITON SQUAMOSUS Linné 1764
Squamous Chiton **fig. 18**
South-east Florida and the West Indies.
Description: Length 2 to 3 inches. Colour yellowish-tan, with irregular brown markings. There are alternating areas of greyish-green, and greyish-white, scales on the girdle. Sculpture: posterior valve with minutely beaded ribs; lateral areas of middle valves with 6 to 8 rows of small beads; central areas smooth except for 5 fine transverse lines.
Occurrence: As for *C. tuberculatus* (the preceding species). It is equally common on rocky shores all around the Island, and is found occasionally just above the high-tide mark. Port Henderson and the general area of Prospect are excellent places to collect this species.

CHITON MARMORATUS Gmelin 1791
Marbled Chiton **fig. 18**
West Indies.
Description: Length 2 to 3 inches. Colour variable: usually dark brown, but sometimes olive or grey, variously marked with lighter areas or longitudinal lines. There are alternating areas of grey and dark green scales on the girdle. The surface of the shell is smooth and polished, with a slightly silky texture.
Occurrence: Relatively common in inter-tidal areas on rocky shores all around Jamaica. It lives on both protected and open stretches of coastline. Our specimens come from Port Henderson and Folly (Port Antonio).

SUBFAMILY ACANTHOPLEURINAE

Genus *Acanthopleura* Guilding 1829

Subgenus *Maugeria* Gray 1857

CAANTHOPLEURA GRANULATA Gmelin 1791
Fuzzy Chiton **fig. 18**
Southern Florida and the West Indies.
Description: Length 2 to 3 inches. Colour brownish; girdle with alter-

nating areas of black and greenish-white. Sculpture indistinct, because the surface of the valves is usually eroded; however, there are a few concentric rows of small beads on the lateral areas of all the valves. The distinctive girdle is thick and fuzzy.

Occurrence: Abundant just below the high-tide line, on rocky coasts all around the Island. It is the most common Jamaican Chiton. Our specimens came from Port Henderson and Discovery Bay.

Other Jamaican Chitons

Order CHITONIDA

Family Cryptoplacidae
Cryptoconchus floridanus Dall 1889
Acanthochitona pygmaea Pilsbry 1893
Acanthochitona spiculosa Reeve 1847
Choneplax lata Guilding 1829

Family Ischnochitonidae
Ischnochiton papillosus C. B. Adams 1845
Callistochiton shuttleworthianus Pilsbry 1893

Family Chitonidae
Chiton viridis Spengler 1797

Subfamily Acanthopleurinae
Tonnicia schrammi Shuttleworth 1856

CLASS SCAPHOPODA

FAMILY SIPHONODENTALIIDAE

Members of this family (with a few exceptions) are small, simple, deep water shells, some of which occur at more than 2,000 feet.

A few delicate species are commonly found in shallow water, in sandy or muddy areas around Jamaica, and their shells are sometimes present in beach drift. Because of their small size – less than 10 mm. – we have not described any of them here, but they are listed at the end of this section.

FAMILY DENTALIIDAE

The shells of this family vary in size from minute, translucent, needle-like forms, to thick, heavy, brightly coloured species, more than 5 inches long.

They have a world-wide distribution in both deep and shallow water, and at least 16 species are believed to occur around Jamaica. The four largest are described here.

Genus *Dentalium* Linné 1758

Subgenus *Graptacme* Pilsbry and Sharp 1897

DENTALIUM EBOREUM Conrad 1846 Ivory Tusk **pl. 1**
South-east U.S. and the West Indies.
Description: Length 1 to 2½ inches. Colour ivory white, sometimes with a pinkish blush. Sculpture of about 20 fine longitudinal lines at the tip. The apical notch is deep and narrow, and is located on the convex side.
Occurrence: This is a relatively uncommon species down to about 30 feet, in sandy or muddy areas all around the Island. Dead shells are occasionally washed ashore at Negril, and commercial dredgings in Ocho Rios harbour recently revealed that it occurs there in about 20 feet.

Subgenus *Dentale* Da Costa 1778

DENTALIUM ANTILLARUM Orbigny 1842

Antillean Tusk **fig. 19**

Southern Florida and the West Indies.

Description: Length about 1 inch. Colour opaque white, with weak bands or splotches of translucent grey. Sculpture of 9 longitudinal ribs at the apex, which increase to 24 near the aperture. There are microscopic, transverse lines between the ribs.

Occurrence: Common in about 20 feet in Ocho Rios harbour, and at Negril. Dead (and usually broken) shells are occasionally found in beach drift in sandy areas.

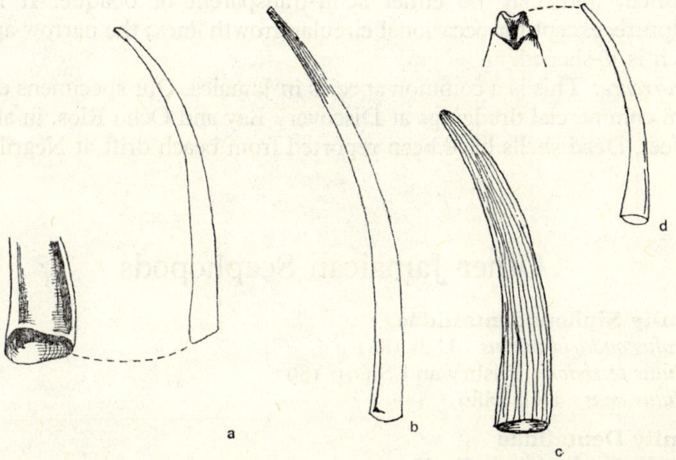

FIG. 19. Four species of tusk shell (*Dentalium*) (after Abbott). a, *Dentalium didymum.* b, *D. semistriolatum.* c, *D. antillarum.* d, *D. liodon*

Subgenus *Bathoxiphus* Pilsbry and Sharp 1897

DENTALIUM DIDYMUM Watson 1879

Flattened Tusk **pl. 1, fig. 19**

Yucatan, Puerto Rico, Barbados and Jamaica.

Description: Length about 1 inch. Colour white; the shell is polished

and porcelainous. Distinguished by a slight, lateral flattening, particularly towards the convex side.
Occurrence: As for *D. antillarum* (the preceding species), but it is more common, and apparently more widely distributed. Dead shells are not uncommon on the beaches at Negril, and they can be dredged in about 6 feet in the Port Royal area of Kingston harbour. This species occurs from shallow water to depths of 2,000 feet.

Subgenus *Laevidentalium* Cossmann 1888

DENTALIUM LIODON Pilsbry and Sharp 1897
Smooth Tusk **fig. 19**
West Indies.
Description: Length about 1 inch. Colour white; the shell is highly polished, and may be either semi-transparent or opaque. It lacks sculpture, except for occasional circular growth lines; the narrow apical notch is V-shaped.
Occurrence: This is a common species in Jamaica. Our specimens come from commercial dredgings at Discovery Bay and Ocho Rios, in about 25 feet. Dead shells have been reported from beach drift at Negril.

Other Jamaican Scaphopods

Family Siphonodentaliidae
Cadulus quadridentatus Dall 1881
Cadulus tetrodon Pilsbry and Sharp 1897
Cadulus acus Dall 1889

Family Dentalidae
Dentalium calamus Dall 1889
Dentalium semistriolatum Guilding 1834. **Fig. 19**
Dentalium ceratum Dall 1881
Dentalium callipeplum Dall 1889

PART III

Collecting, Cleaning and Storing Shells

Collecting

IN a biological sense, molluscs are a notably successful group of animals. Perhaps the most obvious reason for this is the fact that like tortoises and knights of old, most molluscs encase their soft, vulnerable bodies in armoured jackets. But even thick shell is not impregnable against the attack of many predators, and so in their long journey down the corridor of evolution molluscs have learned to supplement their natural defences with a talent for concealment.

In many marine genera this involves deliberate activity like burrowing under the sand, squeezing into coral crevices, or simply seeking shelter beneath a rock. In others, concealment is quite involuntary; the molluscs are camouflaged on the sea bed because the coral and algae which live all around them also flourish on the outside of their shells. Regardless of the method they adopt, the fact is that most live marine molluscs are difficult to find in their natural habitat, and an inexperienced collector can spend hours in an area teeming with living shells without seeing one of them.

The key to productive collecting is to know what to look for, where to look for it, and how to recover it. Much of this knowledge comes only with field experience and exposure to molluscs in their own territory, but there are certain basic guidelines, mainly dictated by common sense, which can save many hours of discouraging effort. This chapter attempts to draw these guidelines.

COLLECTING OUT OF THE WATER

(i) *On beaches*

The simplest way to collect shells is to look for them on a beach, and it is a very good way to start. On a beach of soft sand, with a gradual slope into deeper water, many shells will wash ashore in an acceptable condition. In most cases the sea and its scouring agents will have removed the animal from its shell, and gently cleaned the exterior surface before rolling it up the sand. On a rocky beach, however, shells usually have to run a gauntlet of churning pebbles before they are thrown up above the tide line, and they rarely survive intact.

The most productive time to search a beach is at low tide. The shells washed up by the previous high water are then neatly laid out in a rough line which runs parallel to, and a short distance away from the edge of the sea. If low tide coincides approximately with first light in the morning, so much the better, for the glare reflected from a coral beach in the middle of a cloudless day tires the eyes, and one is inclined to overlook shells which would be seen in a softer light. In addition to this, the very early morning brings out the small land hermit crabs (which live above the high-water mark) to scavenge along the tide line, and occasionally one of these may be wearing an unusual shell.

Many small species, less than one inch long, are difficult to see on a beach at any time, and particularly when their colours blend with the sand. Some of them may still be enveloped by the seaweed in which they were washed ashore. The time and effort involved in searching individually for minute shells on the sand is generally wasted. The most effective method of finding them is to skim off with a trowel the top layer of sand and debris (including the thicker bunches of weed) along the line of the high-tide mark. This can be put in a bucket and carefully sorted later by placing a small quantity at a time in a shallow dish and covering it with water. If the sand is then raked with a pair of tweezers, the larger shells are easily seen; finding the really small ones, some of which may be less than 1/20th of an inch long, requires patience and concentration. The very light species, such as the transparent Bubble Shells (*Bullidae*) and the small Wentletraps (*Epitoniidae*), float to the surface as soon as the water is added, and these can be taken out first with a spoon.

There is another aspect of beach collecting in the West Indies which is surprisingly profitable and often overlooked. This involves the searching of fish pots which have been drawn out of the sea by their fishermen owners, and the beach area immediately around them. These pots are sometimes set in water 100 feet deep, and quite incidentally to their normal function, they trap large hermit crabs. The crabs themselves may have been to still deeper water originally to find their shells, and consequently it is not unusual for shells which normally occur at depths of over 150 feet to be found in fish pots on the beach. To the fishermen, the crabs are destructive and unwelcome, and they are tossed out on the sand before the pots are returned to the sea. As a result, the areas where fishing canoes are drawn up on the beach are often littered with large shells from deeper water, usually with the dead crabs still in place.

But beach collecting has two basic limitations, which explains why one cannot hope to develop a really useful collection by relying on it to the exclusion of other methods. In the first place, shells which have been rolled along the sea floor and eventually washed up out of reach of the waves are rarely found in perfect condition – they have usually suffered some damage, slight as it may seem, and are likely to have been faded by the sun. Secondly, in normal circumstances shells found on a beach have come from relatively shallow water; the deeper water species are rarely washed ashore, and when they are, the turbulent seas which are needed to roll them up from the bottom generally destroy them on the beach. The molluscan by-product of a hurricane is often disappointing.

Nevertheless, beach collecting has provided the nucleus of many fine collections, and occasionally it rewards even advanced collectors with some new species which they have been unable to find anywhere else.

(ii) *In Salt Marsh and Mangrove areas*

A number of locally common species occur above the low-water mark in brackish, salt marsh areas. They include the Melampus snails (*Ellobiidae*), several Nerites (*Neritina*), and the attractive little Horn Shells (*Cerithidea* and *Batillaria*) which are exposed on the mud flats by the retreating tide.

The Horn Shells make no effort to conceal themselves, but the Melampus snails, and some of the Nerites, hide in the daytime under rocks and pieces of damp wood, often some distance from the water. A detailed search of a small area is the most productive method of collecting them.

Salt marshes are not notably attractive places, but they are host to several species which normally occur nowhere else, and, fortunately, these are all relatively common and easily collected.

Mangrove swamps also support a number of interesting species, some of which, like the attractive Angulate Periwinkle (*Littorina angulifera*), live about the high-tide mark, while others are exposed only at low tide, clinging to the mangrove roots. The latter group includes various species of Oyster, Mussel and Chama, and less commonly, several varieties of minute gastropods, hidden in the algae which cover the roots. The West Indian Murex (*Murex brevifrons*), an insatiable predator, often climbs the roots to attack the oysters and is easily collected there. (Fishermen at Falmouth report collecting the beautiful Measled Cowrie (*Cypraea zebra*) from the roots of mangroves in that area, but this needs confirmation.)

Like salt marshes, mangrove swamps are not attractive places, but fortunately one can collect all of the species which occur there above the low-tide line in two or three visits, and the result is well worth the effort.

(iii) *In Rocky areas*

A number of common species live in the inter-tidal zone, or just above the high-water mark, in rocky, wave splashed areas. They include gastropods, bivalves and chitons, and most of them occur in large colonies and are very easily collected.

Search the rock pools for Zebra Nerites (*Puperita pupa*) and Dwarf Periwinkles (*Littorina mespillum*). Most of the other species, such as the several varieties of common Nerite (*Nerita*), make no effort to conceal themselves on the rocks. A later section of this chapter deals with methods for collecting chitons, and the same technique can be applied to the removal of the gastropod *Purpura patula*, which is often difficult to lever out of its crevice in one piece.

COLLECTING UNDER WATER

The majority and most attractive of West Indian shells occur in water less than 20 feet deep, which puts them within reach of the skin diver who does not use self-contained under water breathing apparatus (Scuba). Many of them are found in less than 3 feet at low tide, and can be collected simply by wading with a glass-bottomed box.

Those species which occur at depths greater than 30 or 40 feet are in the province of the Scuba diver. It is dangerous to use Scuba without the requisite training, and inexcusable to dive alone, but used according to accepted practice Scuba is an invaluable collector's asset. It puts within his reach live examples of many of the rare species which are otherwise obtained only by dredging.

There are two important factors to bear in mind when searching for live shells:

(i) Most shells avoid sunlight; only a few species are obliging enough to show themselves obviously when the sun is up,

(ii) Very few shells in their natural state exhibit the bright colours which are often revealed after they are cleaned. They are usually covered by dull coralline or algal growths, and blend well with their backgrounds. The periostracum of many species serves the same purpose. Divers who expect to find the sea floor carpeted with brightly coloured living shells are in for a disappointment. One learns by experience to look for the outline of a shell rather than any revealing splash of colour.

With very few exceptions, marine molluscs which occur below the inter-tidal zone can be found in one of these seven environments –

(i) Under rocks or fragments of broken coral,
(ii) Buried beneath the sand or mud of the sea floor,
(iii) In self-made burrows in coral, clay or wood,
(iv) Deep in the crevices of coral reefs,
(v) Floating on, or below, the surface of the sea,
(vi) Clinging to, or embedded in, marine plants or animals,
(vii) Exposed on the surface of the sea floor.

Productive collecting in these places calls for different techniques, and each merits separate consideration.

(i) *Under rocks and coral fragments*

Collect at low tide whenever possible. Turn over the rock or coral slab (use a crowbar if necessary) and examine first the depression beneath it. Some species will make an immediate attempt to bury themselves, so quick action is necessary. 'Fan' the water above the depression gently with one hand to disturb the sand and uncover those shells already concealed there.

Then search the undersurface of the rock itself, paying attention to crevices and seaweed growths which often conceal small specimens. If the water is shallow and the rock light enough, it is well worth the trouble of lifting it on to dry land and minutely examining it there. Worm shells (*Vermetidae*), and those bivalves which cement themselves to the rock, are sometimes very difficult to remove intact and may require the use of a hammer and small chisel. If their host rock is small enough to be taken home, the job can be done far more safely and efficiently there.

Many of the shells found on or under the rock may contain hermit crabs. These small crustaceans are far more agile than the original molluscan occupants of their shells, and should be transferred quickly to an escape-proof container (a cloth bag with drawstring is not always adequate). They are gregarious creatures, and in shallow water especially on eel grass beds, a large rock may conceal a concentration of several thousand small hermits. Most of the shells will be common species, but in a large colony there are invariably a few interesting specimens from deeper water.

One of the most rewarding places for divers to search is under those broken slabs of coral which fall into the crevices between sections of reef, or are washed over the reef into the protected water just inside it. Many of the choice species which normally live deep within the reef may seek temporary shelter under these slabs. (A short, sturdy piece of metal with a hook on one end is invaluable for lifting loose coral, which often has razor-sharp edges.)

Generally, the heavier the rock or fragment of coral the longer it will have remained in its present position; consequently, more shells will have had an opportunity to adopt it as a permanent home. Small rocks which are constantly shifted by moderate seas are likely to shelter only the most common species.

Finally, remember that it is more productive in the long run to search a small section thoroughly than to attempt to cover a wide area in a cursory way.

(ii) *Buried beneath the sea floor*

A large number of bivalves and gastropods spend much of their lives buried beneath the sea floor.

In the case of the bivalves, there is seldom anything to indicate their exact position under the sand or mud, and consequently they are very difficult to locate. (One exception is the family of little Coquina clams (*Donacidae*), which live on sandy beaches at the point where the waves break, and are momentarily exposed as the water withdraws. The attractive Grey Auger (*Terebra cinerea*) preys on the Coquinas and the two species are often collected together.) Occasionally when a mud flat is exposed by an unusually low tide, some species can be traced by carefully watching for any sudden eruption of water, and then digging immediately beneath it. When the buried bivalves are covered by the sea, however, they are almost impossible to trace.

Fortunately, those gastropods which normally live under the sea floor are far more obliging. Most of them, like the Olives (*Olividae*), and Moon Snails (*Naticidae*), emerge to travel over the surface of the substrate at night, and bury themselves again at the approach of daylight. In doing so, they leave distinctive trails on the surface of the mud or sand. Follow the trails and dig where they disappear into the substrate, and the shells which made them are often uncovered.

Some species, including the voracious Moon Snails, hunt their prey beneath the surface of loose mud or sand, and they are able to move about quite freely in this element without exposing their shells. Very often, however, it is possible to follow their progress by watching for the small, moving mounds of sand which build up just ahead of them and indicate their presence two or three inches below.

When all else fails it is usually worth digging at random in some likely spot, and then passing the uncovered material through a sieve. This may well turn up not only gastropods and bivalves, but also the occasional Tusk Shell.

(iii) *In self-made burrows*

Several species of bivalve burrow deep into certain selected materials and pass the rest of their lives safely entombed there.

The materials most commonly penetrated in Jamaican waters are coral, clay and wood, but some of the smaller species bore into the shells of other living molluscs, and the beautiful Atlantic Thorny Oyster (*Spondylus americanus*) is sometimes unwilling host to several destructive varieties of borer.

The delicate Martesias are found in large colonies in wooden pilings, discarded planks, waterlogged tree trunks and other flotsam. Examine weathered pieces of wood washed up on the beach. Many will bear deep, round holes, at the bottom of which lie the *Martesia* shells safely insulated from damage.

In the case of the Coral Borers, one can recover moveable chunks of dead coral and search for the burrows of the boring Mussels (*Botula, Lithophaga, Gregariella,* etc.). Most of them do not penetrate very deeply into the coral and the tips of their valves may be visible at the entrances to the burrows.

In all cases, take the chunk of wood or coral home, whenever this is possible, and 'operate' on it there to extract the shells. Most borers are delicate and not easily removed from their holes; any clumsy attempt at extraction will probably damage the shell. Where the material is too large to be removed, a small hammer and sharp chisel are necessary. Remember to start chipping some distance from the actual burrow or the wall may cave in and fracture the shell.

Many borer bivalves are surprisingly beautiful, and they are all well worth the effort involved in recovering them from their strange, tubular homes.

(iv) *In the crevices of coral reefs*

A large proportion of the rare and coveted West Indian gastropods occur deep in the crevices of coral reefs. Many of them are not necessarily deep water species, but their peculiar habitat frustrates most collecting methods. They include such choice species as *Conus juliae, Conus granulatus, Murex spectrum,* the delicately beautiful Chocolate-lined Top Shell (*Calliostoma javanicum*), and a number of smaller but equally attractive shells.

Very occasionally, a commercial dredging operation includes the wholesale removal of an entire section of coral reef. This happened in Jamaica at Discovery Bay in 1964, and again in 1967, and on both occasions quantities of rare species were brought to the surface. Searching the debris from a dredging operation of this kind is undoubtedly both the easiest and most productive method of collecting shells which live deep in the reef. Unfortunately, these circumstances do not often arise, and the average collector must rely on other methods.

In general, the species we are referring to are accessible only to divers, and particularly to divers using Scuba. There are two places to search. The first is in the ooze which gathers at the base of the reef, where the shells will probably be dead, but not necessarily damaged. Many species, including those which live deep within the reef fall to its base when they die, and these shells gradually sink into the debris which collects there. In 1966, three enterprising members of the Jamaica Sub-Aqua Club constructed cylindrical sieves, about six feet long and one foot in diameter, from one-eighth inch mesh wire. Using Scuba, they took these sieves down to the bases of a number of coral reefs at depths of about 60 feet, and shovelled handfuls of the debris through the wire mesh. The net result was a large quantity of choice species, many of which normally live deep within the reef. All were dead, but having been cushioned in the soft ooze, most were recovered in perfect condition. This sieving method is highly recommended – but it is hard work, and only for people proficient in the use of Scuba.

The second, and perhaps more obvious method, is to probe the accessible parts of the reef. Here again, to find the rarer species, which occur at depths below 25 feet, Scuba is necessary. The technique involves searching all the accessible crevices below that depth, in the hope that some of the species which usually live deep within the reef will have ventured temporarily to its outer walls. In 1964, a diver casually doing this at Orange Bay, Jamaica, discovered a colony of six *Conus granulatus* deployed in a single hole in the reef at about 50 feet. (The unhappy sequel to this story is that the diver – who had no idea of the significance of his find – took only one specimen, and this was later accidentally destroyed by fire. The colony was never located again.)

One note of caution – shells are not the only marine animals which favour dark crevices; Moray eels are also found there. As a general rule, don't push your hand into a crevice without first looking to see who lives there.

(v) *Floating on, or below, the surface of the sea*

Most people do not associate snails with life on the surface of the sea, yet the exquisite Purple Sea Snails (*Janthinidae*) occur in concentrated masses of millions of individuals far out on the surface of the Atlantic Ocean. A few fathoms below them in even greater numbers, are the Sea

Butterflies (*Pteropoda*), which use their modified mantle flaps to swim through the water.

The *Janthinidae* in the course of their involuntary wanderings, are often washed ashore, but their delicate shells are usually damaged by the waves which sweep them up the beach. Occasionally, living specimens can be intercepted before they reach the shore, and this is the only certain method of obtaining perfect shells. Unfortunately, Jamaica does not receive these molluscs in the same numbers as many other West Indian islands (notably Barbados), but individuals are washed ashore on the north coast in a fairly constant stream during the windy months of the year.

In the case of the Sea Butterflies, whose shells are small, delicate, and inconspicuous, occasional specimens can be found in beach drift on soft, sandy beaches. Dead shells of the common species are more easily obtained by dredging, and the 'sieving operations' referred to in (iv) above produced a number of undamaged *Cavolinidae* (an exquisite family of Pteropods).

The buoyant, chambered shells of the cephalopod *Spirula spirula* are very commonly blown ashore in the West Indies, and in spite of their delicate construction the shells often survive their landings intact, even on rocky beaches. The living animal occurs at depths well outside the range of all but the most sophisticated dredging equipment (unlike many live Pteropods which can be 'trawled' in two or three fathoms).

(vi) *Clinging to, or embedded in, marine organisms*

A number of molluscs maintain a parasitic or symbiotic relationship with other marine organisms. In Jamaican waters, Sea Fans (*Gorgonia*) and Sea Whips (*Plexaura*) are host to several species including the beautiful Cyphomas and Simnias (*Ovulidae*), which normally occur nowhere else. The Sea Fans, and certain sponges, also support various Worm Shells (*Vermetidae*). The Simnias and some Worm Shells adopt the coloration of their host organism, and must be looked for methodically; the spectacular Cyphomas, on the other hand, are not well camouflaged (except on certain *Plexaura*) and can be very easily seen.

The irregular, wood-like base of the Sea Fan often contains two or three different species. The most common of these is the Caribbean Coral Shell (*Coralliophila caribaea*), which seeks shelter in the crevices of the broad base. When the Sea Fans are torn loose from the bottom and cast up on the beach, these shells remain wedged firmly in their crevices, and they perish with their hosts.

Several shallow water species live on blades of eel grass. These include the Modulus shells (*Modulidae*), the Chestnut Turban (*Turbo castanea*), and some Nerites (*Neritidae*). They are usually found on the under-side of the blades, where they feed on the algae which grow there.

Many other marine plants and animals at one time or another may play host to molluscs, and it is interesting to note that a recent Jamaican survey indicates that while brown algae support more individual molluscs, red algae attract a greater variety of species. Green algae run a poor third on both counts.

Some of the rarest shells in the world have been recovered from the stomachs of bottom feeding fish, and this suggests the possibility of making an appropriate arrangement with a local fish processing company.

(vii) *Exposed on the surface of the sea floor*

Some species show themselves fearlessly on the sea floor, even in daylight. Predictably, with a few notable exceptions, they are the larger molluscs whose massive shells are proof against most marauding enemies. They include the Conchs (*Strombidae*), the Helmet Shells (*Cassididae*), and the classic Triton's Trumpet (*Charonia variegata*), but even these species are usually well camouflaged by coral and algal growths, and in the case of the Helmet Shells, partially buried in the sand.

All five local *Strombidae* can be collected occasionally in water less than 3 feet deep, but the Queen Conch (*S. gigas*), the Milk Conch (*S. costatus*), and the Rooster Tail (*S. gallus*), normally prefer deeper water, and are sometimes found at 60 feet. The Atlantic Carrier Shell (*Xenophora conchyliophora*), which is closely related to the Conchs, cements fragments of coral and other solid debris to its shell, and resembles nothing more than a small pile of rubble on the eel grass beds, where it is usually found.

The two most common Helmet Shells (*C. tuberosa* and *C. flammea*), the West Indian Crown Conch (*Melongena melongena*), and the Triton's Trumpet are also shallow water species and can be collected by waders, but the Emperor Helmet (*C. madagascariensis*) is rarely found in water less than 15 feet deep. The larger Murex shells, particularly the West Indian Murex (*M. brevifrons*) and the Apple Murex (*M. pomum*), may also be found on the sea floor in shallow water, both in rocky and sandy areas.

Among the bivalves, various species of Scallop are often found resting on the sea floor. The Zigzag Scallop (*Pecten ziczac*) and the Calico Scallop (*Aequipecten gibbus*) are commonly found on the eel grass beds in Kingston harbour. Several species of Mussel (*Mytilidae*) and

Oyster (*Isognomonidae*) occur in large colonies on wharf pilings, rocks, and general debris in very shallow water; they are all easily seen and collected.

Other sessile bivalves which cement themselves to some firm surface (*Spondylidae* and *Chamidae*, for example), are often covered by algae and inanimate marine debris. The Atlantic Thorny Oyster (*Spondylus americanus*) with its long weed-entangled spines, is particularly well camouflaged and often virtually undetectable – until it snaps its valves shut at a diver's approach. One should watch for this sudden movement of camouflaged bivalves: it is often the only indication of their presence.

There is no particular technique involved in collecting shells from the sea floor, but remember that Scallops can swim and that Helmet Shells often expose only the tops of their shells above the sand. Bear in mind also that a 12-inch Triton's Trumpet will not fit inside a 6-inch collecting bag.

Most molluscs remain fairly placid once they are removed from the water, but the Conchs show their displeasure in no uncertain manner. They use their powerful foot to tumble their shells in every direction, and are quite capable of propelling themselves several yards down a beach and back into the water. If they are left on a hard surface, they will eventually chip the lip of their shells. The most effective way to preserve them from damage, until the animal can be removed from the shell, is to wrap them in a damp towel.

Scallops also attempt to return to the water by vigorously flapping their valves, and the larger species are capable of leaping out of a small boat if handled carelessly.

Divers often encounter large hermit crabs walking busily along the bottom. They are easily captured, but remember that hermits are active animals with powerful claws, which they can extend round the back of their shells to nip the fingers holding them. Most collectors react understandably to this sort of attack by dropping the shell, and many fine specimens have been lost in this way. The solution is to pick up the shell quickly and drop it in the collecting bag before the crab has time to consider his predicament.

EQUIPMENT

One way to guarantee an unsatisfactory day's collecting is to take along all the items of equipment recommended as essential by well meaning 'experts'. As a general rule, the less encumbered one can be when collecting the better.

The following list of basic equipment may be useful. Take at one time only those items you require for the particular type of collecting planned (e.g., don't carry a glass bottomed box if you intend to collect at 60 feet using Scuba equipment).

Collecting bags
Plastic bottle with wide mouth and screw cap
Small, sturdy knife
Trowel and (plastic) bucket
Glass bottomed box (water glass)
Short crowbar
Small hammer and chisel
Small rake
Plastic sieve
Mask, snorkel and fins
Pair of strong gloves
Pair of canvas shoes
Pair of rubber boots
Scuba equipment (if competent to use it)

Rubber boots are invaluable for collecting in shallow water where the bottom is carpeted with the long-spined Sea Urchins (*Diadema*). Contrary to 'expert' opinion, the spines will not penetrate stout rubber.

A small rake can be used effectively on smooth, rocky bottoms, carpeted with thick weed. Use the rake gently and systematically over a pre-selected area.

DREDGING

Dredging for shells is not a pastime reserved only for those fortunate people with large boats, power winches, and plenty of money to spend on the necessary equipment. It is true that dredging in deep water requires these things, but on the other hand, one can dredge quite efficiently in depths down to 150 feet using a small rowing boat and a hand made dredge. And a very large proportion of the really choice West Indian species occur within this depth range.

The simplest type of dredge consists of a rectangular steel frame (measuring about 24 by 8 inches) with a pair of metal bridles which are tied together. A fine-meshed nylon fishing net is sewn to the frame. The leading edges, at the mouth of the frame, are sharp and bent outwards, so that the dredge will bite lightly into the bottom. (See fig. 20).

There are various popular modifications of this design, one of which substitutes copper screening for the nylon net. The tow line may be of $\frac{1}{2}$-inch manila rope, or preferably, $\frac{1}{4}$-inch nylon cord. The ends of the metal bridles should be tied together with a small cord which will break

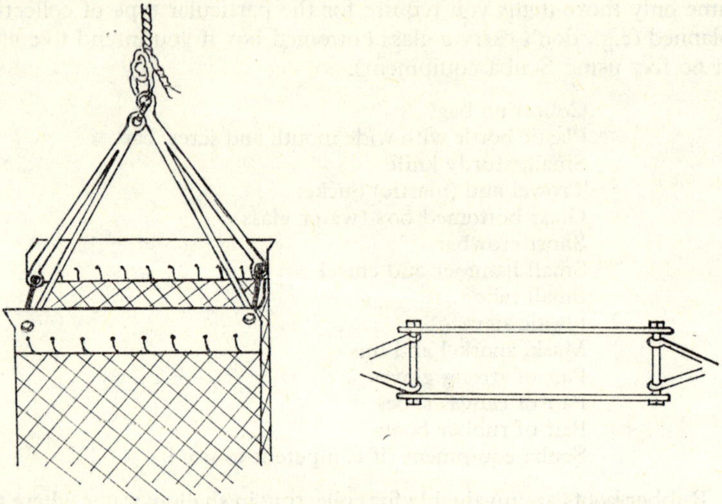

FIG. 20. Diagram of a small hand dredge. *Left,* side view, with netting attached (note small safety cord securing the bridles). *Right,* front view showing attachment of bridles to the frame (which measures about 24 × 8 inches). (After Abbott)

loose and free the dredge, should it snag on some underwater obstacle. The dredged material should be carefully sieved.

Dredging can be a most productive way of collecting shells, but it requires an intelligent approach and knowledge of its limitations. A dredge will operate only on a relatively smooth sea floor of sand, mud or gravel. It is useless on a reef, or in an area where there is plenty of coral or other large obstacles. Reconnaissance of a promising area is time well spent and can obviate a great deal of frustration and disappointment.

Shrimp trawlers which operate in the Caribbean, and are based at several West Indian islands, bring up in their nets many rare deeper water shells. Collectors can make arrangements with the crew of these vessels to save the shells for them, but one must be prepared to pay for this service.

TRAPPING

The fact that many fine deep water shells can be found in the pots set by fishermen suggests the possibility of setting one's own traps.

The type of pot used by Jamaican fishermen is, patently, a most efficient trap for shells. The large size is unnecessary for this purpose, however, and a pot of the common design but measuring roughly 4 by 3 by 1 feet is probably ideal. Bait the pots with decaying meat or fish, and this will attract the carnivorous gastropods. Trapping the vegetarians, and stray swimming bivalves, is a matter of chance, but occasional specimens do wander into pots from time to time.

NIGHT COLLECTING

Most gastropods emerge from their places of concealment at night to feed, and collecting after dark can be most rewarding. All those species which in the daytime conceal themselves beneath the mud or sand of the sea floor, come to the surface at night. They include the Olives (*Olividae*), Moon Shells (*Naticidae*), and Bubble Shells (*Tectibranchia*), and with the aid of a powerful torch and a glass bottomed box, many of them can be collected in 2 or 3 feet of water. (Both sides of Kingston harbour are excellent for this kind of collecting at low tide on calm nights.) Night diving in deeper water is equally profitable, but this is definitely the preserve of experienced divers only.

COLLECTING CHITONS

Chitons must be collected in a prescribed way to ensure that they can be attractively preserved later. When a chiton is removed from the surface on which it is resting, it immediately rolls itself into a tight ball; it can be forcibly straightened out only at the cost of destroying the shell.

The solution is to drop it immediately into a container of sea water, and then allow it to unroll and attach itself to a previously prepared length of smooth, flat wood (a piece the size and shape of a 12-inch ruler is adequate). When it has done this, remove the wood and bind the chiton tightly to it with insulating tape. It can later be immersed in fresh water or 70% alcohol, or frozen in the ice compartment of a refrigerator; after it has died the soft parts can be removed, and the shell will thereafter retain its natural shape.

The initial job of removing the chiton from the surface on which it is resting requires the use of a knife blade. Slip the blade under the mollusc's foot and flip it over. If this can be done quickly, and before the animal has a chance to tighten its hold, there should be no difficulty. However, if the chiton senses danger before the blade can be inserted, it will usually be impossible to remove it intact. In this case leave it alone for about five minutes. This will allow it to relax, and another attempt can then be made with the knife blade.

SALT WATER AQUARIA

The living animals (i.e. the soft parts) of many marine molluscs are even more attractive than their shells, and some of the shell-less Nudibranchs which are usually ignored by collectors, seem to have been compensated for lack of a shell by the most exquisite decoration. The diversity of molluscan behaviour has already been described.

Fortunately, molluscs make excellent subjects for study in an aquarium. They have a high level of tolerance for such factors as temperature and salinity variation, and are subject to few ailments. It is hardly surprising, therefore, that many collectors sooner or later decide to study living molluscs at close quarters. There are many excellent books on marine aquaria, and the following paragraphs deal with the subject only in broad outline.

Equipment

Essential requirements are:

(i) A glass tank, preferably without metal fixtures, and with a minimum capacity of 20 gallons,
(ii) A filter and an aerator, either separate or in combination,
(iii) Sea water (formulae for producing this artificially are available in most books on marine aquaria, and at many pet shops which stock fish),
(iv) An arrangement of rocks, sand, weed, etc. to reproduce as accurately as possible the natural habitat of the molluscs to be kept.

Setting-up procedure

Ensure that the tank is perfectly clean and that it is set on a base strong enough to bear the additional weight of the water. Collect sea water from the cleanest possible area, preferably from a boat in deep water. Avoid collecting it from contaminated harbour areas, or where there is a high concentration of fresh water. If necessary, filter the water before pouring it into the tank.

Arrange the filter/aerator in the tank, and then pour in the water to about three-quarters of the tank's capacity. Introduce the rocks, sand, weed, etc., and make certain that the algae which normally grow from the substrate have their roots securely buried in it. (A substrate depth of about 3 inches is usually ideal.) Start the filter/aerator, and when the water is clear put the live molluscs into the tank; they will quickly discover their particular type of habitat. The sand dwelling species are apt to bury themselves immediately; the Conchs (*Strombidae*) will initially create chaos by flinging themselves all over the bottom and churning up the carefully arranged substrate; and some of the carnivorous gastropods, particularly the *Fasciolariidae*, will lose no time in attacking and eating the other molluscs, maliciously starting with one's most prized specimens. This last consideration merits some thought. Don't mix gastropods which are known to attack other molluscs with less aggressive species if you value the latter; the results of doing this are prompt and depressing.

There are no concrete rules governing the number of molluscs which may be kept in a specified volume of water. The aim is that this should be consistent with reproducing in the aquarium the molluscs' natural habitat as closely as possible. Overcrowding precludes this, and no doubt produces un-typical behaviour. As a general rule when deciding on the number: 'If in doubt, leave out.'

When salt water evaporates, it leaves behind its salt content. Evaporation loss must be made up, therefore, by distilled water – not more sea water. A glass cover over the tank will retard evaporation, but many collectors prefer to have the tank left open, and are prepared to add distilled water more often. The level of the water should not be allowed to fall appreciably.

Feeding

Many collectors make no attempt to feed their shells, some apparently in the belief that all molluscs can exist by abstracting food from the water. As explained in Part I, this is true of most bivalves, but it certainly does not apply generally to gastropods.

The fact is that although many shells can live for a long time without eating, they will, of course, eventually die of starvation if they cannot find food.

One of the most challenging aspects of keeping an aquarium is determining which species eat what. When this has been ascertained, it is a comparatively easy job (for people with access to the sea) to keep the molluscs well fed.

Some carnivorous gastropods, such as the *Conidae*, are very particular about what they eat, and require it live. Others, like the *Muricidae*, will eat scraps of cooked meat, and *M. woodringi* shows a definite preference for corned beef!

Most vegetarians are harder to please, though some of the Conchs (particularly *S. gigas* and *S. raninus*) are not at all fussy, and will happily demolish large quantities of many varieties of common algae.

Bivalves seem to be able to exist successfully in an aquarium in which gastropods are well fed, without special attention. An aquarium with an excessively efficient filter is probably bad for bivalves, which need minute particles of food to be present in suspension in the water.

A number of species are problem eaters. In their natural state their feeding habits are highly specialised, and if one intends to keep them in an aquarium for any length of time they must be provided with their particular diet. Among these species are the beautiful Cyphomas and Simnias, which normally feed only on Sea Fans or Sea Whips, and they require special attention.

All scraps of meat which remain uneaten after about 24 hours, should be removed from the aquarium before they decompose and contaminate the water.

Reproduction

Most, if not all, molluscs will lay eggs in an aquarium if conditions are favourable. If one wants the eggs to hatch it is advisable to remove them to a separate tank to prevent their being eaten.

Many molluscs are hermaphroditic, so it is not always necessary to have two specimens present in the aquarium for eggs to result. There is no doubt that one of the most satisfying achievements for any collector is to find a very rare specimen, and then produce others from it in an aquarium. It does not often happen, but occasional success compensates for many past failures.

It has been said that one can really appreciate a shell only by seeing it first in its living state. It is certainly true that keeping an aquarium tends to develop a different attitude towards collecting, and often opens up an entirely new field of interest. There are large gaps in our knowledge of many aspects of molluscan behaviour, and there is no reason why some of these cannot be filled by amateur collectors with their own salt water aquaria.

CONSERVATION

Marine life is most vulnerable to two destructive factors – over-collection and water pollution. The latter is by far the more dangerous of the two.

The moment that man interferes with the natural environment, life of some description is endangered. In the case of molluscs, the danger usually comes through the discharge into the sea of pollutants such as industrial waste, oil, or even excessive quantities of fresh water (through drainage projects or the re-direction of rivers), or from interference with the sea bed, generally by the use of suction dredges.

In Jamaica, Kingston harbour provides a notorious example of an area in which the molluscan population has measurably declined in the years between 1960-1970. Several previously common species apparently no longer occur in the harbour. In this case, both factors are responsible – pollution from drainage projects and industrial waste, and extensive dredging on the western side. (The same factors have also killed very nearly all corals in the harbour, and decimated the fish population.)

Unfortunately, when ranged against the economic reasons for this situation – particularly the factors of drainage and land reclamation – the case for the preservation of molluscs in one relatively small area can be made to appear very slight. In a poor country, struggling bravely with the problems of poverty and unemployment, one must be realistic about orders of priority, but it should be remembered that the factors which kill molluscs may also destroy priceless and irreplaceable fish nurseries.

However, the question of over-collection is another matter entirely. If all collectors (using the word in its broadest sense), observed the following rules, the problem possibly would never arise:

(i) Don't collect (live) immature molluscs, which have not had a chance to reproduce themselves,

(ii) Don't collect more than 2 or 3 specimens of one species from the same place,

(iii) Don't turn over a rock and then leave it exposed to the air, for this will inevitably kill all the marine organisms fixed to its underside, including molluscan eggs. If possible, replace all overturned rocks, even underwater (this is asking a lot, but it is something to aim at).

In the West Indies, fortunately the problem as it affects individual species at this time is not at all serious, with the single exception of the case of the Queen Conch (*Strombus gigas*). This species is widely used both as fish bait and food, and in Jamaica, for example, examination of the area around any established fishing village will often fail to produce a single adult Conch. The reason is convincingly demonstrated on the Negril beach. There one can see huge piles of discarded Conch shells originally laid down 20 or 30 years ago. The bottom layers are mostly mature adult shells. But the higher up the pile one looks, the more immature are the shells. The layers, of course, are laid down in chronological order up to the present day. The result of this constant assault on the species, before it can reproduce, has been to make the once abundant Conch very scarce indeed in that entire area, and as a natural consequence the fishermen have to sail farther and farther away every year to find an adequate supply of shells. This situation is repeated many times around the Jamaican coastline.

The solution would seem to lie in a combination of legislation and education, and the sooner the better. The speed with which abundance can turn to scarcity has always taken man by surprise.

Cleaning

REMOVAL OF THE ANIMAL FROM ITS SHELL

The job of removing the soft parts of the animal from its shell is not a very pleasant task by any standards – but it has to be done. Decomposing mollusc stands high on anyone's list of the world's most obnoxious odours. Every collector has his own favourite methods of cleaning his shells, and the following paragraphs describe the more popular and effective ones.

In Gastropods

Removing a gastropod animal from its shell, to which it is connected by internal muscular attachment, is often a more difficult and time consuming job than collecting it. In some species like the long, many-whorled *Terebridae*, complete removal is often impossible. The most commonly used methods are as follows:

(i) *Boiling* – place the shell in a sieve in fresh (or sea) water at room temperature. Bring the water to the boil slowly, and allow to simmer for about 5 minutes (double this for very large specimens and halve it for those under 2 inches). Allow the water to cool, take out the shell and remove the soft parts with a piece of bent wire or a pair of forceps, using a corkscrew motion. A powerful jet of water will often flush out any tissue which may have broken off in the spire. Remove the operculum from the animal, if it has one, and when the shell is finally cleaned, place a plug of cotton wool in the aperture and lightly glue the operculum to it, to hold it in its natural position. An operculate shell is not complete without its operculum.

The boiling method is quick and efficient, but it can check the shell's surface, and will almost certainly do so if the shell is put directly into boiling water. For rare or polished species (Cowries, Olives, etc.), one of the following methods is safer.

(ii) *Removal by Insects* – Ants and insect larvae are efficient but slow removers of animal tissue. Bury the shell lightly in loose earth, or leave it exposed in the shade, where insects can get at it. Fly larvae and ants will set to work to remove the animal remains without damaging the shell. Remember to recover the operculum when the cleaning job is done. Rats and mice are attracted to dead shells, and will carry them off if they can, so take what precautions are necessary to prevent this.

(iii) *Removal in water* – If shells are left in water long enough, the animals will eventually decompose and can then be washed out. There are two drawbacks to this method: it takes a long time, and the decomposing animal secretes a substance which may stain the shell (which necessitates frequent changes of water).

(iv) *Freezing* – Place the shell for about 24 hours in the freezing compartment of a refrigerator. Take it out and allow it to thaw; it is then often possible to remove most of the animal with a pair of forceps. Vigorous rinsing may remove the rest of it. The advantage of this method is that it is relatively quick, and does not check the polished surfaces of the shell.

(v) *Use of Alcohol* – Shells less than ½ inch long can be placed in 70% grain alcohol for 48 hours, then removed and dried. They should then be free from odour, and unharmed by the alcohol. For larger shells in which some animal matter remains near the spire, a small quantity of alcohol can be poured into the aperture and left in the inverted shell for a few days. This often kills the remaining odour.

(vi) *Use of Formaldehyde (or formalin)* – Formaldehyde has a disastrous effect on shells which are placed in it to preserve the animal. The shell itself quickly deteriorates and loses colour. Its use in association with shells is not recommended for any purpose, unless it has first been 'buffered' with baking soda.

In Bivalves

In contrast with the procedure for gastropods, the removal of bivalve animals is very simple. Place the bivalve in fresh water; within 36 hours (12 for the smaller species) it will die, the adductor muscles will relax, and the valves will gape open. The soft parts are then easily removed with a spoon or knife.

The ligament will permanently stiffen shortly after this, and one must decide whether to store the shell with the valves closed or gaping. If they are to be closed, bind them together for about 24 hours, and they will set in this position.

In Chitons

The method for collecting chitons has already been described. When the animal is dead, unbind it from its piece of wood, and with a sharp scalpel blade cut around the margin of the foot and remove the soft parts. Rebind the shell on the wood, and allow it to dry thoroughly. It can then be cut free and will be ready for storage.

In Shells containing Hermit Crabs

Hermit crabs, which are firmly attached to their adopted shells by a hook around the columella, drown in fresh water. Place the crab in a container from which it cannot escape and cover it with fresh water. After some hours the crab may evacuate its shell voluntarily (and can then be liberated); otherwise it will drown in about 24 hours and can be shaken out of the shell. Do not attempt to remove it forcibly from the shell while it remains alive.

Cleaning the Shell externally

The shells of most molluscs – excluding those which bury themselves in the substrate (*Naticidae*, etc.), or habitually cover their shells with their mantles (*Cypraeidae*, etc.) – are sooner or later disfigured by coralline or algal growths. Many species also possess the natural cloak of their periostracum. In both cases, the colour pattern is obscured.

The periostracum is as much a part of the shell as the operculum, and most collectors wish to preserve at least one specimen of each species with the periostracum still in place. For the other specimens, a 30% solution of a chlorine clothes bleach will remove the periostracum in a matter of hours, and will not damage the shell – except for the nuclear whorls. Chlorine will destroy the nuclear whorls very quickly, and it is essential that they should be protected by a coating of wax or plasticine before the shell is immersed in the solution. This is an important consideration, for a shell without its nuclear whorls is considered incomplete. The shell should be removed from the solution as soon as the periostracum has loosened sufficiently to be taken off, and it should then be washed thoroughly in fresh water.

In the case of shells encrusted with coralline and algal growths, cleaning is more difficult. Chlorine will remove much of the algal deposits, but it has no effect on the coralline material (which is mostly calcium carbonate, like the shell itself), and this must be tackled laboriously by hand. A variety of sharp instruments is necessary; old dental probes, scalpel blades, and ice picks are very useful. Some of the deposits may flake off quite easily, but most of them must be removed piece by piece, and in such a way that the underlying shell surface is not damaged. Patience and perseverance are necessary; and in most cases the effort is rewarded by a clean shell. No live collected specimen is so badly encrusted that it cannot be restored to an acceptable condition.

The Use of Acid

The use of hydrochloric (or muriatic) acid in the preparation of shells is anathema to most collectors, for it may destroy the scientific value of the specimen and tends to give the shell an unnatural appearance.

Its use is permissible, perhaps, only in the treatment of old, beach-worn specimens which have resisted all other methods of cleaning. In this case, the acid should be diluted by 50%, and the shell plunged into it for a few seconds at a time, and then immediately held under running water. As a general rule, serious collectors avoid using it.

The Use of Oil

A thin coat of mineral oil on the surface of a dull shell can work an apparent miracle. It immediately restores previously hidden colours, and imparts an attractive sheen to the shell.

It can do no damage, and probably assists to some extent in the preservation of the shell's colour. Its one drawback is the fact that it tends to collect dust, but there are obvious remedies for this.

A periodic smear of oil will also prevent the periostracum of a shell from becoming brittle, and eventually flaking off, and it is worth treating the ligaments of bivalve shells in the same way. Excess oil should be removed before the shell is put away.

Storing and Cataloguing

Perhaps the very first fact about shell collecting that one should learn is that shells are of no scientific value (and greatly reduced commercial value) if they are not accompanied by accurate data. The importance of many of the great collections of the eighteenth and nineteenth centuries was substantially diminished because most collectors at that time did not appreciate this. In some cases, if they were not certain of the place from which a particular shell came they cheerfully invented a locality. (Until quite recently, there was a reflection of this in the shell gallery of the British Museum, where one exclusively Caribbean species was labelled as coming from the Red Sea.)

Adequate data should include the following information, of which (ii) and (v) are by far the most important:

(i) Scientific name of shell,
(ii) Locality collected (as precise as possible),
(iii) Name of collector,
(iv) Date collected,
(v) Full details of the circumstances in which the shell was found, i.e., habitat, depth, incidence, etc.

The simplest way to record this information is in a catalogue subdivided into these headings, and with each entry preceded by a serial number. This number is repeated on the shell in Indian ink. For very small specimens, a label bearing the number should be placed in their container. Each new shell to be entered should be given the next available number in the catalogue regardless of its species or class. An example of an actual catalogue entry is:

Serial No.	*Species*	*Locality*	*Collector*	*Date*	*Details*
1886	*Strombus gallus*	Priory, St Ann, Jamaica	J. M. Humfrey	1.6.67 1300 hrs.	Collected live in 6 ins. water on *thalassia* bed, 400 feet from shore. Tide low. No others seen.

Because of their endless variation in size and shape, shells are unsatisfactory objects to store neatly and methodically. Every collector eventually develops his own methods, and the ideal basic requirements are (i) a wooden cabinet with many shallow drawers, and (ii) small containers (for individual species or lots), which will fit neatly into the cabinet drawers.

It is a good plan to look at a number of different cabinets (in Museums and private collections) before deciding on the pattern you prefer. Containers for individual species can be of many types: for the smallest shells, glass or plastic phials fitted with firm plugs are recommended; for larger shells, small cardboard or plastic boxes are suitable, and they should be of different sizes but of a common design. Plastic ice trays are useful (but not elegant) for gastropods between ½ and 2 inches long, and they are particularly appropriate for cabinets with limited space. Plastic cutlery trays can be used effectively for bivalves of many different sizes. Many other suitable containers will no doubt suggest themselves, and the only requirement is that they should be of uniform shape and colour. It is hardly necessary to add that collecting shells should in no way be dependent on having a suitable cabinet to store them in. A simple cupboard and discarded egg boxes are perfectly adequate, and many fine collections are housed like this. The main object is to keep the shells systematically, free from damage, and with correct and adequate data.

There are four paramount 'don'ts' in the storing of shells, and these are – (i) don't store shells in metal containers because rust will permanently disfigure them, (ii) don't expose them to light for prolonged periods, (iii) don't glue them to cards (or anything else), (iv) don't jumble the data which concerns them.

Labels repeating the information contained in the catalogue are sometimes placed with each lot of shells, but this is really an unnecessary duplication of effort (though it does of course provide information on the shells without reference to the catalogue). The labelling of each cabinet drawer to indicate which families it contains may well be considered adequate.

Shells should be arranged in the cabinet by classes in the first instance, and then within each class by families 'in chronological order', i.e., the most ancient families first, the most recent last. In the *Gastropoda* for example, the first drawer would contain the ancient Slit Shells and Limpets (*Pleurotomariidae* and *Fissurellidae*), and the last drawer the far more recently evolved Bubble Shells and Melampus snails (*Tectibranchia* and *Basommatophora*).

Unlike butterflies and most botanical specimens, shells do not deteriorate if they are properly stored and to many collectors this, no doubt, is one of their major attractions. The larger gastropods in fact are almost indestructible. Before the arrival of Columbus in Jamaica, one end of Warieka Hill, which overlooks the city of Kingston, was an Arawak Indian encampment. To-day there are hundreds of shells of *Strombus pugilis*, *Arca zebra* and *Melongena melongena* (which were staple items of the Indian diet) scattered over several acres of hilltop. Most of them have probably been exposed to the weather for well over 400 years. Yet they still retain not only their original form, but much of their colour. A *Cymatium muricinum* which was recently excavated with the floor of St Pauls church at Port Royal (and which almost certainly arrived there shortly after the great earthquake of 1692 which submerged the church) is almost indistinguishable from any recently dead specimen of the species one can pick up on the beach nearby.

In spite of all this however, shells are subject to two 'ailments' which can rob them of much of their beauty – fading and fungus. Most, if not all, shells fade to some extent immediately after they die. In some species this is almost undetectable; in others it is more obvious. There is compensation however: slight fading in shells with dark colours often makes them more attractive. This does not apply to the fading of the beautiful pink and yellow aperture of the Queen Conch however, and if shells of this species are exposed to direct sunlight for five or six months the colour is liable to disappear completely. The answer is to store all shells away from the light, both natural and artificial, i.e., keep them in a cabinet when they are not in use.

The fungus problem usually applies only to shells which are stored in damp and neglected museum basements. The shells afflicted sprout a white fungus on their external surfaces. If this is allowed to remain it will permanently disfigure them. The most effective treatment is a vigorous scrubbing with a soft brush, using soap and warm water. Shells in cabinets which are kept dry and in fairly regular use are more or less immune to this condition.

Shells require little maintenance once they are properly stored. Like everything else they will collect dust if the cabinet drawers are loose fitting, but an occasional wash and a light smear of mineral oil will keep them indefinitely in pristine condition.

PART IV

The Value of Shells

Shells were valued by man long before the dawn of recorded history. The debris from many of the earliest kitchen middens shows quite clearly that, in littoral areas, primitive tribes relied on molluscs for much of their food, and it is usually in this connection that one thinks of their long association with man. But human ingenuity, aesthetic sense – and in some cases superstition – have put shells to far more diverse and interesting uses.

In Mythology and Religion

According to the Greeks, Aphrodite arose naked from the sea between the gaping valves of a scallop shell. Botticelli's fifteenth-century painting of this singular event is well known. But shells figured in the mythology and religion of civilisations far more ancient than that of Greece, and perhaps even more surprising, their religious significance persists in some countries to this day.

The Great Scallop (*Pecten maximus*), which occurs in European waters, was widely used during the Middle Ages as a sort of cap badge which indicated that the Christian wearer had visited the shrine of St James the Apostle at Compostella in north-western Spain. A very similar Mediterranean species, the Jacob's Scallop (*Pecten jacobaeus*), was also used as a 'been to' symbol by the Crusaders and pilgrims to the Holy Land, and for this reason it appears on the coats of arms of many old European families (including those of Sir Winston Churchill and the Duke of Bedford). In many Roman Catholic churches in Europe to-day, the baptismal font is a single valve of the Giant Clam (*Tridacna gigas*), and in the church of St Sulpice in Paris there is a pair of them which were originally presented to François I by the Republic of Venice.

The Indian Chank Shell (*Xancus pyrum*) has been venerated in association with the worship of the God Vishnu for over a thousand years. The rare left-handed specimens of this shell are considered sacred, and can be found encrusted with jewels and gold leaf in many Hindu temples. The four-armed god is usually represented holding the shell in one of his hands (possibly as a victory trumpet), and they are put to more practical purpose by the priests who use them to dispense medicine to their flock.

The historical reasons for the association between molluscs, mythology and religion are obscure. There is evidence that in very early times gastropods acquired female symbolism, and in some areas may have become pagan symbols of fertility. This explanation can hardly be applied to scallop shells, however. Perhaps it was that the extraordinary beauty of shells, and the almost wasteful quantities in which they naturally occur, could be explained only as a manifestation of divine power – and indeed it is a point of view which is far from obsolete to-day. Whatever the reason, the fact is that shells stirred similar human emotions wherever man took root on the face of the earth.

In Art and Decoration

The pervasive influence of shells on art and architecture is evident in many cultures. As far back as 900 B.C., the beautiful form of the Chrysanthemum Shells (*Spondylidae*) influenced the pottery designs of the Cupisnique culture of northern Peru. Leonardo da Vinci and Rembrandt van Rijn, among many other great artists, studied the structure of gastropod shells. Both Bramante and Michaelangelo made frequent use of the shell motif in their incomparable work. In our own time, the celebrated American architect Frank Lloyd Wright, who introduced new concepts of functional simplicity into architecture, acknowledged the influence on his own ideas of the 'inspired design' of shells.

Shell jewellery probably preceded clothes in some primitive societies. Necklaces and pendants fashioned from shells have been found in some of the most ancient burial sites ever discovered. In limestone caverns in France and Belgium, simple shell jewellery has been recovered which was probably worn by people who lived at the end of the last Ice Age. In the Middle Kingdom period of the Egyptian civilisation, not only shells themselves but also their stylised gold replicas, were threaded to form necklaces and bracelets. In Mexico, the Aztecs attempting to improve on nature, inlaid the nodules of the beautiful Lion's Paw Scallop (*Lyropecten nodosus*) with gold, and wore them as pendants (proving perhaps at the same time that no civilisation holds a monopoly on tastelessness). The wide popularity enjoyed by shell-cameo jewellery in Victorian England is well known.

In the eighteenth and nineteenth centuries, and indeed in some cases long before this, exotic tropical shells were embellished and used as household ornaments in Europe. These Roccoco abominations (the word originally comes from a French phrase for shell-work) were carved, scraped, bathed in acid and in some cases encrusted with jewels and precious metals. Two of the shells most commonly used were the Pearly Nautilus (*Nautilus pompilius*) and the Horned Helmet (*Cassis cornuta*). In nearly every case the shells were mutilated in some way, and it is a reflection on the decline in artistic standards towards the end of the period that many of the finest craftsmen in Europe were guilty of this work.

The popularity of shell jewellery – particularly those pieces which use their shells naturally and simply – has persisted. To-day in Florida and the West Indies, the production of this form of ornament provides a livelihood for hundreds of craftsmen, whose basic designs differ scarcely at all from those produced by their counterparts along the shores of the Mediterranean 5,000 years ago.

As Tools, Utensils and Musical Instruments

One can easily imagine the utilitarian attraction of shells for people who knew nothing of metals. There are shells which, with practically no modification, will serve efficiently as water bottles, plates, cups, spoons and forks. With a little ingenuity they can be shaped into dozens of other domestic items.

In the West Indies, before the arrival of Columbus, the Carib Indians worked the lip of the adult Queen Conch into knives, chisels and even axe heads. Their work is crude but effective. In Australia, the Aborigines still use the Bailer Shells (*Melo*), sliced neatly in half, for the purpose which gave them their common name. Shells have made needles and spearheads, pipes and daggers. In the Philippines, the large transparent shell of the Window Pane oyster (*Placuna placenta*) is used as a substitute for glass in lamp shades, screens, and, as their common name implies, in windows. Shells are still used in the making of buttons (mainly the Indo-Pacific Commercial Trochus, *Trochus niloticus*), and in the production of Mother-of-pearl, which is the inner nacreous layer present in many species, but the development of plastics has mortally injured these two industries which flourished in the nineteenth century.

Shinto priests in some parts of Japan still call the people to prayer by blowing into the truncated spire of a large *Charonia tritonis* – appropriately known to shell collectors as Triton's Trumpet. In the Polynesian Islands, the same shells were once used as war trumpets. A hollow bamboo cane was inserted into an opening high up on the spire, and the joint made air-tight with resin from the breadfruit tree. The noise, which can still be produced by examples in the British Museum, is said to be monotonous but impressive.

In many West Indian islands, the fishing boats' return to shore is heralded by a series of blasts on the modified shell of a Queen Conch, to warn the vendors waiting on the beach that fish is available. In Haiti, these trumpets, with their single rather mournful note, are still used in voodoo ceremonies.

Man realised early in his history that large gastropods, with their hollow, spiral construction, could be used as musical instruments. On the walls of King Minos' palace in Crete there are pictures of nubile priestesses triumphantly blowing shell trumpets. In more recent times there have been a number of brave attempts both in the Pacific and the Caribbean (perhaps inspired by pictures such as these) to form shell 'orchestras', using either Conch or Triton shells of varying sizes to produce different notes. The results have been unkindly compared with a herd of milch cows calling for their offspring, and there is no doubt that these beautiful shells look very much more attractive than they sound.

As Producers of Dye

The 'Tyrian purple' which was worn by some of the Roman and Byzantine Emperors as an exclusive badge of office, was not purple at all. Pliny described it as being usually the colour of congealed blood, and in fact the term seems to have been applied to a wide range of reddish tints. Nevertheless, whatever the exact colour, the dye which produced it came chiefly from a Mediterranean gastropod which is now known as *Murex brandaris*. The dye was prepared at Tyre, and its production and marketing developed into an important industry.

Originally the Tyrians produced the dye, which is secreted by the mollusc as a greyish fluid, by crushing the shell, extracting the animal and then boiling it. The vats which they used, and the piles of discarded shells, are still visible along the Mediterranean shore. It is believed that later on, probably when the supply of molluscs started to run low, they discovered that by pressing the animal gently into its shell the fluid could be extracted without killing it. This apparently resulted in a system of farming the shells, and they would be taken out of the sea, 'milked', and then returned to secrete more fluid. The exact colour of the dye seems to have been varied by a number of admixtures. One of these, according to Pliny, was human urine, but it must be assumed that this trade secret was not known to the Roman aristocracy. One can imagine, however, that it gave some satisfaction to the toiling Tyrian labourers as they sweated under the Mediterranean sun.

The common littoral Wide-mouthed Purpura (*Purpura patula*) of the West Indies produces a fluid very similar to that of *Murex brandaris*. This rather dirty looking liquid with a nauseating smell turns violet on exposure to the air. It is a remarkably fast dye, and until artificial dyes became cheaply available in some of the smaller West Indian islands, it was used on a small scale in much the same way as *M. brandaris*.

As Currency

The use of shells as a medium of exchange helped shape the pattern of civilisation, and may have been one of the more important factors in the development of trade in certain societies. In some areas of West Africa, their replacement by metal coins has only recently been completed, and even in the United States shell money, or 'wampum' as it was called in the east, was still in use by some Indian tribes at the time of the civil war.

By far the most popular shell 'coins' were two small species of Cowry, *Cypraea moneta* (the Money Cowry) and *C. annulus* (the Gold-ringer), both of which are abundant in the Pacific and Indian Oceans. This currency, at one time or another, was accepted over a very wide area of the inhabited world. Marco Polo found Money Cowries in common use in China (where the old hieroglyph for the shell is incorporated in the present character for trade), and they were the standard medium of exchange in much of Africa for many centuries. They have been unearthed in Europe as far north as Scandinavia, and also in the graves of American Indians in the United States. It is an interesting fact that in some countries, when metal coins were eventually introduced to supplant the shell currency, the coins themselves were produced in the familiar shape of the Cowry shells.

In north-eastern America shell money served the same purpose, but the 'coins' were very different. Wampum was of two main types, both produced in the form of cylindrical beads and strung together in belts. The more valuable variety was carved from the purple sections of the large Quahog clam (*Mercenaria mercenaria*); the other came mainly from the white columella of the Knobbed Whelk (*Busycon carica*). In north-western America the equivalent of wampum consisted of small, slender Tusk Shells (*Dentalium pretiosum*), which were threaded on strings and required no tedious preparation.

One of the factors which contributed to the fall of the Money Cowry was its low value. About 100 years ago, the bride-price for a young maiden in Ghana was set at 80,000 cowries. The inconvenience involved in assembling, counting, packaging and transporting to the prospective fathers-in-law shells in quantities of this order eventually became unacceptable. The idea of using a rarer shell as a 'coin' of higher value either did not occur to anyone, or was dismissed as impracticable. This particular disadvantage did not apply in America to Tusk Shell money, however, because it was much more sensibly valued. In the early seventeenth century, the bride-price for young Hupé maidens was only 120 shells (3 strings), and it is interesting to speculate on how long this currency system might have survived had it not been smothered by the wave of European conquest.

As Items of Commercial Value

The collection of shells in a systematic way – either as a hobby, or for scientific purposes – is one of the more recent 'uses' to which they have been put. It is known that Aristotle was interested in shells, made a number of accurate observations about them (and also some which were not), and probably collected them. Excavations at Pompeii have revealed a number of shells (some of which are found only in the Pacific) which seem to have been part of a Natural History collection. There is some evidence that Pliny may have been a member of the Society which owned the collection at the time of the town's destruction. But collecting as a widely popular activity is, in a historical sense, a recent phenomenon.

It began in Europe, strange though it may seem, with the advent of imperialism. In the period between the middle of the seventeenth and the end of the nineteenth centuries, the European colonisers who joined the scramble for territory in the Pacific and the Caribbean, sent back from these areas many exotic shells which had never been seen before in temperate countries. They excited wide interest at a time when the study of natural history was becoming increasingly popular among the wealthy and leisured classes.

The names of many of the great collectors are fittingly immortalised in shell nomenclature. They include the Dutchman Rumphius, the Frenchman d'Argenville, the great Swedish naturalist Carl Linnaeus, the Englishman Martin Lister, the Irishman Hans Sloane (whose work is closely associated with Jamaica), and later the dean of English collectors, Hugh Cuming. These men and many like them, recognised the significance of the magnificent shell fauna which was being discovered in the tropical seas, and made a determined effort not just to collect shells but to describe and name them (or in the case of people like Cuming, to sponsor the work).

One of the predictable results of this interest in shells was that, among collectors, the demand soon exceeded the supply, and many species became objects of considerable commercial value. Dealing in shells evolved into a highly profitable business. In the late eighteenth century, several species fetched more than 100 pounds sterling, and in those days this was a large sum of money. (One of the incidental effects of the appearance of these shells in Europe was their use by artists of the period. There is a Rembrandt etching of a *Conus marmoreus* dated 1650, and now in the Rijksmuseum, Amsterdam, which though technically magnificent does the artist no credit – he forgot to engrave it in reverse, so that the print shows it as a 'left-handed' specimen.)

By the middle of the nineteenth century, the areas of occurrence of many of the once-rare shells were well travelled, and the supply of most of the coveted species at last caught up with the demand. Prices declined sharply, and many dealers went out of business. At the same time, interest in shells which had remained at a high level for almost two centuries, gradually waned. (One London family of the period which did not suffer was that of Marcus Samuel. On the old man's death his children sold his shell business, and with the proceeds founded something far more profitable – the Shell Oil Company.)

Shell collecting as a widely popular hobby remained in decline until the end of the Second World War. Its return to favour at that time is probably due to a combination of circumstances, most important of which was the return to the United States of hundreds of thousands of servicemen from the Pacific theatre. Many brought back in their kit, as souvenirs, shells which they had picked up on the beaches from Hawaii to Japan. At the same time, international air travel, particularly to the islands of the Pacific and the Caribbean became relatively cheap and easy, and many people found themselves simultaneously with more leisure time and more money. Interest in collecting was quickly regenerated, dealers re-appeared, and shells once again acquired substantial commercial value.

Establishing Shell Values

The commercial value of a shell is generally determined by five factors:

(i) Its rarity (this really means its availability, and is not necessarily related to its natural incidence),
(ii) Its beauty,
(iii) Its size,
(iv) Its genus,
(v) Its condition.

The market value of a particular species fluctuates constantly, for its value in the final analysis is what people will pay for it. A wealthy individual who wants a particular shell at a particular time may be willing to pay many times its accepted market price.

Nevertheless, there are certain approximate price scales which are set by dealers and accepted (for the most part) by collectors, and the values ascribed to West Indian species in the following section are based on these scales.

Valuable West Indian Species

The following list of the more valuable species which occur in West Indian waters indicates their current market price (1972). The figures are for perfect, live-collected, adult specimens. All of these shells occur in water shallow enough to be dredged without the use of sophisticated and expensive equipment. Values are given in U.S. dollars.

Perotrochus adansonianus (Adanson's Pleurotomaria)
Value: $400

This is one of the rarest shells in the Caribbean. It belongs to the ancient primitive family of Slit Shells (*Pleurotomariidae*), of which there are only 15 known species, and is described in the Systematic Account of Gastropoda in Part II.

It normally occurs in water deeper than 200 feet, but specimens have been carried into fish pots set in shallower depths.

Perotrochus quoyana (Quoy's Pleurotomaria)
Value: $300

This species is closely related to *P. adansonianus*, and looks much like a smaller, less colourful version of it.

A fine, live specimen of *quoyana* was recently dredged about 1 mile south of Black River, Jamaica in 170 feet. It may well occur at this depth in many other areas along the south coast of the Island.

Turcicula imperialis (Imperial Turcicula)
Value: Unknown; it is unlikely that this species has ever been offered for sale. It must be considered an extremely valuable shell.

This species is described in the Systematic Account of Gastropoda in Part II.

Murex beaui (Beau's Murex)
Value: $60 (for the fully webbed variety)

This species is described in the Systematic Account of Gastropoda in Part II.

Murex hystricinus
Value: $35

This is a small, white, delicately spined Murex which generally occurs in depths over 400 feet. Two specimens were recently dredged about 2 miles off Rocky Point, Jamaica, in 200 feet.

Murex tristichus (Tristichus Murex)
Value: $40

This species is described in the Systematic Account of Gastropoda in Part II. Until recently, it was considered a deep water species, but it is now known to occur within the range of amateur dredging.

Ancistrosyrinx elegans (Elegant Star Turret)
Value: $25

This unusual species closely resembles a large Common Star Turret (*A. radiata*). Two fine specimens were dredged in 165 feet just south of Port Morant, Jamaica.

Gaza superba (Superb Gaza)
Value: $20

This is a beautiful, iridescent, turban shaped species, which usually occurs in water deeper than 250 feet. One live specimen was dredged in about 180 feet just south of Black River, Jamaica.

It was once considered one of the rarest Caribbean species, but in recent years shrimp trawlers have provided a fairly constant supply of them.

Eudolium crosseanum (Crosse's Tun)
Value: $20

This is a small, sturdy, Tun Shell which rarely measures more than 3 inches long, and occurs in deep water throughout the Caribbean. It has been dredged in 180 feet off the Pedro Cays.

Conus granulatus (Glory of the Atlantic Cone)
Value: $40 (for the deep red and dark brown variety)

This species is described in the Systematic Account of Gastropoda in Part II.

Conus centurio (Centurion Cone)
Value: $40

This species is described in the Systematic Account of Gastropoda in Part II.

Conus mazei (Maze's Cone)
Value: $45

This is a slender, graceful species which is rarely seen in private collections. It occurs in several areas around Jamaica in depths of 170 feet or more. It may be the rarest Cone in the northern Caribbean and is described in the Systematic Account of Gastropoda in Part II.

Morum dennisoni
Value: $175

This species is described in the Systematic Account of Gastropoda in Part II. (There are about 20 known specimens, one of which was taken from a fishing trawler in Barbados by the author.)

Spondylus americanus (Atlantic Thorny Oyster)
Value: $20 (for exceptionally fine, colourful examples)

This species is described in the Systematic Account of Pelecypoda in Part II. Values for this shell, which is common in the Caribbean, depend almost entirely on its size, colour and the quality of the spines.

Argonauta argo (Paper Nautilus)
Value: $20

This species is described in the Systematic Account of Cephalapoda in Part II. Beach specimens are very rarely undamaged.

Catalogues

Van Nostrand's Standard Catalog of Shells (published by D. Van Nostrand Coy. Inc.) is an excellent publication, which includes the names and current values of most Caribbean marine species.

Collectors who are interested in knowing the relative rarity of their shells will gain a great deal of information from this catalogue, which is the first of its kind ever attempted.

Stability of Prices

In spite of the high prices commanded by certain species, there are no fortunes to be made by collecting rare shells. Professional shell dealers rely for their living on constant trade in 'standard items', rather than the occasional sale of a valuable specimen.

The reason is simple. The current price for a fine specimen of Adanson's Pleurotomaria (*Perotrochus adansonianus*), for example, is about $400. A collector who discovered the exact area of occurrence of this species might be able to collect 100 specimens, using the appropriate equipment. He would then value his haul at $40,000. In practice, he would be lucky to receive more than a small fraction of this sum, for the influx on the market of 100 specimens of *adansonianus* would destroy the very factor responsible for its current high price – its extreme rarity. Shell 'prospecting' is an unprofitable game.

Few collectors, in any event, give much thought to the commercial value of their shells; the pleasures of collecting take more subtle forms. Robert Louis Stevenson once said: 'It is perhaps a more fortunate destiny to have a taste for collecting shells than to be born a millionaire.' Not many of us will have a chance to compare the two conditions, but Stevenson was a perceptive man, and I like to think that he was right.

Bibliography

The problem with literature on molluscs is that it is so widely scattered. There is much of it (in fact there are more than 60,000 individual papers on the subject), but it is not all easily located.

The best record of what is available is contained in the *Zoological Record* published by the Zoological Society of London, Regents Park, London, N.W.1, and found in most good museum libraries. An excellent classified summary of the more useful material is contained in Chapter XII of *American Seashells* by R. Tucker Abbott (see below).

The following list of publications can be regarded as the summary of a summary, and collectively they helped give birth to this book.

Caribbean

Abbott, R. Tucker, 1954. *American Seashells*, 541 pp., 40 pls. (24 in colour). D. Van Nostrand Co. Inc., Princeton, New Jersey, U.S.A.

This is a very attractive summary of Atlantic and Pacific American marine shells, and includes quite a large number of species also found in the West Indies. It is now being revised.

Abbott, R. Tucker, 1968. *Seashells of North America*, 280 pp., 850 illustrations, Golden press, New York, U.S.A.

Contains illustrations of many Caribbean species, and much useful general information on molluscs. The clarity of the plates is outstanding.

Abbott, R. Tucker, 1958. *The Marine Mollusks of Grand Cayman Island, West Indies*, 138 pp., 5 pls. Monograph 11, Academy of Natural Sciences, Philadelphia, Pa., U.S.A.

A useful taxonomic and ecological study of the shallow water marine molluscs of Grand Cayman Island. Nearly all the species mentioned are widely distributed throughout the West Indies.

Adams, C. B., circa 1850. *Occasional Papers on Molluscs* (edited by Clench and Turner), 404 pp., 49 pls. Published by the Dept. of Mollusks, Museum of Comparative Zoology, Harvard University, Cambridge, Mass., U.S.A.

An interesting collection of papers dealing mainly with Jamaican marine molluscs. Some of the descriptions are in Latin, and the nomenclature is hopelessly out of date, but it describes many of the very small West Indian shells.

Clench, William J. (editor). *Johnsonia, Monographs of the Marine Mollusks of the Western Atlantic*, vols. 1 to 4 and continued. Published by the Dept. of Mollusks, Museum of Comparative Zoology, Cambridge, Mass., U.S.A.

This is a most valuable scientific publication which has so far dealt in detail with some of the most interesting West Indian marine molluscan families. Early papers now require up-dating.

Dall, William H., 1886 and 1889. *Reports on the result of dredging . . . Steamer* Blake.' Report on the Mollusca. Pt. 1 (1886) on Brachiopoda and Pelecypoda; Pt. 2 (1889) on Gastropoda and Scaphopoda. Bulletin Museum of Comparative Zoology, Cambridge, Mass., U.S.A., 12, pp. 171-318, pls. 1-9; Bulletin 18, pp. 1-492, pls. 10-40.

Reports (with illustrations) on hundreds of deep water West Indian marine molluscs, some of which have not been taken again since they were dredged by the *Blake*. A massive and valuable work.

Warmke, Germaine L. and Abbott, R. Tucker, 1961. *Caribbean Seashells*, 346 pp. and 43 pls. Livingston Publishing Co., Norberth, Penn., U.S.A.

Descriptions and illustrations of more than 700 Caribbean marine molluscs. It identifies many of the very small shells which occur throughout the West Indies, and is a most convenient reference work for this purpose.

General

Dance, S. Peter, 1966. *Shell Collecting*, 344 pp., 32 pls., Faber and Faber, 24 Russell Square, London.

An interesting account of man's very old passion for collecting shells. Recommended for collectors who share the passion, and for those who think of shells in commercial terms.

Dance, S. Peter, 1971. *Seashells*, 160 pp., many colour illustrations, The Hamlyn Publishing Group Ltd., London.

A useful and beautifully illustrated survey of the major groups of marine molluscs, with special attention paid to those which are used by man. It is similar to, and complements, *Sea Shells of the World* (see below).

Morton, J. E., 1967. *Molluscs*, 244 pp., and numerous explanatory figs. Hutchinson and Co. Ltd., 178-202 Great Portland Street, London, W.1.

An excellent general book on malacology, and a recommended starting point for anyone who wants to make a serious study of the molluscan animal.

National Geographic Magazine (Journal of the National Geographic Society, Washington D.C.). Issues of July 1949 and March 1969 contain interesting articles on molluscs (by Rutherford Platt and Paul A. Zahl respectively), which are lavishly illustrated in colour.

Wagner, R. J. L. and Abbott, R. Tucker (editors), *Van Nostrand's Catalog of Shells*, 303 pp. and numerous illustrations. D. Van Nostrand Co. Inc., Princeton, New Jersey, U.S.A.

A useful shell catalogue which includes the more familiar Caribbean species and their current market values; a number of molluscan groups are treated in detail. Most useful, perhaps, because it indicates the relative rarity of the shells listed.

Zim, Herbert S. and Abbott, R. Tucker, *Sea Shells of the World*, 160 pp., 790 illustrations in colour. Golden Press, New York, U.S.A.

An excellent little summary of the major molluscan families and their distribution throughout the world. Beautifully and accurately illustrated by hand. A very good place for the novice collector to start.

Index

* denotes a non-Caribbean species referred to in the general text

Abra 283
aequalis 283
lioica 283
Acanthochitona 289
hemphilli 289
pygmaea 293
spiculosa 293
Acanthopleura granulata 292
Acanthopleurinae 292, 293
Acar 210
Acmaea 55
antillarum 55
cubensis 56
jamaicensis 56
leucopleura 56
pustulata 55
pustulata pulcherrima 55
Acmaeidae 37, 55
Acteocinidae 204
Acteonidae 187
Acteon punctostriatus 187
Actophila 195
Adapedonta 271
Adesmacea 279, 283
Adrana 206
Aequipecten 228
gibbus 229 307
lineolaris 228
muscosus 228
Aglossa 200
Alaba incerta 200
Alabina 200
adamsi 200
cerithioides 200
Alvania 199
aberrans 199
auberiana 199
gradata 199
Amaea retifera 200
Americardia 246
guppyi 246
media 246
Amphineura 17, 23, 36, 289
Amusium 222
Amygdalum dendriticum 281
Anachis 147
catenata 202
crassilabris 202
hotessieriana 202
iontha 202
mangeloides 202
obesa 202
pulchella 147
sparsa 202
subcostulata 202
Anadara 211
brasiliana 212
chemnitzi 212
lienosa floridana 211
notabilis 211
ovalis 212
transversa 211
Anadarinae 211
Anchomasa 279
Ancistrosyrinx 181
elegans 182, 331
radiata 181
Angulus 259
Anisomyaria 214
Anodontia 237
alba 237
philippiana 238
Anomalocardia brasiliana 250
Anomiacea 231
Anomia simplex 232
Anomiidae 231
Antigona 248
listeri 248
rigida 248
rugatina 282
Antillophos 150
candei 150
oxyglyptus 150
Aplacophora 36
Aplysia protea 21
Aplysiacea 21, 190, 204
Apolymetis intastriata 265
Arca 207
imbricata 207
zebra 207
Arcacea 207, 281
Archaeogastropoda 47, 99
Architectonica nobilis 79
Architectonicidae 78, 199
**Architeuthis harveyi* 35
Arcidae 207, 281
Arcinae 207
Arcopagia fausta 263
Arcopsis adamsi 281
Arene 60
cruentata 60
tricarinata 198
vanhyningi 60
Argonautidae 288
Argonauta 288
argo 23, 36, 288, 332
hians 23, 288
Argopecten 229
Asaphis deflorata 269
Aspella paupercula 142
Asperiscala 93
Astartacea 282
Astraea 63
caelata 63
phoebia 63
tecta 66

Astraea (contd.)
tuber 63
Atlanta peroni 109
Atlantidae 108
Atrina 220
seminuda 220
serrata 221
Atyidae 189
Atys 204
caribaea 204
guildingi 204
lineata 204
riiseana 204
Australium 63

Bailya 149
intricata 149
parva 149
Balcis 200
conoidea 200
intermedia 200
jamaicensis 200
Bankia 32, 283
caribbaea 283
cieba 283
fimbriatula 283
gouldi 283
Barbatia 210
cancellaria 210
candida 210
domingensis 210
tenera 211
Barnea truncata 279
Basommatophora 30, 195, 205
Basterotia 282
elliptica 282
quadrata 282
Bathoxiphus 295
Batillaria 87, 300
minima 87
Batillariinae 87
Bittium varium 200
Borsonia rouaulti 203
Botula 215, 304
fusca 215
Brachidontes 214
citrinus 214
exustus 215
recurvus 215
Buccinacea 146, 202
Buccinidae 149
Bufonaria 127
Bulla striata 189
Bullata ovuliformis 203
Bullidae 189, 299
Bursa 126
cubaniana 126
corrugata 126
spadicea 127
tenuisculpta 127
thomae 126
Bursidae 125
**Busycon carica* 327

Cabestana 121
Cadulus 296
acus 296
tetrodon 296
quadritentatus 296
Caecidae 199
Caecum 199
clava 200
coronellum 200
floridanum 199
imbricatum 199
insigne 199
nebulosum 200
nitidum 200
plicatum 200
pulchellum 199
Calcar 66
Calliostoma 58
javanicum 58, 304
jujubinum 59
pulchrum 59
sarcodum 59
torrei 198
Callistochiton shuttleworthianus 293
Calloplax janeirensis 290
Callucina 237
Calyptracacea 95
Calyptraea centralis 201
Calyptraeacea 201
Calyptraeidae 95, 201
Cancellaria 169
conradiana 169
reticulata 169
Cancellariidae 169
Cantharus 152
auritulus 153
lautus 153
tinctus 152
Capulidae 98
Capulus intortus 99
Cardiacea 244, 282
Cardiidae 244, 282
Cardiomya 283
ornatissima 283
perrostrata 283
Carditacea 234
Cardita gracilis 234
Carditidae 234
Carinodrillia liella 203
Casmaria atlantica 118
Cassididae 116, 307
Cassis 118
* *cornuta* 323
flammea 119, 307
madagascariensis 119, 307
tuberosa 118, 307
Cavolina 191
gibbosa 205
inflexa 205
longirostris 205
tridentata 191
trispinosa 205
uncinata 205
Cavolinidae 191, 205, 306
Cenchritis 74
Cephalopoda 17, 22, 35, 283
Ceratozona squalida 289
Cerithiacea 76, 199
Cerithidea 86, 300
beattyi 86

costata 86
pliculosa 87
Cerithiidae 87, 200
Cerithiopsis 200
emersoni 200
greeni 200
latum 200
Cerithium 87
algicola 88
auricoma 89
eburneum 88
litteratum 87
muscarum 89
variabile 88
Chama 239
congregata 242
florida 243
macerophylla 239
sarda 242
sinuosa 242
Chamidae 239, 308
Charonia 120
* *tritonis* 325
variegata 120, 307
Cheilea equestris 94
Chicoreus 131
Chione 248
cancellata 248
granulata 249
intapurpurea 249
latilirata 250
paphia 250
pygmaea 249
Chiton 291
marmoratus 292
squamosus 292
tuberculatus 291
viridis 293
Chitonida 289, 293
Chitonidae 291, 293
Chitoninae 291
Chlamys 223
benedicti 226
effluens 226
imbricata 226
nana 281
ornata 223
sentis 223
Choneplax lata 293
Cingulina babylonia 204
Circinae 254
Cirsotrema dalli 91
Cittarium pica 57
Clathrodrillia melonesiana 203
Clavinae 182
Clio pyramidata 205
Clypidella 54
Cochlespirinae 181
Cochliolepis parasitica 199
Codakia 238
costata 239
orbicularis 238
orbiculata 238
pectinella 282
Coleoidea 35, 283
Collisella 55
Colubraria 151
lanceolata 151
obscura 151
Colubrellina 126
Columbariidae 202
Columbarium bartletti 202
Columbella mercatoria 146
Columbellidae 146, 202
Compsodrillia 183
Conidae 30, 171, 313
Conus 171
austini 177
centurio 173, 331
clarki 178
daucus 173
dominicanus 179
granulatus 177, 304, 305, 331
jaspideus 172
jaspideus havanensis 172
juliae 176, 304
marmoreus 329
mazei 178, 331
mus 178
patae 177
ranunculus 30, 176
regius 171
regius citrinus 171
spurius 172
verrucosus 172
Cooperella atlantica 283
Coralliophaga coralliophaga 235
Coralliophila 143
abberans 143
abbreviata 143
caribaea 143, 306
scalariformis 146
Corbula 278
aequivalvis 278
caribaea 278
contracta 283
dietziana 278
**disparilis* 278
Corbulidae 278, 283
Crassatellidae 282
Crassinella 282
guadalupensis 282
lunulata 282
martinicensis 282
Crassispira 182
ebenina 203
fuscescens 182
leucocyma 183
nigrescens 203
Crassispirella 182
Crassostrea rhizophorae 233
Cremides 53
Crenella divaricata 281
Crepidula 31, 34, 98
aculeata 98
convexa 201
glauca 201
plana 98
protea 201
Creseis acicula 205
Crucibulum auricula 95
Cryptoconchus floridanus 293
Cryptoplacidae 289, 293
Ctena 238

Ctenoides 231
Cumingia antillarum 283
Cunearca 212
Cuspidariidae 283
Cuvierina columnella 205
Cyathodonta semirugosa 283
Cyclinella tenuis 254
Cyclostrema cancellatum 61
Cyclostremiscus 76
beaui 76
supranitidos 199
trilix 199
Cylichna bidentata 204
Cymatiidae 120
Cymatium 121
caribbaeum 121
femorale 124
krebsi 123
labiosum 121
muricinum 123
nicobaricum 122
parthenopeum 124
pileare martinianum 122
poulseni 121
rubeculum occidentale 122
vespaceum 123
Cyphoma 108
gibbosum 108
intermedium 201
signatum 108
Cypraea 105
* *anulus* 327
* *aurantium* 13
cervus 106
cinerea 106
* *moneta* 327
mus 106
spurca acicularis 106
zebra 105, 300
Cypraeacea 103, 201
Cypraecassis testiculus 119
Cypraeidae 105, 318
Cyrtopleura costata 279

Daphnella 186
lymneiformis 186
stegeri 186
Daphnellinae 186
Decapoda 283, 288
Dentale 295
Dentaliidae 294
Dentalium 294
antillarum 295
calamus 296
calipeplum 296
ceratum 296
didymum 295
eboreum 294
liodon 296
* *pretiosum* 327
semistriolatum 296
Dentiscala 92
Depressicala 200
nautlae 200
nitidella 200
Detracia bullaoides 196
Diadema (Sea Urchin) 309
Diberius 217
Diodora 49
arcuata 51
cayenensis 50
dysoni 50
jaumei 198
listeri 49
minuta 51
sayi 50
variegata 51
viridula 50
Diodorinae 49
Diplodonta 235, 282
gabbi 282
nucleiformis 282
notata 282
punctata 235
semiaspera 282
Diplodontidae 235, 282
Distorsio 125
clathrata 125
mcgintyi 125
Divaricella quadrisulcata 239
Donacidae 266, 303
Donax 267
denticulatus 267
* *fossor* 266
striatus 267
tumidus 267
Doryteuthis plei 288
Dosinia concentrica 254
Dosinidia 254
Dosiniinae 254
Dreissenacea 234
Dreissenidae 234
Drillia 184
actinocycla 203
aepynota 203
albicoma 203
albinodata 184
cydia 183
eucosmia 203
gundlachi 203
interpleura 184
lissotropis 203
ponciana 203
thea 203
Drupa nodulosa 139

Echininus nodulosus 74
Echinochama arcinella 243
Echinolittorina 74
Ellobiidae 195, 205, 300
Emarginula 48
phrixodes 48
pumila 48
tuberculosa 198
Emarginulidae 48
Emarginulinae 48
Engina turbinella 151
Engoniophos guadalupensis 150
Entemnotrochus 47
Episcynia inornata 199
Epitoniacea 89, 200
Epitoniidae 91, 200, 299
Epitonium 92
albidum 93

candeanum 93
denticulatum 200
echinaticostum 200
foliaceicostum 200
frieli 200
krebsi 200
lamellosum 93
novangliae 94
occidentale 92
sericifilum 200
tiburonense 200
turritellulum 200
unifasciatum 200
venosum 200
Erato maugeriae 201
Eratoidae 103, 201
Erosaria 106
Ervilia 283
nitens 283
rostratula 283
Erycina 282
emmonsi 282
periscopiana 282
Erycinacea 282
Erycinidae 282
Euchelus guttarosea 198
Eudolium crosseanum 331
Eulamellibranchia 234, 282
Eulima 201
auricincta 201
bifasciata 201
patula 201
Eulimidae 200
Eurytellina 260
Euvola 222

Fasciolaria tulipa 156
Fasciolariidae 155, 313
Fasciolariinae 156
Favartia 136
Fenimorea janetae 184
Filibranchia 207, 281
Fissurella 53, 54
angusta 54
barbadensis 53
barbouri 198
fascicularis 54
nimbosa 54
nodosa 53
rosea 53
Fissurellidae 19, 37, 48, 198
Fissurellinae 53
Fossaridae 201
Fossarus orbigny 201
Fraginae 246
Fugleria 211
Fusilatirus cayohuesonicus 157
Fusininae 155
Fusinus 155
closter 155
eucosmius 155

Gastrochaena hians 283
Gastrochaenacea 283
Gastrochaenidae 283
Gastropoda 17, 19, 28, 47
Gaza 198
fischeri 198
superba 198, 331
Glycymerididae 213, 281
Glycymeris 213
decussata 213
pectinata 213
sericata 281
undata 213
Glyphepithema 114
Glyphoturris quadrata rugirima 203
Gorgonia (Sea Fan) 306
Gouldia cerina 282
Graptacme 294
Gregariella 217, 304
coralliophaga 217
Gutturnium 123
Gymnosomata 191
Gyroscala 93

Haminoea 189
antillarum 190
elegans 189
petiti 204
succinea 204
Heliacus 78
bisulcatus 78
cylindricus 78
infundibuliformis 79
Hemitoma 49
emarginata 49
octoradiata 49
Heterodonax bimaculatus 269
Heterodonta 234
Heteropoda 108
Hiatellidae 253
Hipponicacea 94, 201
Hipponicidae 94
Hipponix 95
antiquatus 95
subrufus subrufus 95
Homalopoma linnei 198
Hormomya 215
Hyalina 170
albolineata 203
avena 170
avenacea 203
tenuilabra 171
Hydatina vesicaria 188
Hydatinidae 188
Hydrobiidae 75, 199
Hysteroconcha 253

Ianacus 98
Iphigenia brasiliensis 268
Ischadium 215
Ischnochiton 290
limaciformis 290
papillosus 293
Ischnochitonidae 290, 293
Iselica anomala 201
Isognomon 218
alatus 218
bicolor 218
radiatus 218
Isognomonidae 217, 308
Ithycythara 184
lanceolata 184

Ithycythara (contd.)
parkeri 203

Janthina 90
exigua 90
globosa 90
janthina 90
Janthinidae 28, 89, 305, 306
Jodina 90
Jaspidella jaspidea 165

Krebsia 99
Kurtziella limonitella 185

Labiosa lineata 275
Laevicardiinae 246
Laevicardium 246
laevigatum 246
pictum 247
sybariticum 247
Laevidentalium 296
Larkinia 211
Lamellaria rangi 201
Lamellariidae 201
Lamellibranchia 32
Latirus 156
brevicaudatus 156
infundibulum 156
mcgintyi 156
Leptegouana 170
Leptodrillia splendida 203
Leptopecten bavayi 227
Leucozonia 158
nassa 158
nassa leucozonalis 158
ocellata 158
Lima 230
lima 230
pellucida 231
scabra 231
scabra tenera 231
Limaria 231
Limatula subauriculata 281
Limidae 230, 281
Linatella 121
Lioberus castaneus 216
Liotiinae 60
Lirophora 250
Lithophaga 32, 216, 304
antillarum 216
bisulcata 217
nigra 216
Lithopoma 63
Litiopa melanostoma 200
Littoraria 73
Littorina 72
angulifera 73, 300
meleagris 73
mespillum 73, 301
nebulosa 72
ziczac 72
Littorinacea 72
Littorinidae 72
Lobiger souverbii 191
Loliginidae 288
Lolliguncula brevis 288
Lucapina 51
aegis 52
philippiana 52
sowerbii 52
suffusa 51
Lucapinella limatula 52
Lucina 236
blanda 282
pensylvanica 236
phenax 236
Lucinacea 235, 282
Lucinidae 236, 282
Lucinisca 237
Lunarca 212
Luria 106
Lyonsia 283
beana 283
hyalina floridana 283
Lyonsiidae 283
Lyropecten 227
antillarum 227
nodosus 228, 324

Macoma 264
aurora 264
brevifrons 264
constricta 264
orientalis hendersoni 283
pseudomera 265
tageliformis 264
tenta 265
Macrocallista maculata 254
Macrophragma 82
Mactra 274
alata 274
fragilis 275
Mactracea 274, 283
Mactridae 274
Magilidae 192
Malletiidae 281
Malleus candeanus 218
Mancinella 191
Mangelia 203
aguadillana 203
asarca 203
balteata 203
bartletti 203
biconica 203
elata 203
fusca 203
lavalleana 203
luctuosa 203
melanitica 203
morra 203
quadrata 203
quadrilineata 203
trilineata 203
Mangeliinae 184
Marginella 202
denticulata 202
evadne 202
fusca 203
gracilis 203
haematita 202
interrupte-lineata 202
lactea 202
striata 202
subtriplicata 203

torticula 202
Marginellidae 169, 202
Marmorstroma 61
Martesia 280, 304
cuneiformis 280
striata 280
Mathilda barbadensis 200
Mathildidae 200
Maugeria 292
Mcgintiella 165
Melampus 196
coffeus 196
monile 196
Melaraphe 72
**Melo* 325
Melongena melongena 153, 307
Melongeniidae 153
**Mercenaria mercenaria* 327
Meretricinae 251
Mesodesmatidae 283
Mesogastropoda 72, 198
Microcardium 282
tinctum 282
peramabile 282
Microdochus floridanus 199
Microgaza rotella 198
Micromelo undata 188
Mitra 166
barbadensis 166
fluviimaris 166
nodulosa 166
Mitrella 198
idalina 202
fenestrata 202
lunata 202
nitens 198
Mitridae 165, 202
Mitromorpha biplicata 203
Modiolus americanus 214
Modulidae 85, 306
Modulus 85
carchedonius 85
modulus 85
Monilispira albocincta 185
Monoplacophora 17
Monoplex 124
Mopaliidae 289
Morula 139
Morum 116
dennisoni 116, 332
oniscus 116
Mulinia portoricensis 275
Murex 131
antillarum 133
beaui 135, 330
* *brandaris* 326
brevifrons 31, 132, 300, 307
cabriti 134
cailleti 134
cailleti kugleri 134
cellulosus 136
cellulosus nuceus 136
chrysostoma 134
ciboney 133
hystricinus 330
mcgintyi 136
pomum 29, 131, 307
pulcher 135
recurvirostris rubidus 132
spectrum 131, 304
tristichus 136, 331
tryoni 135
woodringi 133 313
Muricacea 130, 201
Muricidae 130, 201, 313
Muricinae 131
Muricopsis 137
oxytatus 137
philippiana 138
pudicus 138
Musculus lateralis 281
Myacea 278, 283
Myacidae 283
Myrtea pristiphora 282
Mytilacea 32, 281
Mytilidae 214, 281, 307
Mytilopsis domingensis 234

Nassariidae 154, 202
Nassarina metabrunnea 202
Nassarius 154
albus 154
consensa 202
hotessieri 202
nanus 154
vibex 155
Natica 111
canrena 111
cayennensis 114
floridana 114
livida 111
marochiensis 114
menkeana 111
sagraiana 201
Naticacea 109, 201
Naticidae 109, 201, 303, 311, 318
Naticinae 111
Nautiloidea 35
**Nautilus pompilius* 36, 325
Neodrillia 183
Neogastropoda 130, 201
Neosimnia 107
acicularis 107
uniplicata 107
Nerita 67, 301
fulgurans 68
peloronta 67
tessellata 68
versicolor 68
Neritacea 67, 198
Neritidae 67, 307
Neritinae 67
Neritina 69, 300
clenchi 70
meleagris 70
piratica 70
punctulata 70
reclivata 71
virginea 69
Neritrema 73
Nitidella 147
dichroa 202
laevigata 148
nitida 148

Nitidella (*contd.*)
  ocellata 147
Niteoliva 164
Nodilittorina tuberculata 74
Nodipecten 228
Nodiscala 92
Notocorbula operculata 283
Nudibranchia 205
Nucula aegeensis 281
Nuculana acuta 281
Nuculacea 206, 281
Nuculidae 206, 281

Ocenebra intermedia 142
Octopoda 283, 288
Octopodidae 288
Octopus 288
  briareus 288
  joubini 288
  vulgaris 22, 228
Odostomia 204
  canaliculata 204
  gemmulosa 204
  laevigata 204
  seminuda 204
  solidula 204
Oliva 162
  caribaeensis 163
  fusiformis 163
  reticularis 162
  reticularis greenwayi 163
  reticularis olorinella 163
Olivella 164
  acteocina 202
  dealbata 202
  minuta 164
  nivea 164
  perplexa 202
  petiolita 164
  rotunda 165
  verreauxi 164
Olividae 162, 303, 311
Ommastrephidae 288
Opalia 92
  crenata 92
  hotessieriana 200
  pumilio 92
  pumilio morchiana 92
Opisthobranchia 28, 187
Ostrea 233
  equestris 233
  frons 233
  permollis 281
Ostreacea 232, 281
Ostreidae 232, 281
Ovulidae 107, 201, 306
Oxynocidae 204
Oxynoe antillarum 204

Pandora bushiana 283
Pandoracea 283
Pandoridae 283
Papyridea 245
  semisulcata 245
  soleniformis 245
Parastarte triquetra 282
Patellacea 55
Patelliformia 197
Parviturboides comptus 199
Pecten 222
  chazaliei 223
  * *jacobeus* 323
  laurenti 222
  * *maximus* 323
  papyraceus 222
  raveneli 281
  ziczac 222, 307
Pectinacea 221, 281
Pectinidae 32, 221, 281
Pedipes 205
  mirabilis 205
Pelecypoda 17, 21, 32, 206
Periglypta 248
Perotrochus 47
Perotrochus adansonianus 47, 330
Perotrochus quoyana 198, 330
Periscula 203
  lavalleeana 203
  fluctuata 203
  pulcherrima 203
Petaloconchus 82
  erectus 82
  floridanus 83
  irregularis 82
  mcgintyi 83
Petricola lapicida 255
Petricolidae 255, 282
Petricolinae 255
Phacoides 236
  muricatus 237
  pectinatus 236
  radians 237
Phalium 117
  cicatricosum 118
  granulatum 117
Phasianellidae 66, 198
Phenacolepadidae 198
Phenacolepas hamillei 198
Philine sagra 204
Philinidae 204
Philippia krebsi 79
Pholadidae 279, 283
Pholas campechiensis 283
Phyllonotus 131
Pinctada radiata 219
Pinna carnea 220
Pinnidae 220
Pira 196
Pisania pusio 152
Pitar 252
  albida 252
  aresta 252
  circinata 253
  dione 253
  fulminata 252
Pitarinae 252
**Placuna placenta* 325
Planaxidae 84
Planaxis 84
  lineatus 200
  nucleus 84
Pleuromalaxis balesi 199
Pleurotoma minor 202
Pleurotomariacea 47, 198

Pleurotomariidae 47, 198
Plexaura (Sea Whip) 306
Plicatula gibbosa 221
Plicatulidae 221
Pododesmus rudis 232
Polinices 109
  hepaticus 110
  lacteus 109
  nubilus 201
Polinicinae 109
Pollia 152
Polyplacophora 36, 289
Polystira albida 181
Ponocyclus 76
Potamididae 86
Potamidinae 86
Propeamussiidae 281
Propeamussium 281
  dalli 281
  pourtalesianum 281
Prosobranchia 28, 47
Protobranchia 32, 206, 281
Prunum 170
  apicinum 170
  guttatum 170
  pruniosum 203
Psarostola 202
  minor 202
  monilifera 202
Pseudochama radians 243
Pseudostomatella 198
  coccinea 198
  erythrocoma 198
Psilaxis 79
Pteria colymbus 219
Pteriacea 32, 217
Pteriidae 219
Pteropoda 28, 191, 205, 306
Pulmonata 195
Puperita 69
  pupa 69, 301
  tristis 69
Purpura patula 140, 301
Purpurina 139
Pusia 166
  albocincta 167
  cubana 166
  gemmata 167
  hanleyi 202
  histrio 202
  puella 167
  pulchella 167
  staminea 167
Pyramidella 191
  candida 204
  dolobrata 191
  fusca 204
Pyramidellidae 191, 204
Pyrene ovulata 147
Pyrogocythara 203
  coxi 203
  densestriata 203

Quadrans lintea 283

Ranularia 121
Retusa 204
  bullata 204
  candei 204
Retusidae 204
Rhizorus 204
  acutus 204
  oxytatus 204
Rimula aequisculpta 198
Risomurex 139
  muricoides 140
  roseus 139
Rissoacea 75, 199
Rissoidae 199
Rissoina 199
  bryerea 199
  cancellata 199
  chesneli 199
  decussata 199
  fischeri 199
  multicostata 199
  striatacostata 199
  striosa 199
Rupellaria typica 255

Sacoglossa 191, 204
Sanguinolaria 268
  cruenta 268
  operculata 269
Sanguinolariidae 268
Saxicavella sagrinata 283
Scaphella neptunia 202
Scaphopoda 17, 24, 37, 294
Schismope cingulara 198
Scissula 259
Scissurellidae 198
Sconsia striata 117
Scrobiculina 260
Seila adamsi 200
Semele 266
  bellastriata 283
  nuculoides 283
  proficua 266
  proficua radiata 266
  purpurascens 266
Semelidae 265, 283
Sepia 22
Sepioteuthis sepioidea 288
Septa 122
Septibranchia 283
Serpulorbis 83
  decussata 83
  riisei 83
Servatrina 220
Sigatica semisulcata 115
Siliquaria anguillae 84
Siliquariidae 84
Sininae 110
Sinum 110
  maculatum 110
  perspectivum 110
Siphonaria pectinata 197
Siphonariidae 197, 205
Siphonodentaliidae 294, 296
Smaragdia viridis viridemaris 71
Smaragdiinae 71
Solariella amabilis 198
Solecurtus 271
  cumingianus 271

Solecurtus (*contd.*)
sanctaemarthae 271
Solemya occidentalis 281
Solemyacidae 281
Solen obliquus 274
Solenacea 271
Solenidae 271
Spengleria rostrata 283
Sphenia antillensis 283
Spiratella 205
inflata 205
leseuri 205
Spirula spirula 22, 36, 283, 306
Spirolaxis exquisita 199
Spirulidae 283
**Spisula solidissima* 274
Spondylidae 229, 308, 324
Spondylus americanus 33, 229, 304, 308, 332
Stenoplax 290
Sthenoteuthis bartrami 288
Stigmaulax sulcata 115
Stilifer subulatus 201
Stiliferidae 201
Stramonita 140
Strigilla 263
carnaria 263
mirabilis 283
pisiformis 283
rombergi 263, 283
Strombacea 99
Strombidae 21, 100, 307, 312
Strombus 101
costatus 102, 307
gallus 102, 307
gigas 19, 29, 101, 307, 313, 315
pugilis 101
raninus 101, 313
raninus nanus 102
Styliola subula 205
Synaptocochlea picta 198
**Syrinx aruanus* 153

Taeniaturbo 62
Tagelus 269
divisus 269
plebeius 270
Taxodonta 207
Tectarius muricatus 74
Tectibranchia 187, 204, 311
Tectininus 74
Tectonatica pusilla 201
Tegula 57
excavata 58
fasciata 57
lividomaculata 57
Teinostoma 199
clavium 199
obtectum 199
parvicallum 199
Tellina 258
aequistriata 283
alternata 262
candeana 283
caribaea 260
cristallina 283
exilis 260
georgiana 262
guildingii 260
laevigata 258
lineata 262
listeri 259
magna 260
martinicensis 283
mera 259
pauperata 283
punicea 262
radiata 258
radiata unimaculata 258
similis 259
sybaritica 283
vespuciana 283
vitrea 283
Tellinacea 258, 283
Tellinella 259
Tellinidae 258, 283
Terebra 179
cinerea 180, 303
dislocata 180
hastata 180
protexta 180
taurinus 179
Terebridae 179
Teredinidae 283
Teredo 32, 283
dominicensis 283
thomsoni 283
Thais 140
deltoidea 141
haemastoma floridana 141
rustica 140
Thecosomata 191
Thraciidae 283
Thyasira trisinuata 282
Timoclea 249
Tindaria corpulenta 281
Tivela 251
abaconis 252
mactroides 251
Tonna 130
galea 130
maculosa 130
Tonnacea 116
Tonnicia schrammi 293
Tonnidae 127
Trachycardiinae 244
Trachycardium 244
isocardia 245
magnum 244
muricatum 244
Tralia ovula 195
Transennella 251
conradina 251
cubaniana 282
culebrana 282
stimpsoni 282
Trapeziidae 235
Tricolia 67, 198
adamsi 198
affinis 67
bella 198
tessellata 198
thalassicola 198
Tricollinae 67

**Tridacna gigas* 22, 33, 323
Trigoniocardia 282
antillarum 282
ceramidum 282
Trimusculidae 205
Trimusculus goesi 205
Triphora 200
melanesia 200
nigrocincta 200
ornata 200
turris-thomae 200
Triphoridae 200
Triptychus niveus 204
Trivia 103
antillarum 104
leucosphaera 104
nix 104
pediculus 103
quadripunctata 104
suffusa 103
Triviinae 103
Trochacea 56, 198
Trochidae 56, 198
**Trochus niloticus* 325
Trona 105
Truncatella 75
bilabiata 199
caribaeensis 199
pulchella 75
pulchella bilabiata 199
scalaris 199
scalaris clathrus 199
Turbinidae 20, 60, 198
Turbininae 61
Turbo 61
canaliculatus 62
castanea 61, 306
filosus 62
Turbonilla 204
abrupta 204
elegans 204
haycocki 204
interrupta 204
pupoides 204
reticulata 204
unilirata 204
Turcicula imperialis 60, 330
Turridae 181, 203
Turrinae 181
Turritella 77
exoleta 77
variegata 77
Turritellidae 76
Tutufa 127
Typhinae 138
Typhis 138
fordi 138
longicornis 201

Vanikoroidae 201
Vanikoro 201
oxychone 201
sulcata 201
Vasinae 162
Vasum 162
capitellus 162
muricatum 162
Veneracea 247, 282
Veneridae 247, 282
Venerinae 248
Ventricolaria 248
Vermetidae 82, 302, 306
Vermicularia 77
knorri 77
spirata 78
Verticordia ornata 283
Verticordiidae 283
Violetta 90
Vitricythara metria 203
Vitrinella 199
helicoidea 199
multistriata 199
Vitrinellidae 76, 199
Vitta 69
Voluta musica 168
Volutacea 162, 202
Volutidae 168, 202
Volutinae 168
Volvarina 170

Williamia krebsi 205

Xancidae 159
Xancinae 159
Xancus 159
angulatus 159
* *pyrum* 323
Xenophora 99
caribaea 100
conchyliophora 99, 307
Xenophoridae 99

Yoldia perprotracta 206

Zebina browniana 199

Subject Matter Index

Acid, use of 319
Anatomy (of molluscs) 26
Aquaria 312
Ark shells 207
Atlantic Wood-louse 116
Auger shells 179

Baby's Ear shells 110
Baily shells 149
Bibliography 335
Bittersweets 213
Bivalves 17, 21, 32, 206
Bleeding Tooth shell 67
Bonnets 117
Bubble shells 187

Calico Clam 254
Camouflage
  (i) general, 298, 301
  (ii) in squid and cuttlefish, 35
Cap shells 98
Carrier shells 99
Cask shells (*see* Tun shells)
Catalogues 332
Cataloguing 320
Ceriths 87, 200
Chank shells 159, 323
Chitons 17, 23, 36, 289
Chrysanthemum shells (*see* Thorny Oysters)
Classes of shells
  (i) gastropods (snails), 19
  (ii) bivalves (clams), 21
  (iii) squids, octopuses and nautiloids, 22
  (iv) chitons, 23
  (v) tusk shells 24
Cleaning
  (i) gastropods, 316
  (ii) bivalves, 317
  (iii) chitons, 317
Cockles 244, 282
Collecting

  Out of water 298
  (i) On beaches, 298
  (ii) In salt marshes and mangroves, 300
  (iii) In rocky areas, 301

  Under water 301
  (i) Under rocks and coral, 302
  (ii) Beneath substrate, 303
  (iii) In burrows, 304
  (iv) In crevices, 304
  (v) Floating on, or below surface of sea, 305
  (vi) Associated with marine plants or animals, 306
  (vii) Exposed on sea floor, 307

  Collecting Chitons, 311
  Dredging, 309
  Equipment 308
  Night collecting, 311
  Trapping, 310
Common Spirula 22, 36, 283, 306
Conchs 21, 100, 307, 312
Cones 30, 171, 313
Conservation 314
Coral shells 143
Corbula shells 278, 283
Cowries 105, 318
Crown Conch 153, 307
Cup & Saucer shells 201
Currency (shells used as) 327
Cuttlebone 22
Cyphomas 108

Date Mussels 32, 216, 304
Dealers (in shells) 329
Diplodons 235, 282
Dove shells 146, 202
Dredging (*see* Collecting)
Drupes 139
Dye (from shells) 326

Environment
  (i) gastropods (snails), 28
  (ii) bivalves (clams), 28
  (iii) squids, octopuses and nautiloids, 35, 36
  (iv) chitons, 37
  (v) tusk shells, 37
Equipment (*see* Collecting)
Evolution (of molluscs) 17

False Limpets 197, 205
False Mussels 234
Feeding habits
  (i) gastropods (snails), 29
  (ii) bivalves (clams), 33
  (iii) squids, octopuses and nautiloids, 35
  (iv) chitons, 37
  (v) tusk shells, 37
File shells 230, 281
Flamingo Tongue 108
Foreword 6
Frog shells 125

Geographic limits (of Caribbean molluscs) 11, 41
Giant Clam 22, 33, 323
Great Scallop 323
Growth
  (i) gastropods (snails), 30
  (ii) bivalves (clams), 34
  (iii) squids, octopuses and nautiloids, 36

Growth (*contd.*)
(iv) chitons, 37
(v) tusk shells, 38

Helmet shells 116, 307
Horn shells 86, 300

Identification 43

Jacknife Clams 269
Jacob's Scallop 323
Jewel Boxes 239, 308
Jewellery (*see* Ornaments)
Jingle shells 231

Kitten's Paw 221

Latirus shells 156
Limpets
(i) keyhole, 19, 37, 48, 198
(ii) true, 55
Lion's Paw 228, 324
Living molluscs 26
Localities 44
Lucines 236, 282

Macomas 264
Margin shells 169, 202
Martesias 280, 304
Measurements 45
Melampus shells 195, 205, 300
Mitres 165, 202
Modulus shells 85, 306
Moon shells 109, 201, 303, 311, 318
Murex shells 130, 201, 313
Musical instruments (shells as) 325
Mussels 214, 281, 307
Mythology & religion (shells association with) 323

Nassa shells 154, 202
Nautiloids 35
Nerites 67, 307
Nomenclature 42
Nudibranchs 205
Nut Clams 206, 281
Nutmeg shells 169

Octopuses 22, 35, 288
Oil, use of 319
Olives 162, 303, 311
Ornaments (shells as) 324
Oysters 232, 281

Paper Nautilus 23, 36, 288, 332
Parts of the shell 18-24
Pen shells 220
Periwinkles 72
Pheasant shells 67, 198
Phos shells 150
Place names (*see* Localities)
Preface 11
Prices, stability of 329
Purple Sea Snails 28, 89, 305, 306

Razor shells 269
Reproduction
(i) gastropods (snails), 30
(ii) bivalves (clams), 34
(iii) squids, octopuses and nautiloids, 36
(iv) chitons, 37
(v) tusk shells, 38
Respiration
(i) gastropods (snails), 30
(ii) bivalves (clams), 33
(iii) chitons, 37
(iv) tusk shells, 37
Rock shells 140

Sanguin shells 268
Scallops 32, 221, 281
Sea Butterflies 28, 191, 205, 306
Sea Hares 21, 205
Semele shells 265, 283
Shipworms 32, 283
Simnias 107
Slipper shells 31, 34, 98, 201
Slit shells 47, 198, 330
Slit Worm shells 84
Snails 17, 19, 28, 47
Spindle shells 155
Spirula (*see* Common Spirula)
Squids 35, 288
Star Shells 63
Storing shells 320
Sub-division of classes 42
Sundials 78, 199
Surf Clams 274
Systematic account (of West Indian shells) 41

Tegulas 57
Tellins 258, 283
Thorny Oysters 33, 229, 308, 324, 332
Tools (shells used as) 325
Top shells 56, 198
Trapping (*see* Collecting)
Trawlers 310
Tree Oysters 217, 308
Tritons 120
Trivias 103
Tulip shells 156
Tun shells 127
Turban shells 61
Turkey Wing 207
Turrids 181, 203
Turritella shells 76
Tusk shells 17, 24, 37, 294

Utensils (shells as) 325

Valuable West Indian shells 330
Value of shells, The 328
Vase shells 162
Venus shells 247, 282
Vitrinellas 76, 199
Volutes 168, 202

Wampum 38, 327
Wedge shells 266, 303
Wentletraps 91, 200, 299
Whelks 149
Wing Oysters 219
Worm shells 82, 302, 306